Hydrogen Sulfide and its Therapeutic Applications

Hideo Kimura
Editor

Hydrogen Sulfide and its Therapeutic Applications

Springer

Editor
Hideo Kimura
Division of Molecular Genetics
National Institute of Neuroscience
Tokyo
Japan

ISBN 978-3-7091-1549-7 ISBN 978-3-7091-1550-3 (eBook)
DOI 10.1007/978-3-7091-1550-3
Springer Wien Heidelberg New York Dordrecht London

Library of Congress Control Number: 2013948359

Printed on acid-free paper

Springer is part of Springer Science+Business Media (www.springer.com)

Preface

Hydrogen sulfide (H_2S) is a toxic gas that emits an unpleasant smell like rotten eggs. About 20 years ago, the mere presence of a pungent gas was considered as a physiological mediator. Following the discovery of endogenous sulfide in the mammalian brain, we found that H_2S is produced by the enzyme cystathionine β-synthase (CBS) and that it functions as a neuromodulator by enhancing the activity of neurotransmitter receptors in the brain. We found another H_2S-producing enzyme called cystathionine γ-lyase (CSE) in tissues, including vasculature. Another interesting observation was that H_2S could function as a smooth muscle relaxant. Although these enzymes were known to produce H_2S in vitro, H_2S was considered as a by-product of the metabolic pathways or as a marker for the evaluation of enzyme activity. Similar to CBS and CSE, 3-mercaptopyruvate sulfurtransferase (3MST) also produces H_2S, but its activity requires reducing cofactors such as thioredoxin and dihydrolipoic acid. Recently, we identified a fourth pathway of H_2S synthesis from D-cysteine. The enzymes described above are expressed in various tissues, including neurons, glia, vasculature, liver, kidney, pancreas, and the gastrointestinal tract. H_2S performs numerous activities in these organs, including neuromodulation, vascular relaxation, angiogenesis, energy formation in mitochondria, and protection against oxidative stress and ischemia–reperfusion injury. Owing to these beneficial effects of H_2S, H_2S-releasing drugs have been developed, and several of them are under clinical trials.

Ruma Banerjee and colleagues describe how endogenous levels of H_2S are enzymatically regulated for proper functioning, in their chapter "Enzymology of Hydrogen Sulfide Turnover." Three pathways for H_2S production from L-cysteine as a major substrate are known, involving CBS, CSE, and 3MST along with cysteine aminotransferase (CAT), which is identical to aspartate aminotransferase (AAT). Since the H_2S elimination pathway is activated at considerably lower sulfide concentrations, the half-life of H_2S is short. This regulation enables H_2S to function as a signaling molecule. This chapter focuses on the structural enzymology and regulation of H_2S metabolism.

Ken Olson, in his chapter "Hydrogen Sulfide as an Oxygen Sensor," describes that eukaryotic cells can detect O_2 availability and transduce it into physiological signals for the proper delivery of O_2 and regulation of O_2 consumption. Tissue H_2S concentrations are inversely related to O_2 concentrations. This reciprocal relationship between H_2S and O_2 indicates the similarity between the effect of hypoxia and that of H_2S. Since H_2S is used by mitochondria to produce ATP along with O_2 consumption, mitochondria can be the site of O_2 sensing. Therefore, the O_2-dependent metabolism of H_2S may be an effective O_2-sensing mechanism. The determination of this mechanism at the subcellular level will be an additional evidence to strongly support this hypothesis.

In the chapter "Multiple Roles of H_2S in Inflammation: A New Class of Therapeutics," Philip Moore and colleagues describe the ability of H_2S to promote the resolution of an inflammatory response and the potential of H_2S-targeting drugs for the treatment of inflammation. They show the importance of H_2S donor concentrations and the timing of their application in an ongoing inflammatory response, which consists of a sequence of processes: initiation, sustaining, and resolving inflammation. Each phase of the inflammatory response includes blood vessel dilatation, adhesion and migration of leukocytes, edema and pain, and the target of H_2S changes depending on the phase of inflammation. The interaction between endogenous and exogenous H_2S is also an intriguing problem that needs to be solved.

In the chapter "Signaling Mechanisms Underlying the Hydrogen Sulfide Effects: Identification of Hydrogen Sulfide 'Receptor,'" Yi-Chun Zhu has summarized a mechanism for the structural modification of proteins to change their activity by H_2S by using a representative targeting molecule – vascular endothelial growth factor receptor (VEGFR), which induces angiogenesis by H_2S stimulation. H_2S breaks the cysteine disulfide bond in VEGFR, whereas S-sulfhydrated cysteine, which is shortly observed, is immediately attacked by a second HS^- and reduced to cysteine. Breaking the cysteine disulfide bond is a reducing reaction, whereas S-sulfhydration is an oxidizing reaction. Considering the fact that H_2S is a reducing molecule, the reduction of cysteine disulfide bond by H_2S is a plausible modification compared to S-sulfhydration.

John Wallace and his colleague in their chapter "Therapeutic Applications of Hydrogen Sulfide" describe the development of drugs, which release appropriate amounts of H_2S at the proper target for a desirable period, on the basis of their functional mechanism rather than their observed effects. H_2S-based experimental drugs for arthritis, inflammatory bowel disease, oxidative stress-induced injury, and even cancer chemoprevention have been developed, and some of them are already entering clinical trials. Companies that have developed H_2S-based compounds aiming at the therapy of various disorders are also outlined in this chapter.

Jin-Song Bian and his colleague describe the biological function of H_2S in both health and disease, with special emphasis on its protective effects on tissues or organs in cardiovascular, central nervous, and renal systems in the chapter entitled "Hydrogen Sulfide: Physiological and Pathophysiological Functions." Owing to the antioxidant, anti-inflammatory, and anti-apoptotic effects, manipulation of the H_2S system is implicated in the therapeutic benefits to patients suffering from various diseases. Research efforts have focused on the development of slow-releasing H_2S donors that mimic endogenous release of H_2S and the selective inhibitors of H_2S-producing enzymes.

Fumito Ichinose, in his chapter entitled "Biological Effects of H_2S Inhalation and Its Therapeutic Potential," describes the beneficial effect of H_2S inhalation in various pathological conditions. The induction of a suspended animation-like metabolic state characterized by reduced energy consumption and hypothermia was observed in rodents, but not in larger animals. The beneficial effects against inflammation, ischemia–reperfusion injury, neurodegenerative diseases, and acute lung injury have a different mechanism from the induction of animation-like state because these effects are observed without any accompanied reduction in metabolism. However, a clear demonstration of these effects is required in larger animals to promise their therapeutic potential. The beneficial effects of inhaled H_2S are critically reviewed in this chapter.

Evgeny Nudler and his colleague in their chapter "H_2S as a Bacterial Defense Against Antibiotics" describe that H_2S produced by bacterial orthologs of the mammalian enzymes CBS, CSE, and 3MST has a role in the defense system against antibiotics. Homologs of nitric oxide synthase exist in a limited number of gram-positive species, whereas those of H_2S-producing enzymes are widely conserved in most bacterial species. Comprehensive understanding of H_2S mechanism provides a new target for drug development against infectious diseases.

In the final chapter, "Hydrogen Sulfide-Mediated Cellular Signaling and Cytoprotection," Hideo Kimura describes H_2S-producing pathways, including a novel pathway with D-cysteine as the substrate. The D-cysteine pathway produces 80 times greater quantity of H_2S than that produced from L-cysteine in the kidney, and it effectively protects the organ from ischemia–reperfusion injury. It provides a new therapeutic approach to renal diseases. He also describes his recent finding that polysulfides, which are derived from H_2S, are found in the brain and activate transient receptor potential ankyrin 1 (TRPA1) channels in glia more potently than parental H_2S. Polysulfides may be responsible for some activities that have previously been ascribed to H_2S.

This book summarizes the recent progress in the study of H_2S, its functions in organisms ranging from bacteria to mammals, and its therapeutic applications. H_2S plays an important role in both physiological and pathophysiological conditions. Although the functional forms of H_2S, HS, and S^{2-} have not been determined, H_2S-derived polysulfides have been added as another active form of this molecule. By understanding the biochemical nature of these molecules, as well as their mechanisms of action, the physiological function and therapeutic potential of H_2S and related molecules will be unveiled.

Tokyo, Japan Hideo Kimura

Contents

Enzymology of Hydrogen Sulfide Turnover

1

Nicole Motl, Pramod Yadav, and Ruma Banerjee

Abstract

Hydrogen sulfide is a biological signaling molecule that is produced by organisms ranging from bacteria to man. Since it is also toxic at high concentrations, strategies exist for its efficient removal and consequent maintenance of low steady-state levels in mammals. Enzymes in the sulfur metabolic network responsible for H_2S biogenesis include the cytoplasmic cystathionine β-synthase and γ-cystathionase enzymes of the transsulfuration pathway as well mercaptopyruvate sulfurtransferase that is both mitochondrial and cytoplasmic. The precursors for H_2S are the amino acids cysteine and homocysteine and a cysteine derivative i.e. mercaptopyruvate. H_2S is cleared via a mitochondrial sulfide oxidation pathway that begins with sulfide quinone oxidoreductase and includes a persulfide dioxygenase, rhodanese and sulfite oxidase. The major oxidation products of H_2S are thiosulfate and sulfate. This chapter focuses on the structural enzymology of H_2S biogenesis and oxidation, emphasizing recent advances in the field.

Keywords

Hydrogen sulfide • Sulfide oxidation • Signaling • Enzymology and regulation

Nicole Motl and Pramod Yadav are equal contributors.
This work was supported in part by the National Institutes of Health (HL58984).

N. Motl • P. Yadav • R. Banerjee (✉)
Department of Biological Chemistry, University of Michigan Medical School, Ann Arbor, MI 48109-0600, USA
e-mail: rbanerje@umich.edu

H. Kimura (ed.), *Hydrogen Sulfide and its Therapeutic Applications*,
DOI 10.1007/978-3-7091-1550-3_1,

Abbreviations

AdoMet	S-adenosylmethionine
CAT	Cysteine aminotransferase
CBS	Cystathionine β-synthase
CSE	γ-cystathionase
ETHE1	Persulfide dioxygenase
H_2S	Hydrogen sulfide
MST	Mercaptopyruvate sulfurtransferase
SQR	Sulfide quinone oxidoreductase

1.1 Introduction to Sulfide Metabolism

Long known as a toxic gas, sulfide like cyanide, targets cellular respiration by reversible inhibition of cytochrome c oxidase. Cells therefore evolved strategies for handling the twin challenges of averting toxicity problems while exploiting the signaling potential of hydrogen sulfide (H_2S), which elicits profound physiological effects (Kimura 2010). Hence, a "safe" window must exist within which low intracellular concentrations of H_2S are maintained and allowed to transiently spike to allow passage of a signal. The concentration width of this window could span three orders of magnitude since steady-state intracellular concentrations of H_2S are estimated to be in the 15–30 nM range (Furne et al. 2008; Levitt et al. 2011; Vitvitsky et al. 2012) while mammalian mitochondrial ATP production is abolished at 50 μM H_2S (Bouillaud and Blachier 2011). The steady-state concentrations of H_2S is a product of both the metabolic flux through the sulfide biogenesis and sulfide oxidation pathways (Vitvitsky et al. 2012) (Fig. 1.1).

Three enzymes in the mammalian sulfur metabolic network catalyze H_2S biogenesis (Kabil and Banerjee 2010; Singh and Banerjee 2011). Two of these enzymes reside in the cytosolic transsulfuration pathway and are cystathionine β-synthase (CBS) and γ-cystathionase (CSE). Both are versatile and catalyze H_2S production in a variety of reactions starting from cysteine and/or homocysteine (Chen et al. 2004; Chiku et al. 2009; Singh et al. 2009). The third enzyme, mercaptopyruvate sulfur transferase (MST) (Nagahara et al. 1998; Shibuya et al. 2009), is both cytosolic and mitochondrial in location and transfers the sulfur atom from mercaptopyruvate to an acceptor from where it can be subsequently released as H_2S. MST works in conjunction with the PLP-dependent enzyme, aspartate/cysteine aminotransferase (CAT), which catalyzes the conversion of cysteine to 3-mercaptopyruvate. Recently, D-amino acid oxidase has been reported to convert D-cysteine to 3-mercaptopyruvate, providing an alternative supply route for the MST substrate (Shibuya et al. 2013).

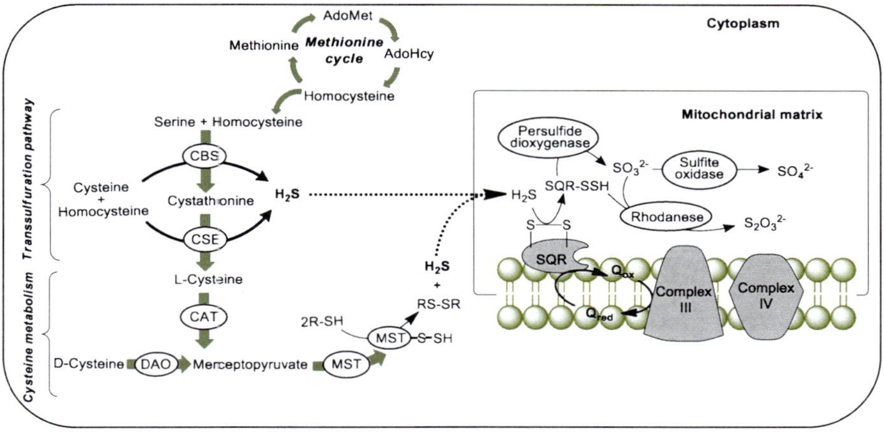

Fig. 1.1 Scheme showing pathways of biogenesis and catabolism of H_2S. DAO is D-amino acid oxidase

Sulfide oxidation occurs in the mitochondria and connects sulfur metabolism to the electron transfer chain and thereby, to both ATP and reactive oxygen species production (Fig.1.1). Half maximal inhibition of cytochrome *c* oxidase in cell extracts occurs at concentrations of ~0.3 μM versus ~20 μM H_2S needed to inhibit cellular respiration in intact cells (Bouillaud and Blachier 2011). To avoid intracellular sulfide build-up and consequent toxicity, the sulfide oxidation pathway is activated at considerably lower (~10–20 nM) sulfide concentrations (Bouillaud and Blachier 2011). The high sensitivity of the sulfide oxidation pathway to H_2S, suggests that the duration of H_2S-based signaling is likely to be short (Bouillaud and Blachier 2011). Sulfate and thiosulfate are the major products of the mammalian mitochondrial sulfide oxidation pathway, which comprises four enzymes: sulfide quinone oxidoreductase (SQR), a persulfide dioxygenase (ETHE1) that is the product of the *ethe1* gene, rhodanese, and sulfite oxidase (Hildebrandt and Grieshaber 2008). This chapter focuses on the structural enzymology and regulation of H_2S metabolism.

1.1.1 Enzymology of H_2S Biogenesis

1.1.1.1 Cystathionine β-Synthase

Structural Organization of CBS. CBS is a multidomainal protein, which in its role in the transsulfuration pathway, catalyzes the condensation of homocysteine and serine to produce cystathionine (Banerjee et al. 2003). Homocysteine is a redox-active nonprotein amino acid that is produced by hydrolysis of S-adenosylhomocysteine, a product of S-adenosylmethionine (AdoMet)-dependent methylation reactions. Homocysteine is either recycled in the methionine cycle via the action of methionine synthase or committed to cysteine synthesis by the action of CBS (Fig. 1.1).

Mutations in CBS are the most common cause of homocystinuria, an autosomal recessive disorder, characterized by severely elevated plasma homocysteine levels (Kraus et al. 1999). CBS deficiency affects multiple organ systems including the ocular, skeletal, cardiovascular and the central nervous system (Mudd et al. 1964).

In addition to the canonical transsulfuration reaction, CBS catalyzes H_2S-producing reactions in which serine is substituted by cysteine (Chen et al. 2004; Singh et al. 2009). These reactions involve β-replacement of cysteine by homocysteine, cysteine or water to generate cystathionine, lanthionine or serine, respectively in addition to the common product, H_2S.

CBS is a homodimeric enzyme. The full-length human enzyme is prone to aggregation and exists in multiple oligomeric states ranging from 2- to 16-mers (Sen and Banerjee 2007). The predominance of the 4-mer in the aggregated mixture has led to confusion in the literature about CBS being a homotetramer. Each monomer is organized into an N-terminal heme-binding domain, a central PLP-binding catalytic core and a C-terminal AdoMet-binding regulatory domain, which contains a tandem repeat of CBS domains. The latter, named after this protein, is found in all three domains of life and refers to a $\beta - \alpha - \beta - \beta - \alpha$ secondary structure motif often associated with energy sensing that binds ATP or AMP (Bateman 1997). In CBS, it binds the allosteric activator, AdoMet, which renders transsulfuration flux sensitive to cellular methyl donor status. Thus, when AdoMet levels are low, sulfur is spared and utilized via the methionine cycle to support AdoMet synthesis. In contrast, when AdoMet levels are plentiful, sulfur metabolism is directed towards cysteine synthesis via activation of CBS.

The crystal structures of full-length CBS from *Drosophila melanogaster* (dCBS, Fig. 1.2a) and a truncated variant of human CBS (hCBS) lacking the C-terminal regulatory domain have been reported (Meier et al. 2001; Taoka et al. 2002; Koutmos et al. 2010). The N terminal heme domain spans residues 1–70 in hCBS and 1–45 in dCBS while the middle PLP domain spans residues 71–411 in hCBS and 46–350 in dCBS. A 27 residue-long linker (351–377) visible in the structure of dCBS connects the PLP and C-terminal domains, which based on sequence alignment, is predicted to extend between residues 382–411 in hCBS.

Catalytic Mechanism of CBS. The active site of CBS is lined with residues that position the substrates and cofactor for catalysis. PLP is covalently linked to an active site lysine (K119 in hCBS and K88 in dCBS) and its phosphate moiety is enveloped by a glycine-rich loop (T_{260}-G_{259}-G_{258}-T_{257}-G_{256}, hCBS numbering) (Fig. 1.2b). N149 and S349 form hydrogen bonds with the exocyclic oxygen and pyridine nitrogen, respectively (Meier et al. 2001; Taoka et al. 2002). Replacement of the corresponding serine in yeast CBS (yCBS) by alanine results in complete loss of activity, while replacement with aspartate results in an ~80-fold lower activity despite the mutants having only two-fold lower PLP occupancy (Quazi and Aitken 2009). Fluorescence and resonance Raman studies demonstrate that the PLP exists in two tautomeric forms: an active ketoenamine and an inactive enolimine (Singh et al. 2009; Weeks et al. 2009; Yadav et al. 2012). N149 helps stabilize the active ketoenamine tautomer. Perturbations in the hydrogen-bonding network in the active site can influence the tautomeric equilibrium, and as described below, is

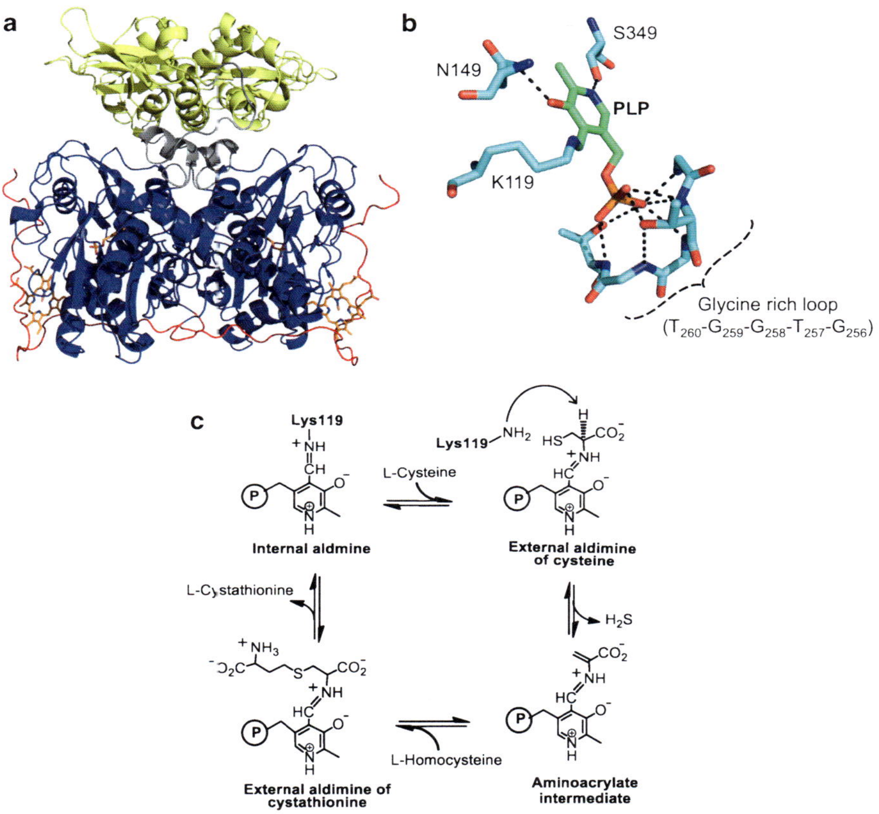

Fig. 1.2 Structure and catalytic mechanism of CBS. (**a**) Structure of full-length dCBS. The protein comprises an N-terminal heme-binding domain (*red*), a PLP domain (*blue*) and a C-terminal AdoMet-binding domain (*yellow*). A linker (*grey*) connects the PLP- and C-terminal domains. (**b**) Close-up of the active site of hCBS. Hydrogen bonds between PLP and amino acids lining the active site are shown as *dotted lines*. Figure 1.2a, b were generated using PDB files 3PC2 and 1 M54, respectively. (**c**) Proposed catalytic mechanism for H_2S generation from cysteine by CBS. The resting enzyme exists as an internal aldimine, which reacts with cysteine to form the external aldimine of cysteine. Elimination of H_2S leads to formation of an aminoacrylate intermediate, which reacts with homocysteine (cysteine or water) to form the external aldimine of the product cystathionine (or lanthionine or serine). The catalytic cycle is completed upon release of product and reformation of internal aldimine

one mechanism by which the heme cofactor exerts its effects on the active site (Singh et al. 2009; Yadav et al. 2012).

The first step in the catalytic cycle (Fig. 1.2c) is binding of serine or cysteine (Fig. 1.2c) to the active site followed by displacement of the lysine residue that forms an internal aldimine with PLP (Banerjee et al. 2003). Abstraction of the α-proton from the resulting external aldimine of serine/cysteine generates a carbanion intermediate. In the crystal structure of dCBS obtained at 1.7 Å resolution, the carbanion intermediate with serine was captured and revealed how the active site stabilizes this reactive species (Koutmos et al. 2010). The Cα atom is clearly sp^2 hybridized in this intermediate and the ε-amino group of K88 is within 2.1, 2.5 and 3.0 Å of the Cα, imino nitrogen of the Schiff base and C4A of PLP, respectively, where the negative charge is predominantly localized. Elimination of H_2O or H_2S from the external aldimine of serine or cysteine, respectively leads to formation of the aminoacrylate intermediate, which was also captured in a crystal structure of dCBS obtained at 1.55 Å resolution (Koutmos et al. 2010). In this structure, the Cα is also clearly sp^2 hybridized and the ε-amino group of K88, which is swung away from Cα, is parked near the oxygen atoms of the phosphate moiety of PLP. The Cβ of the aminoacrylate intermediate is positioned for nucleophilic attack by the thiolate of homocysteine (to generate cystathionine), or cysteine (to generate lanthionine) or by water (to generate serine). The final step preceding product release involves a second transchiffization reaction in which the active site lysine displaces the product and the enzyme returns to its resting internal aldimine state.

Kinetics of H_2S Generation by CBS. Detailed steady-state kinetic analyses of H_2S generation by human and yeast CBS have been reported (Singh et al. 2009). In addition to the canonical β-replacement reaction in the transsulfuration pathway (Eq. 1.1), CBS catalyzes at least three other reactions with a combination of cysteine and homocysteine as substrates, leading to H_2S production (Eqs. 1.2, 1.3, and 1.4). The CBS active site has two pockets for binding amino acids that

$$\text{Serine} + \text{Homocysteine} \rightarrow \text{Cystathionine} + H_2O \tag{1.1}$$

$$\text{Cysteine} + \text{Homocysteine} \rightarrow \text{Cystathionine} + H_2S \tag{1.2}$$

$$\text{Cysteine} \rightarrow \text{Serine} + H_2S \tag{1.3}$$

$$\text{Cysteine} + \text{Cysteine} \rightarrow \text{Lanthionine} + H_2S \tag{1.4}$$

are designated as sites 1 and 2. Site 1 binds the amino acid that forms the external aldimine with PLP while site 2, binds the nucleophilic amino acid (Singh et al. 2009). At saturating substrate concentrations, the specific activity for the β-replacement of cysteine by homocysteine (reaction 2) is ~4-fold higher than for the canonical β-replacement of serine by homocysteine (reaction 1). H_2S generation from one (reaction 3) or two (reaction 4) moles of cysteine results in serine or lanthionine and are ~40- and ~20-fold lower, respectively than H_2S generation from cysteine and homocysteine (Singh et al. 2009).

Within the cell, the efficiency of each CBS-catalyzed reaction is dictated in part by the relative concentrations of the individual substrates and the K_M values, which are high for both yeast and human CBS relative to the intracellular concentrations of their amino acid substrates. The K_M for cysteine for hCBS at site 1 is 6.8 ± 1.7 mM and higher still for site 2 (27.3 ± 3.7 mM). For yCBS, the K_M values for cysteine at sites 1 and 2 are 3.6 ± 1.7 mM and 33 ± 3.7 mM, respectively. Homocysteine only binds to site 2 and the K_Ms for yeast and human CBS are 0.13 ± 0.02 mM and

3.2 ± 1.3 mM, respectively (Singh et al. 2009). Simulations were performed using the steady-state kinetic data for hCBS at physiological concentrations of substrate (560 μM serine, 100 μM cysteine and 10 μM homocysteine). The simulations predicted that ~96 % of H_2S derived from hCBS is via β-replacement of cysteine by homocysteine with the remainder being contributed by the other two reactions (Singh et al. 2009).

Pre-steady state kinetic analysis and characterization of reaction intermediates was first reported for yCBS using stopped flow spectroscopy (Jhee et al. 2001; Taoka and Banerjee 2002; Singh et al. 2011). The heme in hCBS interferes with enzyme-monitored pre-steady state kinetic analysis by masking the PLP absorbance. Kinetic analysis of a heme-less hCBS variant lacking the N-terminal heme domain was limited by its poor stability (Evande et al. 2004). We have recently employed difference UV-visible stopped-flow spectroscopy to characterize intermediates in the hCBS-catalyzed reaction (Yadav and Banerjee 2012), an approach, that we used previously to demonstrate the intermediacy of an aminoacrylate intermediate in dCBS (Koutmos et al. 2010). The rate of aminoacrylate formation is ~2.5-fold faster with serine than with cysteine in the reaction catalyzed by hCBS and product release appears to limit the overall reaction rate (Yadav and Banerjee 2012).

Allosteric Regulation of CBS by Heme. The N-terminal heme domain in CBS is devoid of any secondary structure. The heme is hexa-coordinate and its histidine and cysteine axial ligands were predicted by EPR, extended X-ray absorption fine structure (Ojha et al. 2000) and resonance Raman (Green et al. 2001) spectroscopy. The UV-visible spectrum of ferric CBS exhibits a Soret peak at 428 nm and a broad α/β absorption band centered at 550 nm. In the ferrous state, the Soret peak shifts to 449 nm with concomitant sharpening of the α and β absorption bands at 571 and 540 nm (Taoka et al. 1998). Mutation of either heme ligand diminishes hCBS activity (by ~9-fold) despite full PLP saturation in the C52A and C52S mutants and 75 % saturation in the H65R mutant (Ojha et al. 2002). On the other hand, heme saturation is greatly reduced in the heme ligand mutants (19 % and 40 % in the cysteine and histidine ligand mutants, respectively) compared to wild-type CBS. The C52S or C52A mutants have five-coordinate high-spin heme with a Soret peak blue-shifted from 428 to 415–417 nm in the ferric form and from 449 to 423 nm in the ferrous form (Ojha et al. 2002). In the ferrous-CO state, the Soret peaks of the mutants (at ~422 nm) are virtually identical to that of wild-type enzyme, consistent with the substitution of the cysteine ligand by CO. The Soret peaks in the ferric, ferrous and ferrous-CO states in the H65R mutant are at 424, 421 and 420 nm, respectively, consistent with the presence of a low-spin heme (Ojha et al. 2002).

While the role of the heme domain in CBS has been debated, there is growing evidence that it regulates CBS activity in response to changes in the heme spin-or ligation-state (Taoka and Banerjee 2001; Taoka et al. 2001; Singh et al. 2007). The ferrous heme ligands CO and NO, inhibit CBS activity (Taoka et al. 1999; Taoka and Banerjee 2001). Ferrous-NO CBS is five-coordinate with a broad Soret peak at ~390 nm (Taoka and Banerjee 2001). CBS exhibits nonequivalent binding sites for CO with K_d values of 1.5 ± 0.1 μM and 68 ±14 μM for full-length hCBS and

3.9 ± 2 μM and 50 ± 8 μM for the C-terminal truncated form (Taoka et al. 1999; Puranik et al. 2006). Binding of NO to full-length CBS exhibits a K_d of 30 ± 5 μM for NO (Gherasim et al., unpublished results). Binding of CO to full-length CBS inhibits enzyme activity with a K_i value of 5.6 ± 1.9 μM (Taoka et al. 1999). The redox potential of the heme is −350 ± 4 mV (Singh et al. 2009) and −287 ± 2 mV (Carballal et al. 2008) in full-length and truncated hCBS, respectively. This difference between the full-length and truncated forms suggests that the regulatory C-terminal domain modulates the heme redox potential. We have recently demonstrated that despite the low reduction potential of the CBS heme, reversible regulation by CO can be achieved with physiologically relevant reductants (Kabil et al. 2011). The diflavin oxidoreductases, human methionine synthase reductase and novel reductase 1 reduce ferric CBS and the ferrous-CO species is formed in the presence of NADPH and CO.

Communication between the heme and PLP domains occurs via an α-helix whose N-terminal end leads in from the glycine rich loop harboring the conserved T257 and T260 residues that make contact with the phosphate moiety of PLP. At the C-terminal end of the helix, R266 is involved in a salt-bridge interaction with the heme ligand, C52. Changes in the heme coordination state e.g. by formation of the ferrous-CO species, is predicted to disrupt the salt bridge between C52 and R266, displacing the α-helix, which in turn is propagated to the PLP site shifting the tautomeric equilibrium towards the enolimine (Weeks et al. 2009). Interestingly, replacement of R266 by methionine, a mutation described in homocystinuric patients, also results in the predominance of the inactive enolimine. Similarly, substitutions at T257 and T260 in the PLP domain, stabilize the inactive enolimine tautomer leading to significant loss of CBS activity (Yadav et al. 2012). Interestingly, these CBS mutations impact the H_2S and cystathionine-producing versus H_2O and cystathionine-producing reactions unequally, suggesting that that these two activities can be differentially regulated (Yadav et al. 2012).

Allosteric Regulation of CBS by AdoMet. AdoMet binds to the C-terminal regulatory domain (Scott et al. 2004) and enhances CBS activity ~2–3-fold (Finkelstein et al. 1975). Truncation of the C-terminal domain in hCBS leads to loss of AdoMet-dependent regulation, increases CBS activity and also decreases its propensity for aggregation (Kery et al. 1998). Hence, the regulatory domain exerts an autoinhibitory effect that is alleviated upon AdoMet binding, activating mutations or deletion of the entire domain (Shan and Kruger 1998; Janosik et al. 2001; Evande et al. 2002). The architecture of the regulatory domain and its juxtaposition relative to the catalytic domain was first visualized in the structure of full-length dCBS (Fig. 1.2a). The secondary structures of the two CBS domains are slightly different: β-α-β-β-α-α-β fold in one (spanning residues 416–468 in hCBS) and α-β-α-β-β-α in the second (residues 486–543). The AdoMet binding sites in CBS is predicted to reside in the β-sheet-lined cleft between the two CBS domains.

Other Mechanisms for Regulation of CBS Activity. As a junction enzyme in sulfur metabolism, CBS is the locus of complex regulation. AdoMet is an allosteric regulator, which activates CBS under conditions of methyl group sufficiency (Prudova et al. 2006; Pey et al. 2013). As noted above, gaseous signaling molecules like NO and CO bind to the heme in CBS and inhibits its activity (Taoka et al. 1999; Taoka and Banerjee 2001; Agarwal and Banerjee 2008; Yamamoto et al. 2011; Yadav et al. 2012), which is reversed upon air oxidation (Kabil et al. 2011b). Sumoylation of CBS decreases its activity (Agarwal and Banerjee 2008). K211 in the catalytic core of human CBS appears to be the site of sumoylation. In addition, the C-terminal regulatory domain of CBS is needed for sumoylation. CBS is reportedly inhibited by lanthionine synthase C-like protein 1 (LanCL1) in the presence of glutathione (Zhong et al. 2012). Under oxidative stress conditions, when oxidized glutathione levels rise, inhibition by LanCL1 is alleviated providing a mechanism for increasing transsulfuration flux and consequently, glutathione synthesis.

1.1.1.2 Cystathionine γ Lyase

Structural Organization of CSE. Human CSE (hCSE) is a homotetrameric enzyme that catalyzes the second step in the transulfuration pathway cleaving cystathionine to cysteine, α-ketobutyrate, and ammonia (Fig. 1.3a). The crystal structures of yeast (yCSE) and hCSE are available at 2.6 Å resolution each (Messerschmidt et al. 2003; Sun et al. 2009). Each hCSE monomer comprises a large N-terminal PLP-binding domain (residues 1–263) and a smaller C-terminal domain (residues 264–401) (Fig. 1.3a). The yCSE monomer comprises three domains: an N-terminal domain, which interacts with the active site of a neighboring monomer, a middle catalytic domain and a small C-terminal domain. The yCSE catalytic and C-terminal domains are organized as in hCSE while the N-terminal extension interacts with the active site of the neighboring subunit, is not present in hCSE and comprises an extended loop, an α-helix and a β-strand (Messerschmidt et al. 2003).

Mutations in CSE, inherited as an autosomal recessive disorder, results in cystathioninuria, which is often benign (Wang and Hegele 2003). Cystathioninuria can be secondarily associated with hepatoblastoma, neuroblastoma, poor development, cystic fibrosis and Down's syndrome.

Catalytic Mechanism of CSE. Like CBS, CSE also catalyze various reactions leading to H_2S biogenesis in addition to catalyzing the canonical reaction in the transsulfuration pathway (Chiku et al. 2009; Singh et al. 2009). The PLP is covalently linked to K212 in hCSE via a Schiff base and its mutation to alanine reduces H_2S production ~80-fold compared to wild-type enzyme. Several active site residues are engaged in hydrogen bonding interactions with the PLP (Messerschmidt et al. 2003; Sun et al. 2009); N187 with the pyridine nitrogen and G90, L91, S209 and T211 from one subunit and Y60 and R62 from an adjacent subunit with the phosphate moiety (Fig. 1.3b). Replacement of N187 with alanine or glutamic acid results in complete loss of H_2S production while substitutions of S209 and T211 by alanine result in a modest ~1.5-fold decrease in H_2S production (Huang et al. 2010). Interactions between Y60 and R62 contributed by the N-terminal domain of a neighboring monomer and the phosphate group of PLP helps to stabilize the active site, which is located at the interface between adjacent subunits. Multiple sequence alignment of human,

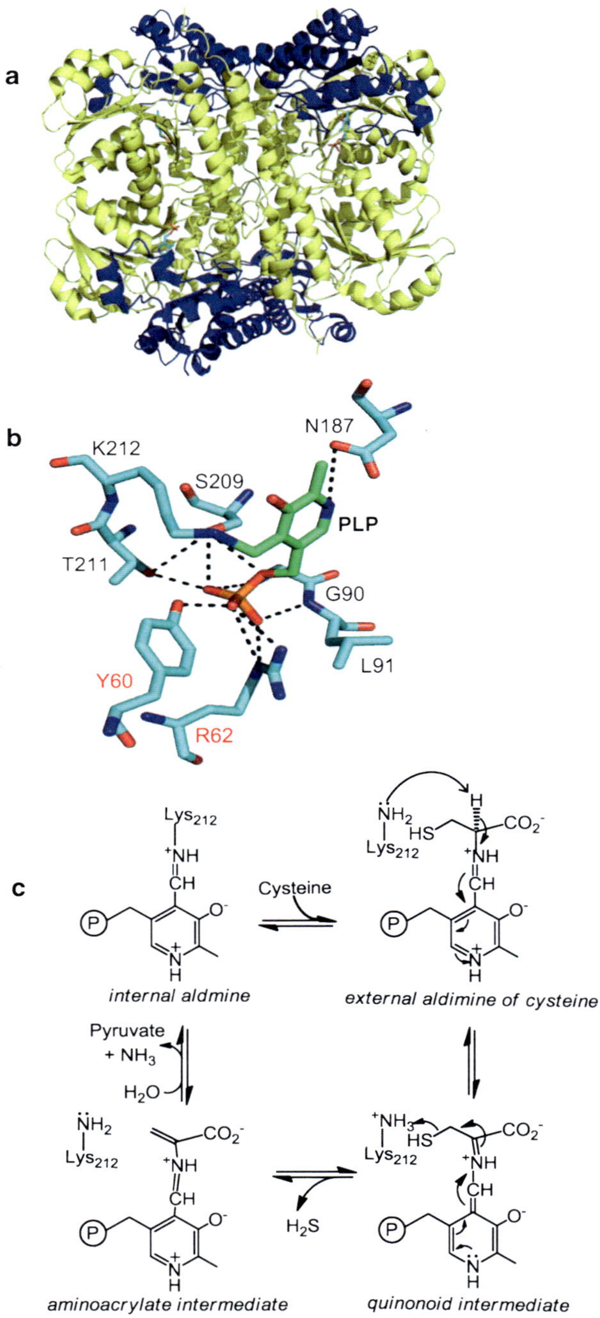

Fig. 1.3 Structure and catalytic mechanism of CSE. (**a**) Structures of tetrameric hCSE comprising an N-terminal catalytic domain (*yellow*) and a C-terminal domain (*blue*). PLP is shown in stick representation (*cyan*). (**b**) Close-up of the active site of hCSE. Hydrogen bonds between PLP and amino acids lining the active site are shown as *dotted lines*. The *red labels* denote residues contributed by a neighboring subunit. Figure 1.3a, b were generated using PDB file 2NMP. (**c**) Proposed catalytic cycle of CSE for H_2S generation. Cysteine reacts with the resting enzyme and forms external aldimine of cysteine. Abstraction of the α-proton of bound cysteine leads to formation of cysteine-ketamine intermediate. Cleavage of the C–β–S bond eliminates H_2S and leads to formation of the aminoacrylate intermediate, which undergoes hydrolysis to give pyruvate and ammonia. In the final step, free PLP rebinds with the K212 to regenerate the resting internal aldimine

yeast, mouse, rat and slime mold CSEs reveals that Y60 and R62 are conserved in all five sequences (Huang et al. 2010). Substitution of Y60 with threonine or alanine results in an ~5–8-fold decrease in hCSE activity, while replacement of R62 with alanine or lysine results in an ~10–36-fold decrease in activity (Huang et al. 2010).

CSE belongs to the γ-family of PLP-dependent enzymes, which catalyze elimination reactions at the γ-carbon. However, in the case of CSE, the specificity is not high and the enzyme catalyzes reactions at both the β- and γ-carbons of the substrate. In the first step, the substrate (i.e. cystathionine or cysteine) forms a Schiff base with the PLP via a transaldimination reaction, freeing the active site K212 residue. In the next step, K212 presumably acts as the general base and abstracts the α-proton from bound substrate. When cystathionine is the substrate, cleavage of the C-γ-S bond is promoted by a second proton abstraction from Cβ resulting in the subsequent elimination of cysteine. Hydrolysis of the resulting imine intermediate yields α-ketobutyrate and ammonia. Alternatively, when cysteine is the substrate, the C-β-S bond is cleaved, releasing H_2S (Fig. 1.3c). A second transaldimination reaction regenerates the resting internal aldimine. UV-visible spectroscopy based pre-steady state kinetic analysis of the reaction catalyzed by yCSE suggests that product release constitutes the rate-limiting step (Yamagata et al. 2003).

The specificity of the hCSE-catalyzed α,γ-elimination versus α,β-elimination is proposed to be governed by the hydrophobicity of the residue at position 339 (Messerschmidt et al. 2003; Huang et al. 2010). This hypothesis was tested by replacing E339 with lysine, alanine and tyrosine, which increases hydrophobicity in the following order: Y>A>K>E. E339K, E339A and E339Y show approximately 1.8-, 3.2- and 7.2-fold increase, respectively in the catalytic efficiency of H_2S production from cysteine as compared to wild-type enzyme (Huang et al. 2010), consistent with the view that enhancing the hydrophobicity of the residue at position 339 in hCSE favors the α,β-elimination reaction.

Kinetics of H_2S Generation by CSE. Detailed steady-state kinetic analysis of hCSE revealed that in addition to the canonical cystathionine cleavage reaction in the transsulfuration pathway (Eq. 1.5), the enzyme can catalyze five distinct H_2S-generating reactions in the presence of cysteine and/or homocysteine (Eqs. 1.6, 1.7, 1.8, 1.9, and 1.10) (Chiku et al. 2009).

$$\text{L-Cystathionine} + H_2O \rightarrow \text{L-cysteine} + \alpha\text{-Ketobutyrate} + NH_3 \tag{1.5}$$

$$\text{L-Cysteine} + H_2O \rightarrow \text{L-Pyruvate} + NH_3 + H_2S \tag{1.6}$$

$$\text{L-Cysteine} + \text{L-Cysteine} \rightarrow \text{L-Lanthionine} + H_2S \tag{1.7}$$

$$\text{L-Homocysteine} + H_2O \rightarrow \text{L-Homoserine} + H_2S \quad (1.8)$$

$$\text{L-Homocysteine} + \text{L-Homocysteine} \rightarrow \text{L-Homolanthionine} + H_2S \quad (1.9)$$

$$\text{L-Homocysteine} + \text{Cysteine} \rightarrow \text{L-cystathionine} + H_2S \quad (1.10)$$

Unlike hCSE (Chiku et al. 2009), rat CSE reportedly utilizes cystine, the oxidized disulfide form of cysteine, as a substrate for H_2S production (Stipanuk and Beck 1982). However, the availability of cystine in the reducing cellular milieu is questionable and the immediate product of the reaction is cysteine persulfide, which is highly unstable. Under maximal velocity conditions, the highest rate of H_2S production is observed for reaction 1.9, i.e. γ-replacement of homocysteine by a second mole of homocysteine while the lowest rate is observed for reaction 1.6, i.e. α,β−elimination of cysteine (Chiku et al. 2009). At physiological substrate concentrations, the highest rate of H_2S production is predicted to occur via the α,β−elimination reaction of cysteine and the lowest rate from the β−replacement of cysteine by a second mole of cysteine (reaction 1.7).

CSE exhibits two substrate-binding sites: site 1 at which the Schiff base is formed between PLP and an amino acid and site 2 where the nucleophilic second amino acid binds. The K_M for cysteine at site 1 (1.7 ± 0.7 mM) is ~1.6-fold lower than for homocysteine (2.7 ± 0.8 mM) while for the K_M for homocysteine at site 2 (5.9 ± 1.2 mM) is ~6-fold lower than for cysteine (33 ± 8 mM). The lower K_M for cysteine versus homocysteine at site 1 together with the higher cellular concentration of cysteine (~100 μM) versus homocysteine (<10 μM), explains the predominance of reaction 1.6 versus 1.9 at physiologically relevant substrate concentrations (Chiku et al. 2009).

The relative contributions of the various CSE-catalyzed H_2S producing reactions has been estimated at physiologically relevant substrate concentrations (5 μM cystathionine, 100 μM cysteine and 10 μM homocysteine) and low (10 μM), moderate (40 μM) and severe (100 μM) hyperhomocysteinemia (Chiku et al. 2009). Simulations predict that under normal and hyperhomocysteinemic conditions, ~87–99.5 % of the total H_2S is derived via α,β-elimination of cysteine and α,γ-elimination reaction of homocysteine, while the remaining three reactions collectively contribute very little. Between the two major H_2S contributing reactions, the α,β-elimination of cysteine predominates (~71 %) at normal homocysteine concentrations while the α,γ-elimination of homocysteine is the major H_2S producer accounting for ~61 % and ~78 % H_2S, respectively under moderate and severe hyperhomocysteinemic conditions (Chiku et al. 2009).

Preliminary pre-steady state kinetic analyses of the yCSE-catalyzed reaction were performed at 5 °C and 30 °C (Yamagata et al. 2003). An aminocrotonate intermediate was detected ($\lambda_{max} = 480$ nm) upon rapid mixing of yCSE with cystathionine but not with L-cysteine. Conversion of the aminocrotonate intermediate to products represents the rate-limiting step in the CSE-catalyzed cleavage of csytathionine (Yamagata et al. 2003).

Regulation of CSE. Unlike CBS, mechanisms for regulating CSE activity are not well understood. H_2S formation by CSE was reported to be upregulated by calmodulin in the presence of 2 mM Ca^{2+} (Yang et al. 2008). However, Ca^{2+}/calmodulin-dependent regulation of purified hCSE has not been observed in our laboratory (Padovani and Banerjee, unpublished results). A recent study on rat CSE supports our observations (Mikami et al. 2013) since the rate of H_2S production was low in the presence of 2 mM cysteine and 0–100 nM Ca^{2+} and increased five-fold when 50 μM PLP was added to the reaction mixture. At higher Ca^{2+} concentrations (0.3–3.0 μM), the rate of H_2S production was ~2.5-fold lower in the presence or absence of PLP compared to that at lower Ca^{2+} concentrations (less than 0.3 μM). Calmodulin (1 μM) had no effect on H_2S production by CSE. These findings suggest that CSE is not regulated by calmodulin and that low Ca^{2+} enhances H_2S production but only in the presence of exogenous PLP while high Ca^{2+} concentrations (300 nM to 3 μM) inhibit CSE (Mikami et al. 2013). While CSE can be sumoylated in vitro (Agarwal and Banerjee 2008), the physiological relevance of this modification is not known. Mammalian CSE has two conserved CXXC motifs. However, their involvement in redox-dependent regulation of CSE activity is not known.

1.1.1.3 Mercaptopyruvate Sulfurtransferase

Structural Organization of MST. MST functions on a catabolic arm of cysteine metabolism and acts downstream of CAT to produce H_2S (Meister et al. 1954; Stipanuk and Beck 1982). CAT, which is identical to aspartate aminotransferase, catalyzes the conversion of cysteine in the presence of α-ketoglutarate to 3-mercaptopyruvate and ammonia. 3-Mercaptopyruvate is a substrate for MST, which transfers the sulfur group to a catalytic cysteine residue forming an enzyme-bound persulfide and releasing pyruvate. In the second half reaction, the sulfane sulfur is transferred to a thiol acceptor e.g. cysteine, homocysteine, dihydrolipoic acid, GSH or thioredoxin and subsequently released as H_2S (Nagahara et al. 2007; Mikami et al. 2011a; Yadav et al. 2013). The MST-bound persulfide can also be transferred to non-thiol acceptors like KCN to form thiocyanate (Nagahara et al. 1999). Recently an alternative to the transamination pathway has been reported for the production of 3-mercaptopyruvate from D-cysteine in a reaction catalyzed by D-amino acid oxidase (Shibuya et al. 2013). The latter, is expressed in multiple tissues and is most abundant in the cerebellum and kidney (Shibuya et al. 2013). Defects in the MST gene are inherited as an autosomal recessive disorder known as mercaptolactate-cysteine disulfiduria (Crawhall et al. 1968). The condition is characterized by excessive excretion of mercaptolactate-cysteine disulfide in the urine, with or without mental retardation. Mercaptopyruvate is oxidized by lactate dehydrogenase to mercaptolactate and reacts subsequent to export, with extracellular cysteine to form the mixed disulfide of mercaptolactate and cysteine (Nagahara and Sawada 2006).

The crystal structures of MST from *E. coli* and *Leishmania major* have been solved at 2.8 Å and 2.1 Å resolution, respectively (Alphey et al. 2003; Spallarossa et al. 2004). Recently, our laboratory has obtained the structure of the product complex of hMST at 2.15 Å resolution with pyruvate and persulfide bound at the active site (Yadav et al. 2013). Both bacterial and human MST comprise two domains connected by a linker, which are structurally similar to the rhodanese domain (Fig. 1.4a). The *Leishmania* MST has an extra ~80 amino acids long domain that shares structural homology with the immunosuppressant FK506-binding protein and to macrophage infectivity potentiator protein, both of which exhibit peptidyl prolyl cis-trans isomerase activity (Alphey et al. 2003). Expression of a C-terminally truncated *Leishmania* MST yields a misfolded protein devoid of catalytic activity (Williams et al. 2003). The C-terminal domain in the *L. major* MST is postulated to be involved in protein folding and protein–protein interactions (Alphey et al. 2003).

Catalytic Mechanism of MST. A conserved cysteine residue (C248 in hMST) plays a key role in the catalytic mechanism of this enzyme (Fig. 1.4b). All three available MST structures exhibit persulfides at the active site cysteine residue, which represents the product of the MST-catalyzed sulfur transfer from 3-mercaptopyruvate. Substitution of the corresponding active site cysteine with serine in rat MST (rMST), results in complete loss of activity (Nagahara and Nishino 1996). Two conserved arginine residues (R188 and R197 in hMST) (Fig. 1.4b) are proposed to be important for proper positioning of the substrate and their replacement with glycine is deleterious for rMST activity. The R187G and R196G mutations increase K_M for 3-mercaptopyruvate by 10- and 60-fold, respectively and decrease k_{cat} 4- and 870-fold, respectively. Based on the *Leishmania* MST crystal structure, a serine protease-like catalytic triad, comprising Ser255-His75-Asp61 was predicted to be a common feature of the MST family. The catalytic triad is proposed to play a role in polarizing the carbonyl bond in the substrate to assist in the nucleophilic attack by the active site cysteine (Alphey et al. 2003). The crystal structure of hMST confirms the presence of a serine protease-like catalytic triad (S250-H74-D63) in the active site (Yadav et al. 2013).

Enzymes belonging to the sulfurtransferase family catalyze the transfer of a sulfur atom from a sulfur donor to a nucleophilic acceptor. Based on steady-state kinetic analysis of bovine MST, the enzyme was proposed to use a sequential mechanism (Jarabak and Westley 1978). Based on the structure of hMST containing a mixture of the product complex (Cys248-SSH and pyruvate) and an unproductive intermediate (3-mercaptopyruvate in a disulfide linkage with Cys248), we have proposed a detailed reaction mechanism (Fig. 1.4c) (Yadav et al. 2013). The conserved arginine residues, R188 and R197, interact via hydrogen bonds with the carboxyl group of mercaptopyruvate while R197 also hydrogen bonds with the carbonyl group of the substrate. H74 in the catalytic triad is proposed to function as a general base, abstracting a proton from S250, which in turn deprotonates C248 for subsequent attack on the sulfur of 3-mercaptopyruvate, leading to transfer of the sulfur atom to form Cys248-persulfide and pyruvate. Following release of pyruvate, an acceptor molecule (e.g. a dithiol or thioredoxin) attacks the sulfane sulfur of the cysteine persulfide regenerating MST and forming a new persulfide on the acceptor. Nucleophilic attack by the second (or resolving cysteine) on the acceptor releases

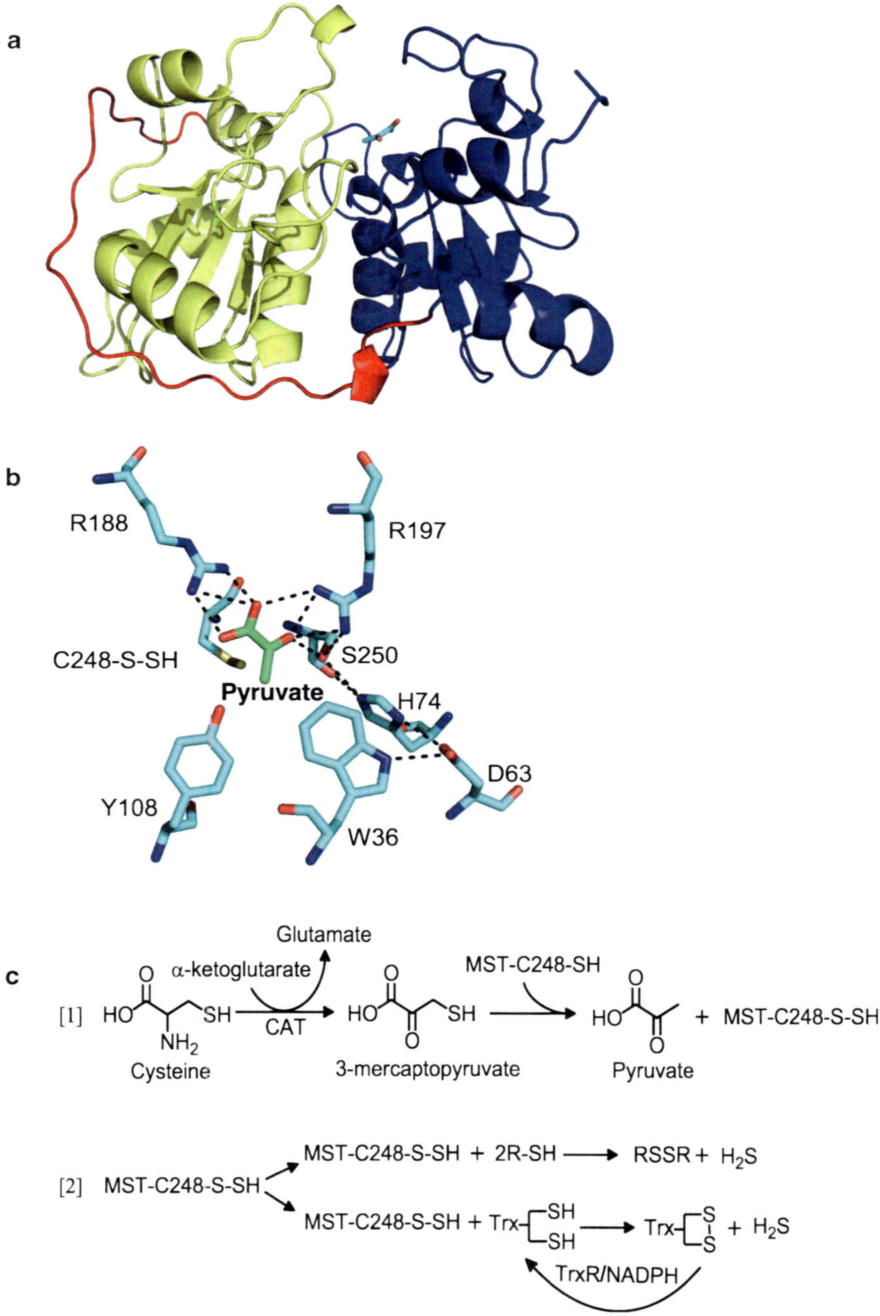

Fig. 1.4 Structure and catalytic mechanism of MST. (**a**) Structure of human MST, which comprises an N-terminal domain (*yellow*) connected to the catalytic domain (*blue*) by a linker (*red*). (**b**) Close-up of the active site of hMST showing a bound persulfide on Cys248 and pyruvate (hydrogen bonds are represented by *dotted lines*). Figure 1.4a, b were generated using PDB file

H_2S and results in oxidized disulfide product. Sulfur transfer from mercaptopyruvate to cyanide generates thiocyanate and pyruvate (Nagahara et al. 1999).

Dihydrolipic acid, thioredoxin, cysteine, homocysteine and glutathione function as acceptors in the in vitro MST assay and a detailed steady state kinetics analysis of H_2S production by MST in the presence of these acceptors have been reported (Yadav et al. 2013). At pH 7.4 and 37 °C kinetic studies revealed that, in the presence of thioredoxin or DHLA, hMST exhibits the highest k_{cat}/K_M values are obtained in the presence of human thioredoxin (520,000 M^{-1} s^{-1}) and dihydrolipoic acid (390 M^{-1} s^{-1}), while the lowest value (12.0 M^{-1} s^{-1}) was obtained with glutathione. Based on kinetic simulation at physiologically relevant concentrations of acceptors, thioredoxin is predicted to couple most efficiently to the MST reaction (Yadav et al. 2013).

Regulation of MST. Regulation of the MST/CAT pathway is poorly understood. Ca^{2+}-dependent regulation of MST has been reported based on the effect of varying Ca^{2+} concentrations on MST/CAT-dependent H_2S production in mouse retinal lysate (Mikami et al. 2011b). H_2S production decreased as Ca^{2+} concentration increased (0–2.9 μM) in the presence of cysteine and α-ketoglutarate (substrates for the CAT/MST pathway) but not when 3-mercaptopyruvate was used, suggesting that Ca^{2+} regulates CAT (Mikami et al. 2011b). Calmodulin is not involved in regulation of MST/CAT-dependent H_2S production (Mikami et al. 2011b). Rat MST has five cysteines and appears to be redox regulated. Of the five cysteines, three (i.e. C154, C247 & C263) are surface exposed (Nagahara and Katayama 2005; Nagahara 2012). An intersubunit disulfide bond forms between C154 and C263 under oxidizing conditions and can be reduced by thioredoxin. Thioredoxin-reduced MST is ~4.6-fold more active than the oxidized form of MST, while pretreatment of MST with DTT results in lower activation (~2.3-fold) (Nagahara et al. 2007). Human MST is a monomer and the cysteine residues that form the intersubunit disulfide in rMST are not conserved in hMST (Yadav et al. 2013). The active site cysteine (i.e. C247 in rMST) acts as another redox-sensitive switch with the potential to regulate MST (Nagahara and Katayama 2005), Treatment of rMST with stoichiometric oxidants (H_2O_2 or tetrathionate) results in inhibition due to formation of cysteine sulfenate at the active site, which can be reversed by reductants such as DTT or thioredoxin (Nagahara and Katayama 2005).

1.1.1.4 The Relative Contributions of CBS, CSE and MST to H_2S Production

It is not readily possible from the available kinetic data collected under varied buffer, pH and temperature conditions, to assess the relative roles of CBS, CSE and MST to H_2S production in different tissues. As a first step towards addressing this question,

Fig. 1.4 (Continued) 4JGT. (**c**) Reaction scheme for CAT/MST-dependent H_2S generation. (*1*) CAT catalyzes the transamination between cysteine and α-ketoglutarate to generate mercapto-pyruvate and glutamate. (*2*) MST catalyzes the sulfur transfer from mercaptopyruvate to an active site cysteine, giving pyruvate and MST-bound persulfide. The latter reacts with thiols or thioredoxin (in the presence of NADPH and thioredoxin reductase) to generate H_2S

our laboratory has initiated kinetic studies in a limited set of tissues (murine liver, kidney and brain) at pH 7.4 and at 37 °C (Kabil et al. 2011; Vitvitsky et al. 2012). A second key piece of information that is needed to evaluate the contributions of the individual enzymes, is their concentrations in a given tissue, which can be obtained using quantitative Western blot analysis (Kabil et al. 2011). Finally, it is essential that sensitive and reliable methods for H_2S detection be used to monitor its formation at physiologically relevant substrate concentrations. Initial studies in our laboratory evaluating total H_2S production in murine tissues indicates that in the presence of 100 μM cysteine, liver exhibits the highest rate of H_2S production (484 ± 271 μmole h^{-1} kg^{-1} tissue) followed by kidney (104 ± 44 μmole h^{-1} kg^{-1} tissue at 0.5 mM cysteine) and then brain (29 ± 7 μmole h^{-1} kg^{-1} tissue) (Kabil et al. 2011). Quantitative Western blot analyses suggest that the expression level of all three H_2S-producing enzymes decreases in the following order: liver>kidney>brain (Kabil et al. 2011; Yadav and Banerjee, unpublished results).

The relative contributions of CBS and CSE to H_2S production at physiologically relevant substrates concentrations (560 μM serine, 100 μM cysteine and varying homocysteine ranging from its normal concentration (10 μM), to those seen under mild (40 μM) and severe (200 μM) hyperhomocysteinemia conditions) have been assessed. The kinetic simulations predict that CSE is the major H_2S producer and accounts for ~97 % of H_2S production in liver at 10 μM homocysteine, with the proportion only increasing under hyperhomocysteinemic conditions (Kabil et al. 2011). A comparable analysis including the contribution of MST is needed to provide a more complete picture of the quantitative significance of CBS, CSE and MST to H_2S production in different tissues. To assess the contribution of the MST/CAT pathway to cysteine-derived H_2S production, the assays need to be conducted in the presence of physiological concentrations of α-ketoglutarate needed by CAT to convert cysteine to 3-mercaptopyruvate.

1.1.2 Enzymology of H_2S Oxidation

1.1.2.1 Microbial Strategies for Sulfide Oxidation

A major reaction of the global sulfur cycle is the oxidation of hydrogen sulfide to sulfate. It is instructive to examine the variations in sulfide oxidation strategies used by microbes where the pathways are much better understood than in man. In microbes, inorganic sulfur compounds such as sulfide, sulfur globules, sulfite, thiosulfate and polythionates are oxidized to sulfate for generation of ATP.

Sulfide and Thiosulfate Oxidation by Phototrophic Bacteria. Green and purple sulfur bacteria utilize both sulfide and thiosulfate as electron donors for photoautotrophic growth (Gregersen et al. 2011; Grimm et al. 2011). Oxidation occurs in a stepwise manner with sulfide being converted initially to elemental sulfur, which is stored as sulfur globules either in the periplasm or on the surface of the outer membrane and utilized when sulfide is limiting. The two major enzymes directly involved in sulfide oxidation are SQR (Eq. 1.11) and flavocytochrome c sulfide dehydrogenase (Eq. 1.12), a heterodimeric flavoprotein comprising a glutathione reductase-like subunit containing an FAD cofactor and a redox-active disulfide and a diheme cyctochrome c subunit (Chen et al. 1994). Flavocytochrome c sulfide dehydrogenase exists in both soluble and membrane bound forms and, like SQR, catalyzes the oxidation of sulfide to polysulfide while reducing the diheme cytochrome via FAD (Chen et al. 1994).

$$S^{2-} + \mathrm{Ubiquinone_{ox}} \rightarrow S^0 + \mathrm{Ubiquinone_{red}} \qquad (1.11)$$

$$S^{2-} + \mathrm{cyt\, c_{ox}} \rightarrow S^0 + \mathrm{cyt\, c_{red}} \qquad (1.12)$$

Sulfur globules are converted to sulfite via the cytoplasmic dissimilatory sulfite reductase pathway (Holkenbrink et al. 2011). The proteins common to bacteria harboring the dissimilatory sulfite reductase pathway include a siroheme or siroamide-containing sulfite reductase, a transmembrane electron transporting protein complex, a putative siroheme amidase and the product of the *dsr*C gene. In addition, a sulfurtransferase complex and an iron-sulfur cluster-containing NADH oxidoreductase, which is proposed to reduce sulfur polysulfides, are also present in some bacteria. The transmembrane complex shuttles electrons released during oxidation of H_2S to sulfite to quinones (Gregersen et al. 2011; Holkenbrink et al. 2011). Sulfite is toxic and is further oxidized to sulfate by one of two pathways. The first is direct oxidation catalyzed by sulfite oxidase (Eq. 1.13) and the second involves oxidation of sulfite to adenosine-5′-phosphosulfate and its subsequent conversion to sulfate liberating either ADP or ATP as the co-product (Eqs. 1.14 and 1.15) (Kappler and Dahl 2001).

$$SO_3{}^{2-} + H_2O \rightarrow SO_4{}^{2-} + 2e^- + 2H^+ \qquad (1.13)$$

$$SO_3{}^{2-} + \mathrm{AMP} \rightarrow \mathrm{APS} + 2e^- \qquad (1.14)$$

$$\mathrm{APS} + \mathrm{PPi\,(or\ Pi)} \rightarrow SO_4{}^{2-} + \mathrm{ATP\,(or\ ADP)} \qquad (1.15)$$

Thiosulfate is oxidized to sulfate by the multi-enzyme sulfur oxidizing (Sox) system (Eq. 1.16). The Sox system is widely distributed among phototrophic and sulfur oxidizing bacteria and also present in some green and purple bacteria

$$^{-}SSO_3 + 5\,H_2O \rightarrow 2SO_4{}^{2-} + 8e^- + 10H^+ \qquad (1.16)$$

(Friedrich et al. 2001; Gregersen et al. 2011; Grimm et al. 2011). It is localized in the periplasm and comprises four proteins: SoxAX, SoxYZ, SoxB and SoxCD. In the first step, thiosulfate is oxidized to a cysteinyl S-thiosulfonate intermediate bound to SoxY, which is then hydrolyzed to sulfate by a dimanganese thiosulfohydrolase, SoxB. The remaining protein-bound sulfane sulfur is oxidized to sulfate by SoxCD, a heterotetramer containing a molybdenum cofactor-containing subunit and a diheme cytochrome c-containing subunit. Electrons released during thiosulfate oxidation are transferred to cytochromes and from there, to the electron transport chain. Green and purple sulfur bacteria do not have a SoxCD component in their Sox system, and are unable to completely oxidize thiosulfate using the Sox system (Sakurai et al. 2010; Gregersen et al. 2011; Grimm et al. 2011). In these bacteria, the remaining sulfane sulfur in the SoxYZ complex is proposed to either be transferred to the bacterial sulfur globule pool or, to be oxidized by the dissimilatory sulfite reductase pathway to sulfite.

Sulfur Oxidation by Acidophilic Bacteria. Acidophilic bacteria such as *Acidithiobacillus ferrooxidans* and *Acidithiobacillus thiooxidans*, utilize a sulfur dioxygenase to oxidize elemental sulfur to sulfite (Eq. 1.17) (Rohwerder and Sand 2003). Thiol-containing proteins are predicted to be involved in mobilization of extracellular octameric elemental sulfur and presentation of sulfane sulfur to the

$$^{1}/_{8}\,S_8 + O_2 + H_2O \rightarrow SO_3^{2-} + 2H^+ \tag{1.17}$$

periplasmic sulfur dioxygenase, which does not accept sulfide or elemental sulfur as substrate (Rohwerder and Sand 2003). The product of this reaction, sulfite is oxidized to sulfate by sulfite oxidase. Two other enzymes, SQR and thiosulfate quinone oxidoreductase are also active in sulfur oxidation in acidophilic bacteria. SQR oxidizes sulfide to elemental sulfur which can react abiotically with sulfite to produce thiosulfate (Rohwerder and Sand 2007). Thiosulfate can be oxidized to tetrathionate by thiosulfate quinone oxidoreductase with the electrons being transferred to quinones.

Sulfur Oxidation by Archaea. Sulfur-oxidizing Achaea use a variation of the oxidation schemes discussed above. Thus, in *Acidianus ambivalens*, sulfur oxidation is initiated by sulfur oxygenase reductase a non-heme iron protein that catalyzes the disproportionation of elemental sulfur and/or polysulfide to sulfide and sulfite (Kletzin 1992; Kletzin et al. 2004; Urich et al. 2006). H_2S is oxidized by SQR to elemental sulfur, which can be utilized by sulfur oxygenase reductase setting up an energy-yielding cycle between the two enzymes and allowing the organism to maximize the energy gained from sulfur compounds (Brito et al. 2009). Sulfite can be further oxidized to sulfate by sulfite oxidase (Kletzin et al. 2004; Brito et al. 2009) or it can react with elemental sulfur to form thiosulfate, which is oxidized to tetrathionate. Alternatively, sulfite can be oxidized to sulfate via the APS pathway (Zimmermann et al. 1999).

In the next section, we review the literature on the mitochondrial sulfide oxidation pathway focusing on the enzymes, SQR, the persulfide dioxygenase (ETHE1) and rhodanese. The conversion of sulfide to sulfite and thiosulfate by this trio of enzymes most closely resembles the components of the sulfide oxidation pathway in acidophilic bacteria with the exception that polysulfide is not generated as an intermediate in the mammalian pathway. Instead, as discussed below, the persulfide product of SQR is transferred to an as yet unidentified acceptor. Sulfite is ultimately oxidized to sulfate by sulfite oxidase in mammals.

1.1.2.2 Sulfide Quinone Oxidoreductase

Structural Organization of SQR. SQR is found in the inner mitochondrial membrane (Theissen et al. 2003) in eukaryotes, which are believed to have acquired the nuclear encode gene from a mitochondrial endosymbiont (Theissen et al. 2003). SQR exists as a dimer or a trimer, with one FAD and one redox active disulfide in each monomer (Marcia et al. 2010). SQR oxidizes sulfide to a protein-bound persulfide and transfers electrons from H_2S to ubiquinone via a bound FAD, coupling sulfide oxidation to the electron transport chain (Fig. 1.1). Hence, sulfide functions as an inorganic substrate for the ATP-generating electron transfer chain (Bouillaud and Blachier 2011).

Several crystal structures of SQRs have been reported, notably from *Acidianus ambivalens* (Brito et al. 2009), *Aquifex aeolicus* (Marcia et al. 2009), and *Acidithiobacillus ferrooxidans* (Cherney et al. 2010). SQR belongs to the family of flavin disulfide reductases that includes glutathione reductase. SQR contains two Rossman fold domains and a C-terminal domain that is important for membrane binding (Fig. 1.5a). The FAD cofactor is found in the first N-terminal Rossman fold domain and can be non-covalently or covalently bound. In the latter case, a thioether linkage exists between a cysteine or a cysteine persulfide and the 8-methylene group of the isoalloxazine ring of FAD (Marcia et al. 2009). The catalytic disulfide is located on the *re* face of the FAD. A conserved glutamate residue in the active site is proposed to serve as a general base for deprotonating H_2S (Cherney et al. 2010). The ubiquinone-binding site is located on the *si* face of the FAD and the majority of residues in contact with the quinone are hydrophobic including F41, P43, G322, Y323, N353, Y411, F394 and F357 (*A. ferrooxidans* numbering) (Cherney et al. 2010). The aromatic ring of the quinone is sandwiched between the benzene rings of F394 and F357. The O2 atom of the flavin electron donor, and the O4 atom of the quinone acceptor are <3 Å apart. Residues Y411 and K391 have been proposed to transfer protons from water for protonation of the reduced quinone (Marcia et al. 2009; Cherney et al. 2010).

Catalytic Mechanism of SQR. The catalytic cycle of SQR is initiated by nucleophilic attack of the sulfide on the disulfide resulting in formation of a persulfide and a cysteine thiolate (Fig. 1.5b) (Brito et al. 2009; Marcia et al. 2009; Cherney et al. 2010, 2012; Jackson et al. 2012). The cysteine thiolate attacks the FAD cofactor forming a C4A adduct (Fig. 1.5c). Nucleophilic attack of an acceptor on the sulfane sulfur results in reformation of the active site disulfide and subsequent two-electron transfer to the FAD results in formation of $FADH_2$. Electron transfer from $FADH_2$ to ubiquinone regenerates FAD.

In bacterial SQRs, multiple rounds of electron transfer occur from sulfide to quinone without release of the persulfide at the end of each catalytic cycle (Brito et al. 2009; Marcia et al. 2009; Cherney et al. 2010; Jackson et al. 2012). Instead, the catalytic cycle repeats until the maximum length of the polysulfide product that can be accommodated in the active site is obtained at which point two consecutive nucleophilic attacks by the sulfide results in product release and reformation of the active site disulfide. Snapshots of these polysulfide species have been observed in structures of prokaryotic SQRs. The structure of *A. ambivalens* SQR revealed a

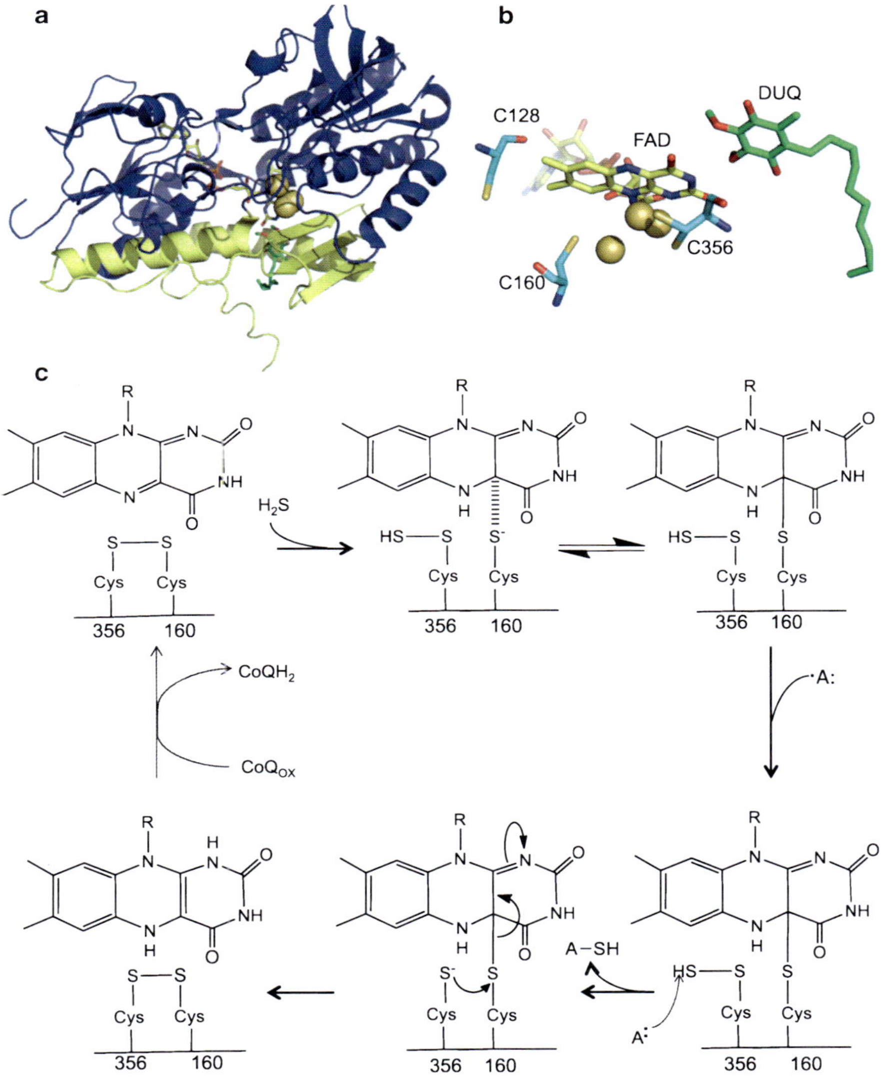

Fig. 1.5 Structure and reaction mechanism of SQR. (**a**) Structure of *A. ferrooxidans* SQR monomer displaying the N-terminal Rossman fold domains (*blue*) and the C-terminal membrane binding domain (*yellow*) and the cofactors, FAD (*yellow*) and DUQ (*green*). Active site sulfurs are shown as gold spheres (**b**) Close-up of the *A. ferrooxidans* SQR active site viewing the *re* face of the FAD isoalloxazine ring and decylubiquinone (*DUQ*) on the *si* face of FAD. Sulfur atoms between the active site cysteine residues are shown as *gold spheres*. Figure 1.5a, b were generated using PDB file 3T31. (**c**) Proposed mechanism of SQR. Reaction of H_2S reduces with the active site disulfide bond results in the formation of a persulfide on one cysteine and a covalent linkage between the other cysteine and FAD. The cysteine bound persulfide is released from the enzyme by an acceptor molecule. Reformation of the active site disulfide results in reduction of FAD to $FADH_2$. Electrons are then transferred from $FADH_2$ to ubiquinone allowing for regeneration of the oxidized active site

trisulfide bridge between the two active site cysteines (Brito et al. 2009), while the *A. ferrooxidans* structure revealed a branched intermediate between the two cysteine residues consisting of five sulfur atoms (Cherney et al. 2010, 2012). In the *A. aeolicus* SQR structures, both linear and cyclic polysulfur intermediates were observed (Marcia et al. 2009). The released product can be soluble polysulfide containing up to ten sulfur atoms (Griesbeck et al. 2002) or an octasulfur ring (Marcia et al. 2009).

In contrast, in mammalian SQRs, the persulfide product is transferred to an acceptor at the end of each catalytic cycle. Although several molecules have been proposed as acceptors of the persulfide product including sulfite and glutathione, the identity of the physiological co-substrate is not known (Jackson et al. 2012). Human SQR utilizes cyanide and sulfite as persulfide acceptors and exhibits k_{cat} values of $82 \pm 6\ s^{-1}$ and $251 \pm 9\ s^{-1}$ in the presence of 1 mM cyanide and 600 μM sulfite respectively (Jackson et al. 2012). In contrast, the k_{cat} in the presence of 1 mM glutathione as an acceptor is $19 \pm 3\ s^{-1}$, which is reportedly identical to the background rate constant observed in the absence of an additional acceptor ($k_{cat} = 18.5 \pm 0.9\ s^{-1}$). The activity of SQR in the absence of an exogenous acceptor was attributed to sulfide acting as an acceptor for the SQR product resulting in formation of hydrogen disulfide (Jackson et al. 2012). It is not known whether ETHE1 can directly accept the persulfide product from SQR.

1.1.2.3 Persulfide Dioxygenase (ETHE1)

ETHE1, a persulfide dioxygenase, is a soluble, mitochondrial matrix enzyme that catalyzes the oxidation of glutathione persulfide to sulfite (Hildebrandt and Grieshaber 2008; Tiranti et al. 2009; Kabil and Banerjee 2012). However, the physiological substrate is not known unequivocally. Persulfide dioxygenase contains a mononuclear non-heme iron in its active site (McCoy et al. 2006; Holdorf et al. 2008). Mutations in persulfide dioxygenase result in ethylmalonic encephalopathy, an autosomal recessive disorder that results in developmental delay hemorrhagic diarrhea, acrocyanosis, petechiae, and progressive neurological failure (Tiranti et al. 2004, 2009; Mineri et al. 2008). The clinical profile of ethylmalonic encephalopathy includes high levels of lactate, high C5 and C4 acylcarnitine levels in blood, increased ethylmalonic acid concentrations in urine and cytochrome c oxidase deficiency in muscle and brain (Burlina et al. 1991; Garcia-Silva et al. 1994; Tiranti et al. 2009). Persulfide dioxygenase deficiency leads to accumulation of thiosulfate and sulfide with the latter inhibiting cytochrome *c* oxidase and possibly other enzymes and accounting in part for the observed pathology (Di Meo et al. 2011).

Both glutathione persulfide (GSSH) and CoA persulfide (CoA-SSH) serve as substrates for human persulfide dioxygenase while cysteine persulfide, glutathione and thiosulfate do not (Kabil and Banerjee 2012). However, the specific activity of the enzyme with CoA-SSH is ~50-fold lower than with GSSH. Interestingly, 5 mM glutathione, a high but physiologically relevant concentration, increases the catalytic efficiency of the persulfide dioxygenase ~3-fold. Despite the similarity between persulfide dioxygenase and glyoxylase II, the former is unable to hydrolyze S-D-lactoylglutathione and other glutathione thioesters like glyoxylase II (Holdorf et al. 2008). This is most likely due to a C-terminal loop covering much of the active site in the dioxygenase making it smaller than the glyoxylase II active site (McCoy et al. 2006; Holdorf et al. 2008).

Structural Organization of Persulfide Dioxygenase. The structure of the *Arabidopsis thaliana* persulfide dioxygenase (McCoy et al. 2006) has been used to model the structure of the human enzyme (Fig. 1.6a) (Kabil and Banerjee 2012). The two proteins share 54 % identity. However, while the *Arabidopsis* persulfide dioxygenase is a dimer, as revealed by gel-filtration chromatography and crystallography (McCoy et al. 2006), the human enzyme is a monomer (Kabil and Banerjee 2012). The *Arabidopsis* persulfide dioxygenase has a typical metallo-β-lactamase fold containing two central β-sheets enclosed by three helices on each side. The active site iron is ligated via two histidine and one aspartate residue, which correspond to H84, H135 and D154 in the human sequence (Fig. 1.6b). The three remaining coordination sites are occupied by water resulting in octahedral coordination, typical for ferrous iron (McCoy et al. 2006; Holdorf et al. 2008).

The geometry of the iron site in persulfide dioxygenase resembles that of the 2-His-1-carboxylate facial triad family of oxygenases (Hegg and Que 1997), a common structural motif that binds mononuclear non-heme Fe^{2+}. The three coordination sites occupied by water are available for binding ligands such as O_2, substrates and/or cofactors, and allow the enzymes to tune the reactivity of the iron center (Koehntop et al. 2005; Bruijnincx et al. 2008). With three coordination sites occupied by solvent, the metal center is not reactive towards dioxygen. Substrate binding displaces solvent molecules and results in formation of a five-coordinate metal center. Replacement of neutral solvent molecules with an anionic ligand can decrease the Fe(III)/Fe(II) redox potential rendering the iron center more susceptible to oxidation by dioxygen (Koehntop et al. 2005). Hence, substrate binding primes the iron center for dioxygen binding and protects the enzyme from auto-inactivation (Koehntop et al. 2005).

The substrate-binding site is predicted to comprise residues M226, R163, Y197 and F166 in the human sequence based on structural alignment with the active site of glyoxylase II, which provides insights into how the glutathione persulfide might interacts with persulfide dioxygenase. The glyoxylase II residues Y145 and K143 are engaged in hydrogen bonds with the glutamate portion of glutathione and Y175 forms a hydrogen bond with the cysteine portion of glutathione (McCoy et al. 2006). The corresponding residues in human persulfide dioxygenase are F166, R163 and Y197. R163 is predicted to be positioned similarly to K143 and to interact with the glutamate portion of the GSSH substrate. In contrast, F166, which replaces Y145 in the persulfide dioxygenase structure, is not expected to engage in a hydrogen bonding interaction with the substrate. Y197 like Y175 in the *A. thaliana* protein, might interact with the cysteine portion of GSSH, while the sulfane sulfur of GSSH would be coordinated to the iron center (Kabil and Banerjee 2012).

Catalytic Mechanism of Persulfide Dioxygenase. Kinetic characterization of human persulfide dioxygenase and two variants mimicking missense mutations described in ethylmalonic aciduria patients, T151I and D196N, has been described

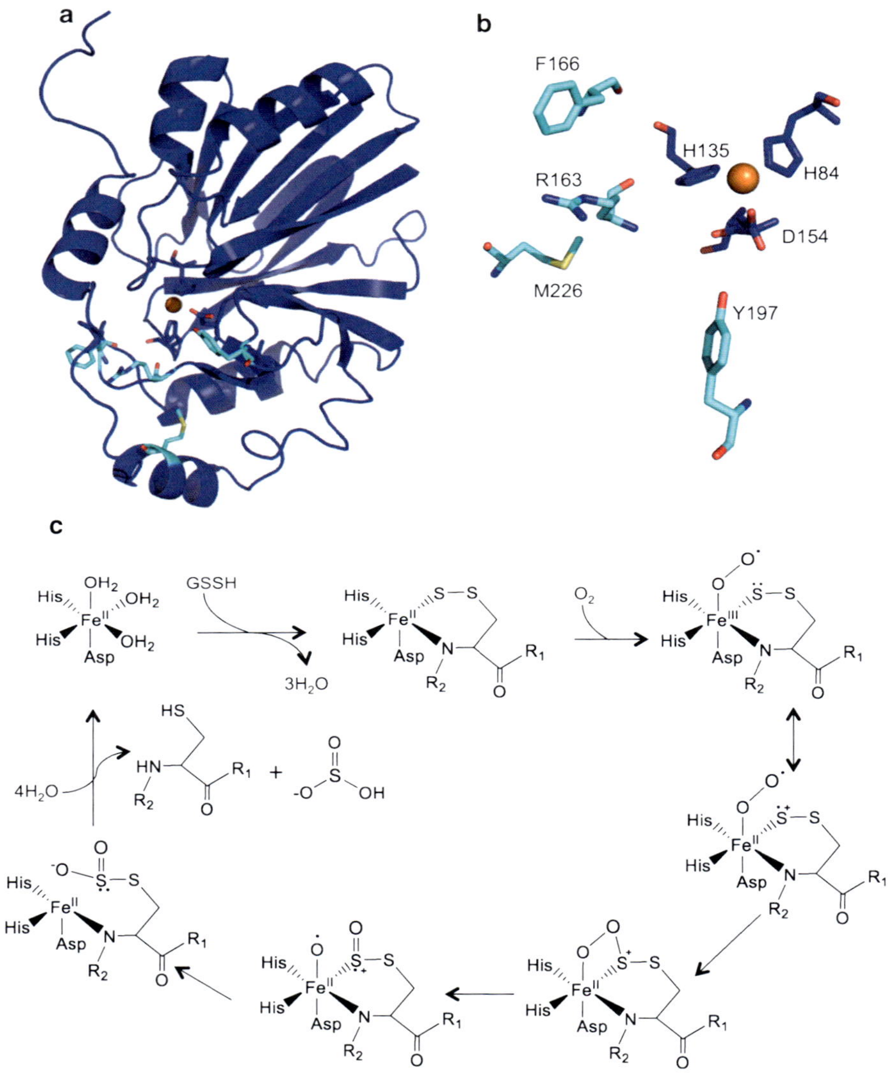

Fig. 1.6 Structure and catalytic mechanism of persulfide dioxygenase. (**a**) Structure of *A. thaliana* persulfide dioxygenase monomer. The mononuclear non-heme iron is shown in *orange* and active site residues are shown in *cyan*. (**b**) Close up of the active site of *A. thaliana* persulfide dioxygenase. The mononuclear non-heme iron is coordinated by the 2His-1Asp facial triad residues (*blue*) and other active site residues (*cyan*). Figure 1.6a, b were generated using PDB file 2GCU. (**c**) Proposed catalytic mechanism of persulfide dioxygenase for sulfite generation. Binding of the GSSH substrate displaces coordinated water and promotes binding of oxygen to the iron center to form the Fe (III) superoxo intermediate. Resonance allows for partial radical cation character of the coordinated sulfur leading to recombination and formation of a cyclic peroxo-intermediate. Cleavage of the O-O bond results in a sulfoxy-cation and an iron-bound activated oxygen atom which is transferred to the sulfur to sulfoxy-cation. Subsequent hydrolysis yields sulfite, and GSH is displaced from the active site upon water binding to the metal center. R_1 and R_2 represent residues glutamate and glycine in the GSSH substrate

(Kabil and Banerjee 2012). In the presence of GSSH, the V_{max} for human persulfide dioxygenase is 113 ± 4 μmol min^{-1} mg^{-1} protein, which corresponds to a k_{cat} of 47 s^{-1}at 22 °C. Glutathione (5 mM) decreases the K_M for GSSH ~1.4 fold and increases k_{cat} an ~2.2-fold yielding an ~3-fold increase in k_{cat}/K_M. The physiological relevance of modulation of the persulfide dioxygenase activity by GSH is not known. The T152I mutant does not affect the K_M for GSSH while the k_{cat} is diminished ~4-fold and correlates with an ~2.5 fold lower iron content compared to wild-type enzyme. T152 is located on the same loop as D154, a predicted iron ligand and participates in a hydrogen bonding interaction with the backbone of L156, which may be necessary for correct positioning of D154. The T152I mutation would result in loss of the interaction with L156 possibly leading to repositioning of D154 (Kabil and Banerjee 2012).

The impact of the D196N mutation is to increase the K_M for GSSH ~2-fold while leaving the k_{cat} unaltered. D196 is located on an internal loop distal from the active site and is proposed to hydrogen bond with F200 and H198 to stabilize the loop. The D196N mutation likely destabilizes this loop. Several other pathogenic missense mutations have been described in human persulfide dioxygenase and include Y38C, L55P, T136A, R163Q, R163W, C161Y, T164K and L185R (Tiranti et al. 2004, 2006; Mineri et al. 2008). Many of these mutations are predicted to be located near the active site and may disrupt metal or substrate binding and/or destabilize the enzyme.

The reaction mechanism proposed for persulfide dioxygenase is adapted from that described for cysteine dioxygenase (McCoy et al. 2006) and is also based on the general reaction mechanism of mononuclear non-heme iron oxygenases (Koehntop et al. 2005; Bruijnincx et al. 2008). In the proposed mechanism (Kabil and Banerjee 2012), solvent is displaced from the iron upon GSSH binding (Fig. 1.6c). The sulfane sulfur and a nitrogen atom from GSSH coordinate to the iron center resulting in a five-coordinate iron species, which is primed for O_2 binding. Binding of O_2 results in formation of a Fe (III) superoxo intermediate in which the coordinated sulfur acquires a partial radical cation character via resonance. Recombination of the Fe(III) superoxo species with the coordinated sulfur atom leads to formation of a cyclic peroxo-intermediate and is followed by homolytic cleavage of the O-O bond resulting in a sulfoxy-cation species and a metal-bound activated oxygen atom. Transfer of the activated oxygen to the sulfur and hydrolysis yields sulfite. Finally, release of GSH followed by rebinding of solvent to the active site completes the catalytic cycle (Kabil and Banerjee 2012).

1.1.2.4 Rhodanese

Rhodanese is a widely distributed protein found in Archaea, bacteria and eukaryotes (Cipollone et al. 2007). It is a mitochondrial matrix protein that catalyzes the transfer of a sulfur atom from a sulfur donor to an acceptor (Eqs. 1.18 and 1.19). Despite extensive studies on rhodanese, its precise physiological function is still not known. Historically, rhodanese was thought to be involved in

$$SO_3^{2-} + RSSH \rightarrow S_2O_3^{2-} + RSH \tag{1.18}$$

$$S_2O_3^{2-} + CN^- \rightarrow SO_3^{2-} + SCN^- \quad (1.19)$$

cyanide detoxification due to its ability to convert cyanide and thiosulfate to thiocyanate (Westley 1973; Cipollone et al. 2007). In fact, thiosulfate along with sodium nitrite is administered to treat acute cyanide poisoning. Cells that are routinely exposed to gaseous and dietary intake of cyanide, such as the epithelial cells surrounding bronchioles, hepatocytes that are proximal to the liver's blood supply and proximal tubule cells in kidney, have the highest rhodanese levels (Sylvester and Sander 1990). However, the activity of rhodanese is confined to the mitochondrial matrix, where thiosulfate enters with low efficiency suggesting that a different sulfur source than thiosulfate might be used for clearing cyanide (Westley et al. 1983). Alternatively, rhodanese might play a role in sulfur metabolism (Westley et al. 1983; Hildebrandt and Grieshaber 2008), particularly for mitochondrial thiosulfate production by transfer of the sulfane sulfur from a donor to sulfite (Hildebrandt and Grieshaber 2008).

Structural Organization of Rhodanese. Bovine liver rhodanese, which shares 89 % sequence identity with the human enzyme, has been characterized extensively. This protein is folded into two globular domains of equal size, with each containing a five stranded parallel β-sheet enclosed by two α-helices on one side and three α-helices on the other (Fig. 1.7a) (Hol et al. 1983). While both domains have a similar fold, the sequence homology is low with only 16 % sequence identity between them. The N-terminal domain is inactive due to replacement of the catalytic cysteine residue with an aspartate and plays a role in forming the active site. The catalytic C247 residue in bovine liver rhodanese is located at the bottom of a pocket formed between the two domains. The walls of the pocket comprise hydrophobic and hydrophilic regions (Fig. 1.7b). The hydrophobic regions consists of residues F212, F106, Y107, W35, and V251, and the hydrophilic region comprise residues D180, S181, R182, R186, E193, R248, K249, and T252. The persulfide intermediate is stabilized by hydrogen bonds from the backbone amides of R248, K249, V251 and the hydroxyl group of T252. Residues R186 and K249 located at the entrance of the active site pocket and their side chains are positioned to participate in binding and positioning of thiosulfate through ionic interactions and to polarize the sulfane sulfur for nucleophilic attack by the C247 thiol. It is possible that the hydrophobic region lining the active site is important for binding other substrates that contain aromatic or hydrophobic residues (Hol et al. 1983).

Catalytic Mechanism of Rhodanese. The sulfurtransferase reaction catalyzed by rhodanese occurs via a double displacement mechanism and involves formation of a stable persulfide intermediate (Hol et al. 1983; Cipollone et al. 2007a). The reaction is initiated by the nucleophilic attack of the C247 thiolate on the sulfane sulfur of thiosulfate resulting in formation of an enzyme-bound persulfide intermediate (Fig. 1.7c). Next, sulfite is released and followed by binding of a sulfur acceptor, which attacks the sulfane sulfur of the persulfide intermediate. In the case of cyanide, the product of this sulfur transfer reaction is thiocyanate (Hol et al. 1983). Release of thiocyanate completes the catalytic cycle.

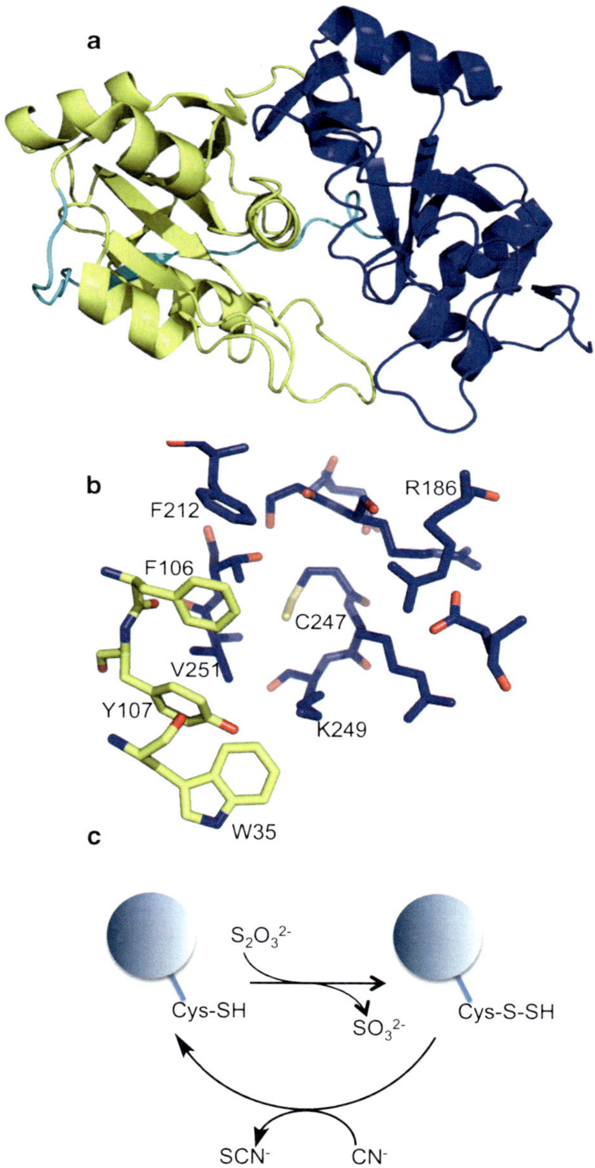

Fig. 1.7 Structure and reaction mechanism of rhodanese. (**a**) Structure of *Bos taurus* rhodanese. The protein comprises two equal sized N-terminal (*yellow*) and C-terminal (*blue*) globular domains. The linker region between the two domains is shown in *cyan*. (**b**) Close up of the rhodanese active site. The active site is located at the interface of the two domains. N- and C-terminal domain residues contribute to a hydrophobic patch (*yellow* and *blue*), while C-terminal domain residues contribute to a hydrophilic region (*blue*). Figure 1.7a, b were generated using PDB file 1RHD. (**c**) Catalytic mechanism of rhodanese for thiocyanate generation from thiosulfate. Rhodanese catalyzes the sulfur transfer from thiosulfate to an active site cysteine, resulting in sulfite and a rhodanese-bound persulfide intermediate. The latter reacts with a sulfur acceptor, in this example cyanide, to generate thiocyanate

Rhodanese also displays sulfurtransferase activity between a persulfide donor and sulfite acceptor (Eq. 1.18). The K_M values for cyanide and thiosulfate for bovine rhodanese are 0.087 ± 0.009 mM and 16.2 ± 1.6 mM, respectively (Hildebrandt and Grieshaber 2008). Similar values have also been reported for rat and lugworm rhodanese (Hildebrandt and Grieshaber 2008). In contrast, the K_Ms for GSSH and sulfite are considerably lower (61.3 ± 17.8 μM and 21.8 ± 3.6 μM respectively), suggesting that the persulfide transferase activity of rhodanese might be more relevant physiologically than cyanide detoxification.

1.1.2.5 Mitochondrial Sulfide Oxidation: Unanswered Questions

Endogenously produced H_2S must be regulated to maintain low intracellular concentrations. In mammals, steady-state levels of H_2S are governed by flux through the synthetic pathways (transsulfuration and CAT/MST) and the sulfide oxidation pathway (Vitvitsky et al. 2012). In some prokaryotes, sulfide oxidation is essential for ATP generation. Similarly, low concentrations of H_2S (0.1–1 μM), can stimulate mammalian mitochondrial ATP production and serve as an inorganic source of ATP (Goubern et al. 2007; Bouillaud and Blachier 2011; Modis et al. 2013). An important unanswered question regarding the sulfide oxidation pathway is its organization. Since SQR oxidizes sulfide, it is reasonable to propose that it catalyzes the first step in the pathway. The ambiguity arises thereafter since the persulfide acceptor of SQR is not known. Both persulfide dioxygenase and rhodanese can oxidize persulfide forming either sulfite or thiosulfate, respectively. However, the co-substrate for rhodanese is sulfite, which is derived from the persulfide dioxygenase-catalyzed reaction. This dependence of rhodanese on the product of the dioxygenase suggests that oxidation of H_2S proceeds through SQR, the dioxygenase and then rhodanese. The products of this pathway thus configured are thiosulfate and sulfate, which is derived by oxidation of sulfite catalyzed by sulfite oxidase. The presence of two routes for sulfite oxidation in the mitochondria is paralleled in microbes where multiple sulfite oxidation routes co-exist in the same organism, e.g. the oxidation of sulfite by sulfite oxidase or via the adenosine-5′-phosphosulfate reductase pathway (Kappler and Dahl 2001).

Variants of persulfide dioxygenase fused to a rhodanese domain are found in certain bacteria (Tiranti et al. 2009). The occurrence of fused persulfide dioxygenase/rhodanese variants suggests that their proximity enhances utilization of sulfite produced by one active site and consumed by the other. However, this order of the pathway is brought into question by clinical data on ETHE1 deficient patients and ETHE1 knockout mice in which sulfite levels are greatly diminished, as expected, but thiosulfate and H_2S levels are elevated (Tiranti et al. 2009). Elevated thiosulfate in the absence of persulfide dioxygenase activity suggests that an alternative route for sulfite synthesis exists, which supports production of thiosulfate by rhodanese (Kabil and Banerjee 2012). One branch of the cysteine catabolic pathway is initiated by cysteine dioxygenase that oxidizes cysteine to cysteinesulfinate, which is further metabolized to 3-sulfinylpyruvate by CAT. 3-sulfinylpyruvate is unstable and decomposes to form pyruvate and sulfite. We have posited that the persulfide product of SQR is preferentially utilized by rhodanese under conditions of persulfide dioxygenase deficiency, explaining the observed accumulation of thiosulfate under these conditions (Kabil and Banerjee 2012). However, while cysteine catabolism is up regulated under conditions of cysteine excess (Stipanuk et al. 2004), it is unclear how this pathway responds to persulfide dioxygenase deficiency and its role in sulfite production under these conditions warrants investigation.

In this context, understanding the fate of the persulfide product of SQR is pertinent. Bacterial polysulfide products of SQR are stored in sulfur globules until further oxidation to sulfate. However, sulfur globules have not been reported in higher organisms and in fact, eukaryotic lugworm, rat, and human SQRs require a persulfide acceptor for catalytic turnover under in vitro conditions (Hildebrandt and Grieshaber 2008; Jackson et al. 2012). The persulfide product bound to the SQR active site might be either oxidized directly by persulfide dioxygenase or rhodanese or transferred to a small molecule carrier such as GSH with the resulting GSSH serving as substrate for the persulfide dioxygenase and rhodanese. Sulfane sulfur acceptors that support the activity of SQR include sulfite, cyanide, sulfide and glutathione. Human SQR has been proposed to utilize sulfite as the physiological acceptor since it displays a 4- or 13-fold higher catalytic efficiency in the presence of either cyanide or sulfite versus glutathione (Jackson et al. 2012). Based on these results, it has been proposed that sulfite is the physiological acceptor of SQR's persulfide product resulting in the formation of thiosulfate (Jackson et al. 2012). The product of persulfide transfer to sulfite is thiosulfate, which is however not a substrate for persulfide dioxygenase (Kabil and Banerjee 2012). Of the persulfide donors that have been tested as substrates for persulfide dioxygenase only CoA-SSH exhibits activity in addition to GSSH (Kabil and Banerjee 2012). Hence, the mechanism by which the persulfide product of SQR is transferred to the persulfide dioxygenase and rhodanese is an important unanswered question in the field.

References

Agarwal N, Banerjee R (2008) Human polycomb 2 protein is a SUMO E3 ligase and alleviates substrate-induced inhibition of cystathionine β-synthase sumoylation. PLoS One 3:e4032. doi:4010.1371

Alphey MS, Williams RA, Mottram JC, Coombs GH, Hunter WN (2003) The crystal structure of Leishmania major 3-mercaptopyruvate sulfurtransferase. A three-domain architecture with a serine protease-like triad at the active site. J Biol Chem 278(48):48219–48227

Banerjee R, Evande R, Kabil O, Ojha S, Taoka S (2003) Reaction mechanism and regulation of cystathionine beta-synthase. Biochim Biophys Acta 1647(1–2):30–35

Bateman A (1997) The structure of a domain common to archaebacteria and the homocystinuria disease protein. Trends Biochem Sci 22:12–13

Bouillaud F, Blachier F (2011) Mitochondria and sulfide: a very old story of poisoning, feeding and signaling? Antioxid Redox Signal 15:379–391

Brito JA, Sousa FL, Stelter M, Bandeiras TM, Vonrhein C, Teixeira M, Pereira MM, Archer M (2009) Structural and functional insights into sulfide:quinone oxidoreductase. Biochemistry 48(24):5613–5622

Bruijnincx PC, van Koten G, Klein Gebbink RJ (2008) Mononuclear non-heme iron enzymes with the 2-His-1-carboxylate facial triad: recent developments in enzymology and modeling studies. Chem Soc Rev 37(12):2716–2744

Burlina A, Zacchello F, Dionisi-Vici C, Bertini E, Sabetta G, Bennet MJ, Hale DE, Schmidt-Sommerfeld E, Rinaldo P (1991) New clinical phenotype of branched-chain acyl-CoA oxidation defect. Lancet 338(8781):1522–1523

Carballal S, Madzelan P, Zinola CF, Grana M, Radi R, Banerjee R, Alvarez B (2008) Dioxygen reactivity and heme redox potential of truncated human cystathionine beta-synthase. Biochemistry 47(10):3194–3201

Chen ZW, Koh M, Van Driessche G, Van Beeumen JJ, Bartsch RG, Meyer TE, Cusanovich MA, Mathews FS (1994) The structure of flavocytochrome c sulfide dehydrogenase from a purple phototrophic bacterium. Science 266(5184):430–432

Chen X, Jhee KH, Kruger WD (2004) Production of the neuromodulator H_2S by cystathionine beta-synthase via the condensation of cysteine and homocysteine. J Biol Chem 279(50):52082–52086

Cherney MM, Zhang Y, Solomonson M, Weiner JH, James MN (2010) Crystal structure of sulfide: quinone oxidoreductase from Acidithiobacillus ferrooxidans: insights into sulfidotrophic respiration and detoxification. J Mol Biol 398(2):292–305

Cherney MM, Zhang Y, James MN, Weiner JH (2012) Structure-activity characterization of sulfide:quinone oxidoreductase variants. J Struct Biol 178(3):319–328

Chiku T, Padovani D, Zhu W, Singh S, Vitvitsky V, Banerjee R (2009) H_2S biogenesis by cystathionine gamma-lyase leads to the novel sulfur metabolites, lanthionine and homolanthionine, and is responsive to the grade of hyperhomocysteinemia. J Biol Chem 284:11601–11612

Cipollone R, Ascenzi P, Visca P (2007a) Common themes and variations in the rhodanese superfamily. IUBMB Life 59(2):51–59

Cipollone R, Frangipani E, Tiburzi F, Imperi F, Ascenzi P, Visca P (2007b) Involvement of Pseudomonas aeruginosa rhodanese in protection from cyanide toxicity. Appl Environ Microbiol 73(2):390–398

Crawhall JC, Parker R, Sneddon W, Young EP, Ampola MG, Efron ML, Bixby EM (1968) Beta mercaptolactate-cysteine disulfide: analog of cystine in the urine of a mentally retarded patient. Science 160(826):419–420

Di Meo I, Fagiolari G, Prelle A, Viscomi C, Zeviani M, Tiranti V (2011) Chronic exposure to sulfide causes accelerated degradation of cytochrome c oxidase in ethylmalonic encephalopathy. Antioxid Redox Signal 15(2):353–362

Evande R, Boers GHJ, Blom HJ, Banerjee R (2002) Alleviation of intrasteric inhibition by the pathogenic activation domain mutation, D444N, in human cystathionine beta-synthase. Biochemistry 41:11832–11837

Evande R, Ojha S, Banerjee R (2004) Visualization of PLP-bound intermediates in hemeless variants of human cystathionine beta-synthase: evidence that lysine 119 is a general base. Arch Biochem Biophys 427(2):188–196

Finkelstein JD, Kyle WE, Martin JL, Pick AM (1975) Activation of cystathionine synthase by adenosylmethionine and adenosylethionine. Biochem Biophys Res Commun 66(1):81–87

Friedrich CG, Rother D, Bardischewsky F, Quentmeier A, Fischer J (2001) Oxidation of reduced inorganic sulfur compounds by bacteria: emergence of a common mechanism? Appl Environ Microbiol 67(7):2873–2882

Furne J, Saeed A, Levitt MD (2008) Whole tissue hydrogen sulfide concentrations are orders of magnitude lower than presently accepted values. Am J Physiol Regul Integr Comp Physiol 295(5):R1479–R1485

Garcia-Silva MT, Campos Y, Ribes A, Briones P, Cabello A, Santos Borbujo J, Arenas J, Garavaglia B (1994) Encephalopathy, petechiae, and acrocyanosis with ethylmalonic aciduria associated with muscle cytochrome c oxidase deficiency. J Pediatr 125(5 Pt 1):843–844

Goubern M, Andriamihaja M, Nubel T, Blachier F, Bouillaud F (2007) Sulfide, the first inorganic substrate for human cells. FASEB J 21(8):1699–1706

Green EL, Taoka S, Banerjee R, Loehr TM (2001) Resonance raman characterization of the heme cofactor in cystathionine beta-synthase. Identification of the Fe-S(Cys) vibration in the six-coordinate low-spin heme. Biochemistry 40(2):459–463

Gregersen LH, Bryant DA, Frigaard NU (2011) Mechanisms and evolution of oxidative sulfur metabolism in green sulfur bacteria. Front Microbiol 2:116

Griesbeck C, Schutz M, Schodl T, Bathe S, Nausch L, Mederer N, Vielreicher M, Hauska G (2002) Mechanism of sulfide-quinone reductase investigated using site-directed mutagenesis and sulfur analysis. Biochemistry 41(39):11552–11565

Grimm F, Franz B, Dahl C (2011) Regulation of dissimilatory sulfur oxidation in the purple sulfur bacterium allochromatium vinosum. Front Microbiol 2:51

Hegg EL, Que L Jr (1997) The 2-His-1-carboxylate facial triad–an emerging structural motif in mononuclear non-heme iron(II) enzymes. Eur J Biochem 250(3):625–629

Hildebrandt TM, Grieshaber MK (2008) Three enzymatic activities catalyze the oxidation of sulfide to thiosulfate in mammalian and invertebrate mitochondria. FEBS J 275(13):3352–3361

Hol WG, Lijk LJ, Kalk KH (1983) The high resolution three-dimensional structure of bovine liver rhodanese. Fundam Appl Toxicol 3(5):370–376

Holdorf MM, Bennett B, Crowder MW, Makaroff CA (2008) Spectroscopic studies on Arabidopsis ETHE1, a glyoxalase II-like protein. J Inorg Biochem 102(9):1825–1830

Holkenbrink C, Barbas SO, Mellerup A, Otaki H, Frigaard NU (2011) Sulfur globule oxidation in green sulfur bacteria is dependent on the dissimilatory sulfite reductase system. Microbiology 157(Pt 4):1229–1239

Huang S, Chua JH, Yew WS, Sivaraman J, Moore PK, Tan C-H, Deng L-W (2010) Site-directed mutagenesis on human cystathionine-γ-lyase reveals insights into the modulation of H_2S production. J Mol Biol 396(3):708–718

Jackson MR, Melideo SL, Jorns MS (2012) Human sulfide: quinone oxidoreductase catalyzes the first step in hydrogen sulfide metabolism and produces a sulfane sulfur metabolite. Biochemistry 51(34):6804–6815

Janosik M, Kery V, Gaustadnes M, Maclean KN, Kraus JP (2001) Regulation of human cystathionine beta-synthase by S-adenosyl-L- methionine: evidence for two catalytically active conformations involving an autoinhibitory domain in the C-terminal region. Biochemistry 40(35):10625–10633

Jarabak R, Westley J (1978) Steady-state kinetics of 3-mercaptopyruvate sulfurtransferase from bovine kidney. Arch Biochem Biophys 185(2):458–465

Jhee KH, Niks D, McPhie P, Dunn MF, Miles EW (2001) The reaction of yeast cystathionine beta-synthase is rate-limited by the conversion of aminoacrylate to cystathionine. Biochemistry 40(36):10873–10880

Kabil O, Banerjee R (2010) The redox biochemistry of hydrogen sulfide. J Biol Chem 285:21903–21907

Kabil O, Banerjee R (2012) Characterization of patient mutations in human persulfide dioxygenase (ETHE1) involved in H_2S catabolism. J Biol Chem 287:44561–44567

Kabil O, Vitvitsky V, Xie P, Banerjee R (2011a) The quantitative significance of the transsulfuration enzymes for H_2S production in murine tissues. Antioxid Redox Signal 15:363–372

Kabil O, Weeks CL, Carballal S, Gherasim C, Alvarez B, Spiro TG, Banerjee R (2011b) Reversible heme-dependent regulation of human cystathionine beta-synthase by a flavoprotein oxidoreductase. Biochemistry 50(39):8261–8263

Kappler U, Dahl C (2001) Enzymology and molecular biology of prokaryotic sulfite oxidation. FEMS Microbiol Lett 203(1):1–9

Kery V, Poneleit L, Kraus JP (1998) Trypsin cleavage of human cystathionine beta-synthase into an evolutionarily conserved active core: structural and functional consequences. Arch Biochem Biophys 355(2):222–232

Kimura H (2010) Hydrogen sulfide: from brain to gut. Antioxid Redox Signal 12(9):1111–1123

Kletzin A (1992) Molecular characterization of the sor gene, which encodes the sulfur oxygenase/reductase of the thermoacidophilic Archaeum Desulfurolobus ambivalens. J Bacteriol 174(18):5854–5859

Kletzin A, Urich T, Muller F, Bandeiras TM, Gomes CM (2004) Dissimilatory oxidation and reduction of elemental sulfur in thermophilic archaea. J Bioenerg Biomembr 36(1):77–91

Koehntop KD, Emerson JP, Que L Jr (2005) The 2-His-1-carboxylate facial triad: a versatile platform for dioxygen activation by mononuclear non-heme iron(II) enzymes. J Biol Inorg Chem 10(2):87–93

Koutmos M, Kabil O, Smith JL, Banerjee R (2010) Structural basis for substrate activation and regulation by cystathionine beta-synthase domains in cystathionine beta-synthase. Proc Natl Acad Sci USA 107:20958–20963

Kraus JP, Janosik M, Kozich V, Mandell R, Shih V, Sperandeo MP, Sebastio G, de Franchis R, Andria G, Kluijtmans LA, Blom H, Boers GH, Gordon RB, Kamoun P, Tsai MY, Kruger WD, Koch HG, Ohura T, Gaustadnes M (1999) Cystathionine beta-synthase mutations in homocystinuria. Hum Mutat 13(5):362–375

Levitt MD, Abdel-Rehim MS, Furne J (2011) Free and acid-labile hydrogen sulfide concentrations in mouse tissues: anomalously high free hydrogen sulfide in aortic tissue. Antioxid Redox Signal 15:373–378

Marcia M, Ermler U, Peng G, Michel H (2009) The structure of Aquifex aeolicus sulfide:quinone oxidoreductase, a basis to understand sulfide detoxification and respiration. Proc Natl Acad Sci USA 106(24):9625–9630

Marcia M, Ermler U, Peng G, Michel H (2010) A new structure-based classification of sulfide: quinone oxidoreductases. Proteins 78(5):1073–1083

McCoy JG, Bailey LJ, Bitto E, Bingman CA, Aceti DJ, Fox BG, Phillips GN Jr (2006a) Structure and mechanism of mouse cysteine dioxygenase. Proc Natl Acad Sci USA 103(9):3084–3089

McCoy JG, Bingman CA, Bitto E, Holdorf MM, Makaroff CA, Phillips GN Jr (2006b) Structure of an ETHE1-like protein from Arabidopsis thaliana. Acta Crystallogr D Biol Crystallogr 62(Pt 9):964–970

Meier M, Janosik M, Kery V, Kraus JP, Burkhard P (2001) Structure of human cystathionine beta-synthase: a unique pyridoxal 5′- phosphate-dependent heme protein. EMBO J 20(15):3910–3916

Meister A, Fraser PE, Tice SV (1954) Enzymatic desulfuration of beta-mercaptopyruvate to pyruvate. J Biol Chem 206(2):561–575

Messerschmidt A, Worbs M, Steegborn C, Wahl MC, Huber R, Laber B, Clausen T (2003) Determinants of enzymatic specificity in the Cys-Met-metabolism PLP-dependent enzymes family: crystal structure of cystathionine gamma-lyase from yeast and intrafamiliar structure comparison. Biol Chem 384(3):373–386

Mikami Y, Shibuya N, Kimura Y, Nagahara N, Ogasawara Y, Kimura H (2011a) Thioredoxin and dihydrolipoic acid are required for 3-mercaptopyruvate sulfurtransferase to produce hydrogen sulfide. Biochem J 439(3):479–485

Mikami Y, Shibuya N, Kimura Y, Nagahara N, Yamada M, Kimura H (2011b) Hydrogen sulfide protects the retina from light-induced degeneration by the modulation of Ca2+ influx. J Biol Chem 286(45):39379–39386

Mikami Y, Shibuya N, Ogasawara Y, Kimura H (2013) Hydrogen sulfide is produced by cystathionine gamma-lyase at the steady-state low intracellular Ca(2+) concentrations. Biochem Biophys Res Commun 431(2):131–135

Mineri R, Rimoldi M, Burlina AB, Koskull S, Perletti C, Heese B, von Dobeln U, Mereghetti P, Di Meo I, Invernizzi F, Zeviani M, Uziel G, Tiranti V (2008) Identification of new mutations in the ETHE1 gene in a cohort of 14 patients presenting with ethylmalonic encephalopathy. J Med Genet 45(7):473–478

Modis K, Coletta C, Erdelyi K, Papapetropoulos A, Szabo C (2013) Intramitochondrial hydrogen sulfide production by 3-mercaptopyruvate sulfurtransferase maintains mitochondrial electron flow and supports cellular bioenergetics. FASEB J 27(2):601–611

Mudd SH, Finkelstein JD, Irreverre F, Laster L (1964) Homocystinuria: an enzymatic defect. Science 143:1443–1445

Nagahara N (2012) Regulation of mercaptopyruvate sulfurtransferase activity via intrasubunit and intersubunit redox-sensing switches. Antioxid Redox Signal [Epub ahead of print]

Nagahara N, Katayama A (2005) Post-translational regulation of mercaptopyruvate sulfurtransferase via a low redox potential cysteine-sulfenate in the maintenance of redox homeostasis. J Biol Chem 280(41):34569–34576

Nagahara N, Nishino T (1996) Role of amino acid residues in the active site of rat liver mercaptopyruvate sulfurtransferase. J Biol Chem 271(44):27395–27401

Nagahara N, Sawada N (2006) The mercaptopyruvate pathway in cysteine catabolism: a physiologic role and related disease of the multifunctional 3-mercaptopyruvate sulfurtransferase. Curr Med Chem 13(10):1219–1230

Nagahara N, Ito T, Kitamura H, Nishino T (1998) Tissue and subcellular distribution of mercaptopyruvate sulfurtransferase in the rat: confocal laser fluorescence and immunoelectron microscopic studies combined with biochemical analysis. Histochem Cell Biol 110(3):243–250

Nagahara N, Ito T, Minami M (1999) Mercaptopyruvate sulfurtransferase as a defense against cyanide toxication: molecular properties and mode of detoxification. Histol Histopathol 14(4):1277–1286

Nagahara N, Yoshii T, Abe Y, Matsumura T (2007) Thioredoxin-dependent enzymatic activation of mercaptopyruvate sulfurtransferase. An intersubunit disulfide bond serves as a redox switch for activation. J Biol Chem 282(3):1561–1569

Ojha S, Hwang J, Kabil O, Penner-Hahn JE, Banerjee R (2000) Characterization of the heme in human cystathionine beta-synthase by X-ray absorption and electron paramagnetic resonance spectroscopies. Biochemistry 39:10542–10547

Ojha S, Wu J, LoBrutto R, Banerjee R (2002) Effects of heme ligand mutations including a pathogenic variant, H65R, on the properties of human cystathionine beta syntase. Biochemistry 41:4649–4654

Pey AL, Majtan T, Sanchez-Ruiz JM, Kraus JP (2013) Human cystathionine beta-synthase (CBS) contains two classes of binding sites for S-adenosylmethionine (SAM): complex regulation of CBS activity and stability by SAM. Biochem J 449(1):109–121

Prudova A, Bauman Z, Braun A, Vitvitsky V, Lu SC, Banerjee R (2006) S-Adenosylmethionine stabilizes cystathionine beta-synthase and modulates redox capacity. Proc Natl Acad Sci USA 103(17):6489–6494

Puranik M, Weeks CL, Lahaye D, Kabil O, Taoka S, Nielsen SB, Groves JT, Banerjee R, Spiro TG (2006) Dynamics of carbon monoxide binding to cystathionine beta-synthase. J Biol Chem 281(19):13433–13438

Quazi F, Aitken SM (2009) Characterization of the S289A, D mutants of yeast cystathionine beta-synthase. Biochim Biophys Acta-Proteins Proteomic 1794(6):892–897

Rohwerder T, Sand W (2003) The sulfane sulfur of persulfides is the actual substrate of the sulfur-oxidizing enzymes from Acidithiobacillus and Acidiphilium spp. Microbiology 149(Pt 7): 1699–1710

Rohwerder T, Sand W (2007) The sulfane sulfur of persulfides is the actual substrate of the sulfur-oxidizing enzymes from Acidithiobacillus and Acidiphilium spp. Microbiology 149:1699–1710

Sakurai H, Ogawa T, Shiga M, Inoue K (2010) Inorganic sulfur oxidizing system in green sulfur bacteria. Photosynth Res 104(2–3):163–176

Scott JW, Hawley SA, Green KA, Anis M, Stewart G, Scullion GA, Norman DG, Hardie DG (2004) CBS domains form energy-sensing modules whose binding of adenosine ligands is disrupted by disease mutations. J Clin Invest 113(2):274–284

Sen S, Banerjee R (2007) A pathogenic linked mutation in the catalytic core of human cystathionine beta-synthase disrupts allosteric regulation and allows kinetic characterization of a full-length dimer. Biochemistry 46(13):4110–4116

Shan X, Kruger WD (1998) Correction of disease-causing CBS mutations in yeast. Nat Genet 19(1):91–93

Shibuya N, Tanaka M, Yoshida M, Ogasawara Y, Togawa T, Ishii K, Kimura H (2009) 3-Mercaptopyruvate sulfurtransferase produces hydrogen sulfide and bound sulfane sulfur in the brain. Antioxid Redox Signal 11:703–714

Shibuya N, Koike S, Tanaka M, Ishigami-Yuasa M, Kimura Y, Ogasawara Y, Fukui K, Nagahara N, Kimura H (2013) A novel pathway for the production of hydrogen sulfide from D-cysteine in mammalian cells. Nat Commun 4:1366

Singh S, Banerjee R (2011) PLP-dependent H_2S biogenesis. Biochim Biophys Acta 1814:1518–1527

Singh S, Madzelan P, Banerjee R (2007) Properties of an unusual heme cofactor in PLP-dependent cystathionine beta-synthase. Nat Prod Rep 24:631–639

Singh S, Madzelan P, Stasser J, Weeks CL, Becker D, Spiro TG, Penner-Hahn J, Banerjee R (2009a) Modulation of the heme electronic structure and cystathionine beta-synthase activity by second coordination sphere ligands: the role of heme ligand switching in redox regulation. J Inorg Biochem 103:689–697

Singh S, Padovani D, Leslie RA, Chiku T, Banerjee R (2009b) Relative contributions of cystathionine beta-synthase and gamma-cystathionase to H_2S biogenesis via alternative trans-sulfuration reactions. J Biol Chem 284(33):22457–22466

Singh S, Ballou DP, Banerjee R (2011) Pre-steady-state kinetic analysis of enzyme-monitored turnover during cystathionine beta-synthase-catalyzed H_2S generation. Biochemistry 50:419–425

Spallarossa A, Forlani F, Carpen A, Armirotti A, Pagani S, Bolognesi M, Bordo D (2004) The "rhodanese" fold and catalytic mechanism of 3-mercaptopyruvate sulfurtransferases: crystal structure of SseA from Escherichia coli. J Mol Biol 335(2):583–593

Stipanuk MH, Beck PW (1982) Characterization of the enzymic capacity for cysteine desulphhydration in liver and kidney of the rat. Biochem J 206(2):267–277

Stipanuk MH, Hirschberger LL, Londono MP, Cresenzi CL, Yu AF (2004) The ubiquitin-proteasome system is responsible for cysteine-responsive regulation of cysteine dioxygenase concentration in liver. Am J Physiol Endocrinol Metab 286(3):E439–E448

Sun Q, Collins R, Huang S, Holmberg-Schiavone L, Anand GS, Tan CH, van-den Berg S, Deng LW, Moore PK, Karlberg T, Sivaraman J (2009) Structural basis for the inhibition mechanism of human cystathionine gamma-lyase, an enzyme responsible for the production of H_2S. J Biol Chem 284(5):3076–3085

Sylvester M, Sander C (1990) Immunohistochemical localization of rhodanese. Histochem J 22(4):197–200

Taoka S, Banerjee R (2001) Characterization of NO binding to human cystathionine [beta]-synthase: possible implications of the effects of CO and NO binding to the human enzyme. J Inorg Biochem 87(4):245–251

Taoka S, Banerjee R (2002) Stopped-flow kinetic analysis of the reaction catalyzed by the full length yeast cystathionine beta synthase. J Biol Chem 277:22421–22425

Taoka S, Ohja S, Shan X, Kruger WD, Banerjee R (1998) Evidence for heme-mediated redox regulation of human cystathionine beta-synthase activity. J Biol Chem 273(39):25179–25184

Taoka S, West M, Banerjee R (1999) Characterization of the heme and pyridoxal phosphate cofactors of human cystathionine β-synthase reveals nonequivalent active sites. Biochemistry 38(9):2738–2744

Taoka S, Green EL, Loehr TM, Banerjee R (2001) Mercuric chloride-induced spin or ligation state changes in ferric or ferrous human cystathionine beta-synthase inhibit enzyme activity. J Inorg Biochem 87:253–259

Taoka S, Lepore BW, Kabil O, Ojha S, Ringe D, Banerjee R (2002) Human cystathionine beta-synthase is a heme sensor protein. Evidence that the redox sensor is heme and not the vicinal cysteines in the CXXC motif seen in the crystal structure of the truncated enzyme. Biochemistry 41(33):10454–10461

Theissen U, Hoffmeister M, Grieshaber M, Martin W (2003) Single eubacterial origin of eukaryotic sulfide: quinone oxidoreductase, a mitochondrial enzyme conserved from the early evolution of eukaryotes during anoxic and sulfidic times. Mol Biol Evol 20(9):1564–1574

Tiranti V, D'Adamo P, Briem E, Ferrari G, Mineri R, Lamantea E, Mandel H, Balestri P, Garcia-Silva MT, Vollmer B, Rinaldo P, Hahn SH, Leonard J, Rahman S, Dionisi-Vici C, Garavaglia B,

Gasparini P, Zeviani M (2004) Ethylmalonic encephalopathy is caused by mutations in ETHE1, a gene encoding a mitochondrial matrix protein. Am J Hum Genet 74(2):239–252

Tiranti V, Briem E, Lamantea E, Mineri R, Papaleo E, De Gioia L, Forlani F, Rinaldo P, Dickson P, Abu-Libdeh B, Cindro-Heberle L, Owaidha M, Jack RM, Christensen E, Burlina A, Zeviani M (2006) ETHE1 mutations are specific to ethylmalonic encephalopathy. J Med Genet 43(4):340–346

Tiranti V, Viscomi C, Hildebrandt T, Di Meo I, Mineri R, Tiveron C, Levitt MD, Prelle A, Fagiolari G, Rimoldi M, Zeviani M (2009) Loss of ETHE1, a mitochondrial dioxygenase, causes fatal sulfide toxicity in ethylmalonic encephalopathy. Nat Med 15(2):200–205

Urich T, Gomes CM, Kletzin A, Frazao C (2006) X-ray structure of a self-compartmentalizing sulfur cycle metalloenzyme. Science 311(5763):996–1000

Vitvitsky V, Kabil O, Banerjee R (2012) High turnover rates for hydrogen sulfide allow for rapid regulation of its tissue concentrations. Antioxid Red Signal 17(1):22–31

Wang J, Hegele RA (2003) Genomic basis of cystathioninuria (MIM 219500) revealed by multiple mutations in cystathionine gamma-lyase (CTH). Hum Genet 112(4):404–408

Weeks CL, Singh S, Madzelan P, Banerjee R, Spiro TG (2009) Heme regulation of human cystathionine beta-synthase activity: insights from fluorescence and Raman spectroscopy. J Am Chem Soc 131(35):12809–12816

Westley J (1973) Rhodanese. Adv Enzymol Relat Areas Mol Biol 39:327–368

Westley J, Adler H, Westley L, Nishida C (1983) The sulfurtransferases. Fundam Appl Toxicol 3(5):377–382

Williams RA, Kelly SM, Mottram JC, Coombs GH (2003) 3-Mercaptopyruvate sulfurtransferase of Leishmania contains an unusual C-terminal extension and is involved in thioredoxin and antioxidant metabolism. J Biol Chem 278(3):1480–1486

Yadav PK, Banerjee R (2012) Detection of reaction intermediates during human cystathionine beta-synthase-monitored turnover and H_2S production. J Biol Chem 287:43464–43471

Yadav PK, Yamada K, Chiku T, Koutmos M, Banerjee R (2013) Structure and kinetic analysis of H_2S production by human mercaptopyruvate sulfurtransferase. J Biol Chem 288(27): 20002–20013

Yadav PK, Xie P, Banerjee R (2012) Allosteric communication between the pyridoxal 5′-phosphate (PLP) and heme sites in the H_2S generator human cystathionine beta-synthase. J Biol Chem 287(45):37611–37620

Yamagata S, Yasugahira T, Okuda Y, Iwama T (2003) Conversion of the aminocrotonate intermediate limits the rate of gamma-elimination reaction catalyzed by L-cystathionine gamma-lyase of the yeast Saccharomyces cerevisiae. J Biochem 134(4):607–613

Yamamoto T, Takano N, Ishiwata K, Suematsu M (2011) Carbon monoxide stimulates global protein methylation via its inhibitory action on cystathionine beta-synthase. J Clin Biochem Nutr 48(1):96–100

Yang G, Wu L, Jiang B, Yang W, Qi J, Cao K, Meng Q, Mustafa AK, Mu W, Zhang S, Snyder SH, Wang R (2008) H_2S as a physiologic vasorelaxant: hypertension in mice with deletion of cystathionine gamma-lyase. Science 322(5901):587–590

Zhong WX, Wang YB, Peng L, Ge XZ, Zhang J, Liu SS, Zhang XN, Xu ZH, Chen Z, Luo JH (2012) Lanthionine synthetase C-like protein 1 interacts with and inhibits cystathionine beta-synthase: a target for neuronal antioxidant defense. J Biol Chem 287(41):34189–34201

Zimmermann P, Laska S, Kletzin A (1999) Two modes of sulfite oxidation in the extremely thermophilic and acidophilic archaeon acidianus ambivalens. Arch Microbiol 172(2):76–82

Hydrogen Sulfide as an Oxygen Sensor

2

Kenneth R. Olson

Abstract

Eukaryotic cells depend upon oxygen (O_2) for their survival and elaborate mechanisms have evolved in multicellular animals, especially vertebrates, to monitor the availability of environmental O_2, the efficiency of O_2 extraction from the environment, ensure adequate O_2 delivery to tissues and even to regulate cellular metabolism when O_2 availability is compromised. In vertebrates, specialized O_2 "sensing" cells have developed to carry out many of these processes. Although all O_2 sensing cells ultimately couple low Po_2 (hypoxia) to physiological responses, how these cells actually detect hypoxia, i.e., the "O_2 sensor" remains controversial. We have recently proposed that hydrogen sulfide (H_2S) through its O_2-dependent metabolism is a universal and phylogenetically ancient O_2 sensing mechanism. This hypothesis is based on a variety of experimental evidence including; (1) the effects of exogenous H_2S mimic hypoxia, (2) H_2S production and/or metabolism is biochemically coupled to O_2, (3) tissue H_2S concentration is inversely related to Po_2 at physiologically relevant Po_2s, (4) compounds that inhibit or augment H_2S production inhibit and augment hypoxic responses, (5) H_2S acts upon effector mechanisms known to mediate hypoxic responses, (6) H_2S was key to the origin of life and the advent of eukaryotic cells and the reciprocal relationship between O_2 and H_2S has been inexorably intertwined throughout evolution. The evidence for H_2S-mediated O_2 sensing is critically examined in this review.

Keywords

Hypoxia • Cardiovascular • Chemoreceptors • Mitochondria

K.R. Olson (✉)
Indiana University School of Medicine - South Bend, 1234 Notre Dame Avenue, South Bend, IN 46617, USA
e-mail: kolson@nd.edu

H. Kimura (ed.), *Hydrogen Sulfide and its Therapeutic Applications*,
DOI 10.1007/978-3-7091-1550-3_2,

Abbreviations

3MP	3-Mercaptopyruvate
3-MST	3-Mercaptopyruvate sulfur transferase
AMPK	AMP-activated protein kinase
AOA	Amino-oxyacetate
AOX	Alternative oxidase
Asp	Aspartic acid
CAT	Cysteine aminotransferase
CBS	Cystathionine β-synthase
CDO	Cysteine dioxygenase
CO	Carbon monoxide
CSE	Cystathionine γ-lyase
Cys	Cysteine
DAO	D-amino acid oxidase
DHLA	Dihydrolipoic acid
EC_{50}	Effective concentration for half-maximal activity
ETHE1	Mitochondrial sulfur dioxygenase
Gly	Glycine
GSH	Reduced glutathione
GSSG	Oxidized glutathione
H_2O_2	Hydrogen peroxide
H_2S	Hydrogen sulfide
HA	Hydroxylamine
HIF	Hypoxia-inducible factor
HPC	Hypoxic pulmonary vasoconstriction
HSD	Hypoxic systemic vasodilation
IK_{Ca}	Intermediate conductance potassium channel
K_{ATP}	Adenosine triphosphate sensitive potassium channel
KCl	Potassium chloride
K_i	Inhibition constant
K_v	Voltage-gated potassium channels
NEB	Neuroepithelial bodies
NEC	Neuroepithelial cells
NO	Nitric oxide
O_2^-	Superoxide
PASMC	Pulmonary artery smooth muscle cells
pB	Pre-Bötzinger respiratory group
PKCε	Protein kinase C epsilon
PLP	Pyridoxal 5′phosphate
Po_2	Partial pressure of oxygen
PPG	Propargyl glycine

Rde	Rhodanase
RI	Ischemia reperfusion injury
ROS	Reactive oxygen species
R-SO	Sulfenyl
$S_2O_3^{2-}$	Thiosulfate
SO	Sulfur oxidase
SO_3^{2-}	Sulfite
SO_4^{2-}	Sulfate
SQR	Sulfide:quinone oxidoreductase
ST	Sulfur transferase
TASK	Acid-sensitive potassium channel
TR	Thiosulfate reductase, a.k.a. rhodanase
TRD	Thioredoxin reductase
Trx	Thioredoxin
α-Kg	α-ketoglutarate

2.1 Introduction

Specialized O_2 sensing tissues in vertebrates are strategically placed to monitor ambient O_2, O_2 transport in blood, and to match blood flow with ventilation or tissue demand. Neuroepithelial cells (NEC) are present on the external surfaces of fish gills and monitor water Po_2 (Milsom and Burleson 2007). This is especially important for these aquatic vertebrates because, compared to air, water has considerably less O_2 (1/30), slower rates of O_2 diffusion (Krogh's diffusion coefficients 1/200,000) and higher viscosity (60 times). Even more problematic, aquatic O_2 levels can vary both temporally (minutes to seasons) and spatially within meters (Bickler and Buck 2007). In lungs of newborn mammals, cells similar to neuroepithelial-like cells are found in clusters (called neuroepithelial bodies, NEB) near airway bifurcations and here they may be important in the transition from the relatively hypoxic uterine environment during and shortly after birth (Kemp et al. 2002). External O_2 sensors other than NEB are relatively uncommon in terrestrial vertebrates. Presumably this is because atmospheric O_2 is relatively constant (21 %) and internal O_2 sensors appear to be able to accommodate changes in O_2 availability (such as in borrows or with increasing altitude) if needed. Arterial O_2 sensors that monitor blood Po_2 and O_2 delivery are found in fish as vascular-facing NEC and in higher vertebrates as the type I glomus cells of carotid and aortic bodies. (It is perhaps no coincidence that the first gill arch of fish is the homolog of the mammalian carotid body.) Mammalian adrenal medullary cells and homologous chromaffin cells that line systemic veins in fish secrete catecholamines in response to hypoxemia (Nurse et al. 2006; Perry et al. 2000) and may monitor tissue O_2 extraction. Perhaps the best characterized O_2 sensing tissues are the blood vessels themselves. It has generally been accepted that hypoxia relaxes systemic vessels thereby matching tissue perfusion to metabolic demand, whereas hypoxia contracts pulmonary vessels to match ventilation to perfusion (Sylvester et al. 2012). However, recent studies have shown that many systemic vessels in non-mammalian vertebrates are contracted by hypoxia (Russell et al. 2008) and hypoxia relaxes pulmonary vessels in diving mammals to prevent pulmonary hypertension that would otherwise occur during a prolonged dive (Olson et al. 2010). Although the vascular response to hypoxia is intrinsic to vascular smooth muscle (Madden et al. 1992) there appears to be considerable plasticity in the functional organization of this response. These atypical responses have been key to evaluating H_2S-mediated mechanisms for acute O_2 sensing as described in this chapter.

There is considerable controversy concerning the actual mechanism with which these cells detect O_2 levels or availability and then transduce this into physiologically relevant signals and of the numerous proposed O_2 sensing mechanisms none have received unanimous support. Our work has suggested that the O_2 dependent metabolism of hydrogen sulfide (H_2S) is an effective and efficient mechanism of H_2S sensing. This review presents the evidence supporting our hypothesis.

2.2 Mechanism(s) of O_2 Sensing

Despite the apparent ubiquity of O_2 sensing tissues in vertebrates there has been little consensus on the mechanism that specifically measures Po_2 or O_2 concentration. Potassium channels have long been a likely candidate (Weir and Archer 1995) although it is now believed that they operate downstream of the actual sensing mechanism. The various theories of O_2 sensing mechanisms have been extensive reviewed Sylvester et al. (2012). Because mitochondria account for most of a cell's O_2 consumption they are central to most O_2 sensing theories and implicit in the three most prevalent theories, the redox hypothesis, the reactive oxygen species (ROS) hypothesis and the energy state/AMPK (AMP-activated protein kinase) hypothesis. In the redox hypothesis, hypoxia suppresses mitochondrial oxidative phosphorylation which further reduces the cytosol and decreases ROS production. Voltage-gated potassium (K_v) channels that were tonically kept open during normoxia by ROS now close and the resulting cell depolarization opens voltage-gated calcium channels and the influx of calcium produces contraction. Essentially the opposite occurs in the ROS hypothesis where hypoxia increases mitochondrial production of ROS, namely superoxide (O_2^-) and probably more importantly hydrogen peroxide (H_2O_2). The ROS thus produced activate a variety of intracellular signaling cascades that also increase intracellular calcium concentration. In the energy state/AMPK hypothesis, hypoxia decreases mitochondrial ATP production which increases the AMP to ATP ratio and activates AMP kinase. The resulting production of cyclic ADP ribose then brings about an increase in intracellular calcium and contraction. It should be noted, however, that vascular smooth muscle and endothelial cells can and do derive their energy from glycolysis and even though they respond to hypoxia, their ATP levels do not appear to be affected (Dromparis and Michelakis 2013). Other O_2 sensing mechanisms such as heme oxygenase (which generates the gasotransmitter carbon monoxide, CO), plasma membrane bound NADPH oxidase or a yet identified hemoprotein or mitochondrial complex III and nitric oxide (NO) have also been described for various tissues (Evans et al. 2011; Gonzalez et al. 2010; Haldar and Stamler 2013; Waypa and Schumacker 2010; Wolin et al. 2010).

2.3 H_2S Oxidation as an O_2 Sensing Mechanism

We (Olson et al. 2006) originally proposed that the O_2-dependent metabolism of endogenously generated, and biologically active H_2S functioned as an efficient O_2 sensing mechanism that initiated hypoxic pulmonary vasoconstriction (HPC) and hypoxic systemic vasodilation (HSD). This hypothesis appears to fulfill the criteria for an O_2 sensing mechanism (Olson 2011) in that; (1) the effects of exogenous H_2S mimic hypoxia, (2) H_2S production and/or metabolism is biochemically coupled to O_2, (3) tissue H_2S concentration is inversely related to Po_2 at physiologically relevant Po_2s, (4) compounds that inhibit or augment H_2S production inhibit and augment hypoxic responses, (5) H_2S acts upon effector mechanisms known to mediate hypoxic responses, (6) H_2S-mediated O_2 sensing has an evolutionary precedent and a phylogenetic history. The following sections describe how these criteria have been met.

2.3.1 The Effects of Exogenous H_2S Mimic Hypoxia

2.3.1.1 Cardiovascular System

Mechanical responses to hypoxia are identical to those produced by H_2S in all vessels (Fig. 2.1). Relaxation of rat thoracic aorta and portal vein was one of the first "physiological" effects of H_2S identified (Hosoki et al. 1997) and this mimics the well-known hypoxic vasodilation of mammalian systemic vessels. Similar hypoxic and H_2S vasodilations have been observed in all classes of vertebrates from the most primitive hagfish and lamprey to mammals (cf Russell et al. 2008; Dombkowski et al. 2005). This response has now been observed in over 20 mammalian studies. H_2S also dilates the mouse ductus arteriosus (Baragatti et al. 2013) which would be expected to keep this vessel patent in the relative hypoxic intrauterine environment. H_2S vasodilation of mammalian systemic vessels has also been observed in perfused organs (Cheng et al. 2004) and in vivo (Derwall et al. 2011; Leffler et al. 2010; Mustafa et al. 2009; Yan et al. 2004; Yang et al. 2008). Consistent with hypoxic pulmonary vasoconstriction, H_2S constricts isolated bovine pulmonary arteries (Olson et al. 2006, 2010), increases vascular resistance in perfused fish gills (Skovgaard and Olson 2012) and perfused rat lungs (Madden et al. 2012) and increases pulmonary arterial blood pressure in vivo (Derwall et al. 2011; Sowmya et al. 2010).

Hypoxic responses of non-mammalian vertebrates are considerably more variable in that systemic vessels can be either dilated or contracted by hypoxia and often these effects are multiphasic. Nevertheless, in over 30 animals from all

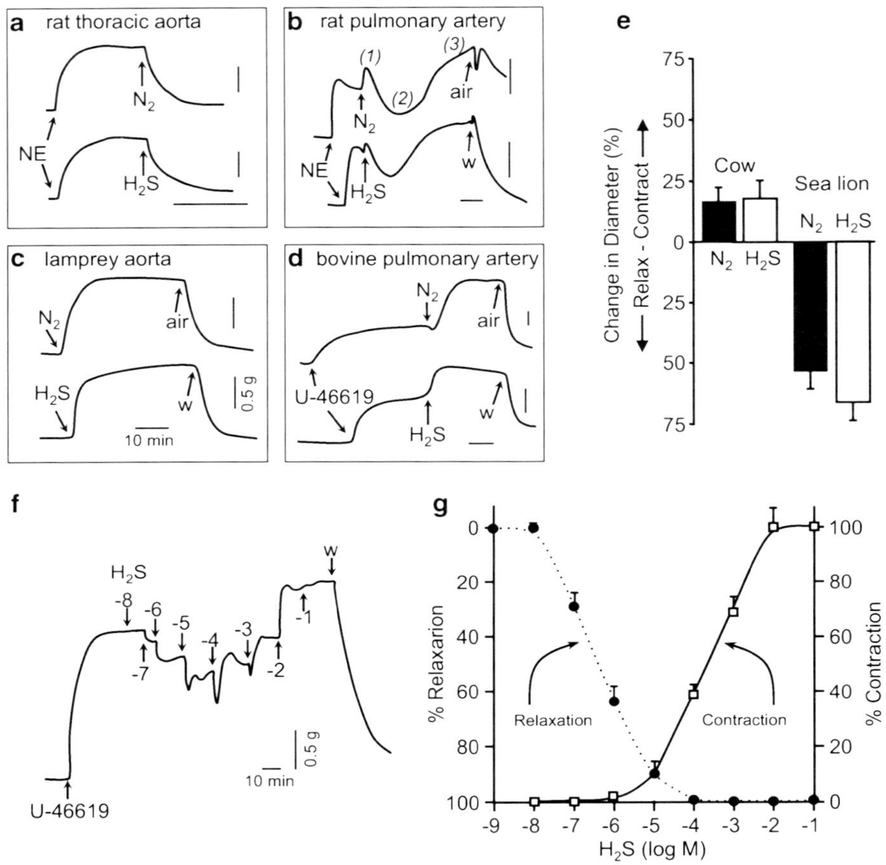

Fig. 2.1 Hypoxia (95 % N_2/5 % CO_2 **a**, **b**, **d**, **e**; or 100 % N_2, **c**) and H_2S (300 μmol/l) produce identical responses in conductance (>500 μm diameter) vessels from rat thoracic aorta (**a**), rat pulmonary artery (**b**), lamprey dorsal aorta (**c**) and bovine pulmonary artery (**d**). Vessels pre-contracted with 1 μmol/l norepinephrine (NE) or U-46619 (thromboxane A_2 mimetic; 0.1 μmol/l); *air* aeration with room air; *w* wash; *1*, *2*, *3*, tri-phasic response. *Horizontal time bar* in **a–d** = 10 min, vertical tension scale = 0.5 g. (**e**), hypoxia (Po_2 ~ 50 mmHg) and H_2S (3×10^{-4} M) produce identical contractions of cow resistance (<500 μm diameter) pulmonary arteries while both stimuli relax sea lion resistance pulmonary arteries. Mean ± SE; *N* (animals,vessels) = cow (9,9), sea lion (3,5). (**f**, **g**), H_2S has two dose-dependent effects on pre-contracted (0.1 μmol/l U-46619) bovine pulmonary arteries. H_2S appears to produce a dose-dependent relaxation between 10^{-8} and 10^{-5} mol/l, whereas above 10^{-5} mol/l it produces a dose-dependent constriction. The EC_{50} for relaxation ($5.5 \pm 1.8 \times 10^{-7}$ M) is significantly ($p < 0.05$) different from the EC_{50} for contraction ($3.7 \pm 1.5 \times 10^{-4}$ M). (**f**) single trace of cumulative doses, *arrows* indicate log M [H_2S]; *w* wash; (**g**) average (+SE) of eight vessels, *filled circles* denote relaxation and *open squares* contraction (values extrapolated where curves overlap ~ 10^{-5} M) (**a–d**, **f**, **g** Adapted from Olson et al. 2006, with permission; **e** adapted from Olson et al. 2010, with permission)

vertebrate classes, the hypoxic responses of both systemic and respiratory vessels are consistently mimicked by H_2S (summarized in Olson and Whitfield 2010). This includes the signature multiphasic contraction-relaxation-contraction of rat pulmonary arteries (Fig. 2 1b) and the unique hypoxic and H_2S-mediated dilation of sea lion pulmonary resistance vessels (Fig. 2.1e). It is also evident that many of these multiphasic responses such as the initial relaxation followed by constriction in bovine pulmonary arteries are the results of separate, but dose-dependent effects of H_2S (Fig. 2.1f), a hint of which can also be seen during the onset of hypoxia (Fig. 2.1d). Hypoxia and H_2S also relax non-vascular smooth muscle of fish urinary bladder and the gastrointestinal tract and in the gastrointestinal tract both of these stimuli produce a unique and transient increase in spontaneous contraction frequency and amplitude prior to the onset of the inhibitory effects (Dombkowski et al. 2006, 2011). H_2S also relaxes human corpus cavernosum and urinary bladder smooth muscle (d'Emmanuele di Villa Bianca et al. 2009; Fusco et al. 2012).

Involvement of H_2S has also been shown in other hypoxia-related responses. In the cardiovascular system this includes angiogenesis, ischemia reperfusion injury (RI) and pre- and post-conditioning against RI (Bian et al. 2006; Cai et al. 2007; Calvert et al. 2010; Lavu et al. 2010; Liu et al. 2010; Pan et al. 2006; Papapetropoulos et al. 2009; Szabo and Papapetropoulos 2011; Wang et al. 2010a, b; Yong et al. 2008). H_2S also contributes to hypoxia-induced radiation resistance (Zhang et al. 2011) and it is a cryogenic mediator of hypoxia-induced anapyrexia (Kwiatkoski et al. 2012).

2.3.1.2 Respiration

Intravascular injection or inhalation of H_2S at low concentrations has long been known to mimic hypoxemia by stimulating respiration in many mammals (Beauchamp et al. 1984; Reiffenstein et al. 1992; Haggard and Henderson 1922; Haouzi et al. 2009 [in mice but not rats], 2011; Haouzi 2012; Van de Louw and Haouzi 2012) and birds (Klentz and Fedde 1978). This may have both a central and peripheral component mediated through the carotid bodies. Intracerebroventricular injection of H_2S produces a K_{ATP} channel-mediated dose dependent bradycardia and hypotension (Liu et al. 2011) akin to the hypoxic diving reflex. In more specific studies it was observed that H_2S increases discharge frequency from the pre-Bötzinger (pB) dorsal inspiratory respiratory group and it may initially produce transient inhibition of the pB by stimulating the nearby parafacial respiratory group (Chen et al. 2013; Hu et al. 2008). H_2S stimulates peripheral chemoreceptors (neuroepithelial cells) in the fish gill (Olson et al. 2008) and mammalian carotid body (Li et al. 2010; Makarenko et al. 2012; Peng et al. 2010) and it stimulates the mammalian adrenal medulla (Peng et al. 2010; Zhu et al. 2012) and homologous fish chromaffin cells (Perry et al. 2009). H_2S may also contribute to the sequela of events in which heart failure, hypertension and renal failure activate the carotid body leading to breathing instability and increased sympathetic nerve activity (Schultz et al. 2012).

2.3.2 H_2S Production and/or Metabolism is Biochemically Coupled to O_2

2.3.2.1 Biosynthesis

Pathways for H_2S synthesis and metabolism are shown in Fig. 2.2. L-cysteine and L-homocysteine account for most H_2S production through the activity of four enzymes, cystathionine β-synthase (CBS), cystathionine γ-lyase (CSE aka CGL) and sequential catalysis by cysteine aminotransferase (CAT) and 3-mercaptopyruvate sulfurtransferase (3-MST). H_2S may be synthesized in the cytosol as CBS and CSE and CAT/3-MST are cytosolic enzymes, or it may be synthesized in the mitochondria as CAT/3-MST are present there as well (Kamoun 2004). 3-MST is especially abundant in the mitochondrial matrix (Mikami et al. 2011a) where it can take advantage of threefold higher cysteine concentration than in the cytosol (Fu et al. 2012). CSE can also be translocated to the mitochondria by a variety of stress-related stimuli (Fu et al. 2012). Initially it was believed that CBS was predominantly found in the brain and CSE in the cardiovascular system (reviewed in Kimura 2010), although a broader distribution is becoming evident, i.e., CBS has been identified in vascular endothelium, CAT and 3-MST in vascular endothelium and brain and MST, but not CAT, in vascular smooth muscle (Kimura 2010; Olson et al. 2010). CBS and CSE have also been identified in human plasma (Bearden et al. 2010). H_2S can be generated from D-cysteine, however, this pathway appears limited to the brain and kidney where it protects the former from oxidative stress and the latter from re-perfusion injury; it may be of limited function elsewhere (Shibuya et al. 2013). CBS, CSE and CAT are pyridoxal 5′phosphate (PLP)-dependent, enzymes. *S*-adenosylmethionine allosterically activates CBS (Stipanuk 2004). CBS contains a heme group that can be inhibited by presumably physiological levels of carbon monoxide (CO; inhibition constant, $K_i = 5.6\ \mu M$) and although it is also inhibited by nitric oxide (NO), the K_i is so high (320 μM) that it is of questionable physiological significance; O_2 does not affect CBS activity (Banerjee and Zou 2005). CSE and both cytosolic and mitochondrial CAT activity are inhibited by calcium, independent of calmodulin (Mikami et al. 2011b, 2013).

2.3.2.2 Metabolism (Inactivation)

Mathematical models of H_2S diffusion out of cells versus intracellular metabolism suggest that most H_2S is inactivated intracellularly (Olson 2013), and this is supported by the considerable body of evidence that mitochondria efficiently oxidize H_2S (reviewed in Olson 2012c). Mitochondrial enzymes, sulfide:quinone oxidoreductase (SQR), 3-MST, rhodanase (Rde), thiosulfate reductase (TR), sulfur dioxygenase (ETHE1) and sulfite oxidase (SO) ultimately oxidize H_2S to sulfate (SO_4^{2-}) which is then excreted. Sulfite (SO_3^{2-}) and thiosulfate ($S_2O_3^{2-}$) are intermediates. SQR is bound to the inner mitochondrial membrane and it is closely associated with the respiratory chain "supercomplex"(Hildebrandt 2011) which provides a hint at its O_2 sensing function as the mitochondrion is a lead candidate for the site of O_2 sensing (Sylvester et al. 2012). Not surprisingly, 3-MST,

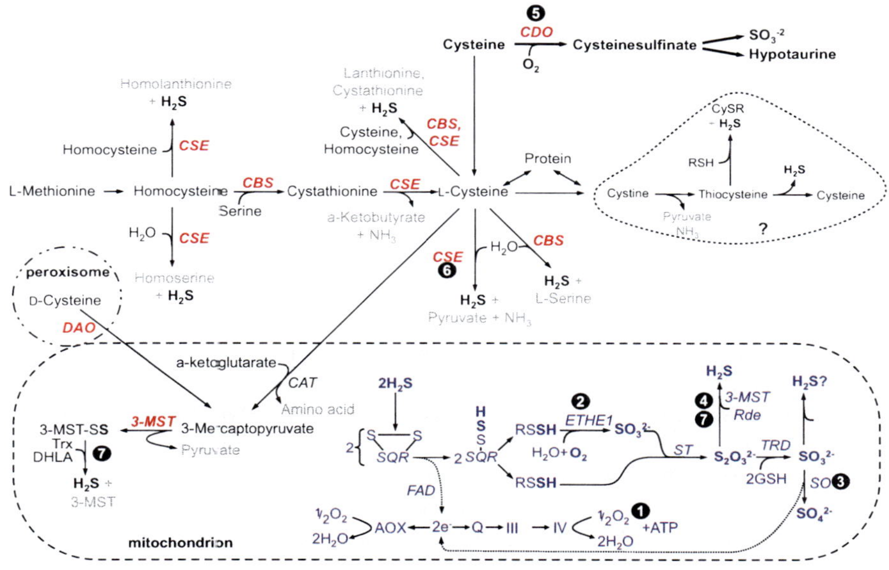

Fig. 2.2 Pathways for H_2S production and degradation. H_2S is synthesized from homocysteine or cysteine by the cytosolic enzymes cystathionine β-synthase (*CBS*), cystathionine γ-lyase, or the tandem action of cysteine aminotransferase (*CAT*) and 3-mercaptopyruvate sulfur transferase (3-*MST*) in both the cytosol and mitochondria. Other potential mechanisms for H_2S biosynthesis have been described in invertebrates (*dotted enclosure*; Julian et al. 2002) but have yet to be confirmed in mammals. 3-mercaptopyruvate can also be produced from D-cysteine in the peroxisome by D-amino acid oxidase (*DAO*). Oxidation of H_2S in the mitochondria (*blue letters*) is initiated by interaction with the enzyme sulfide:quinone oxidoreductase (SQR) producing two SQR persulfides that are then transferred to a mobile sulfide carrier (RSSH), one of which is oxidized to sulfite (SO_3^{2-}) by mitochondrial sulfur dioxygenase (*ETHE1*). Sulfur transferase (*ST*) transfers the other persulfide to form thiosulfate ($S_2O_3^{2-}$). H_2S may be regenerated from thiosulfate by 3-MST or rhodanase (*Rde*), or the thiosulfate may be oxidized to sulfate (SO_4^{2-}) by the sequential actions of thioredoxin reductase (*TRD*) in combination with glutathione (*GSH*) and sulfur oxidase (*SO*). Oxidation of H_2S donates electrons to the respiratory chain (Q, III, IV) that has been shown in invertebrates and mammals to result in ATP production and O_2 consumption. An alternative oxidase (*AOX*) that oxidizes H_2S without concomitant ATP production has also been observed in invertebrates. H_2S can also be regenerated from thiosulfate by 3-mercaptopyruvate sulfur transferase (*3-MST*) or rhodanase (*Rde*) in the presence of other reducing disulfides such as thioredoxin (*Trx*) or dihydrolipoic acid (*DHLA*). *Circled numbers* indicate actual or potential hypoxia-sensitive sites (see text for details) (Modified from Olson et al. (2012c), with permission)

rhodanase and thiosulfate reductase are abundant in the mitochondrial matrix and to a lesser extent the intermembrane space (Koj et al. 1975). H_2S oxidation begins with H_2S binding to the highly conserved Cys-Cys disulfide bridge of SQR. The sulfide is oxidized to elemental sulfur forming SQR persulfide (sulfane sulfur) with the now-reduced SQR cysteine (SQR-SSH). Two H_2S and two SQR are involved, one persulfide sulfur is transferred to sulfur dioxygenase (SDO) where it is oxidized to sulfite while sulfur from the second persulfide is transferred from the SQR to sulfite by sulfur transferase (ST) producing thiosulfate. One electron from each of the two H_2S are fed via the quinone pool (Q) into the respiratory chain (Marcia et al. 2010) and they ultimately reduce O_2 at complex IV. SQR is bound to the inner mitochondrial membrane and it is believed that sulfur is shuttled from SQR by a mobile carrier such as glutathione (GSH), dihydrolipoate, thioredoxin or even sulfite (Hildebrandt and Grieshaber 2008; Jackson et al. 2012; Theissen and Martin 2008). One of the mobile persulfides is oxidized to sulfite by ETHE1 which consumes molecular O_2 and water in the process. Metabolism of H_2S through SQR appears ubiquitous in tissues although the brain may be an exception (Hildebrandt 2011; Lagoutte et al. 2010; Linden et al. 2011; but see Ackermann et al. 2011). The capacity of cells to oxidize sulfide appears to be greater than the estimated rate of sulfide production (Bouillaud and Blachier 2011; Furne et al. 2008). Thus it is expected that intracellular H_2S concentrations are very low under normoxic conditions.

Under normoxic conditions, most thiosulfate is further metabolized to sulfate by thiosulfate reductase (TR) and sulfite oxidase (SO). Elimination of one sulfur atom as sulfate is accompanied by four atoms of O_2. Even though sulfur excretion as thiosulfate would conserve O_2, apparently there is little need for this when is O_2 plentiful and thiosulfate excretion in vertebrates is generally low. In fact, it is not clear why most vertebrates, especially terrestrial ones, would bother with this pathway at all. However, as described below, it may be an important avenue for regeneration of H_2S during hypoxia. Details of H_2S biosynthesis and metabolism can be found in (Kabil and Banerjee 2010; Kimura 2010; Olson 2012c; Stipanuk 2004; Stipanuk et al. 2009).

2.3.2.3 H_2S Oxidation via Electron Transport

In our original mechanism of H_2S-mediated O_2 sensing (Olson et al. 2006) we proposed that H_2S was constitutively produced in the cytosol through transsulfurration and oxidized in the mitochondria. As the amount of O_2 available to the mitochondria determined the rate of H_2S oxidation, and hence H_2S concentration, this was the "O_2 sensor." Although this is still likely an integral mechanism, it is now evident that there are a number of other mechanisms with which O_2 can influence H_2S concentration and thereby contribute to O_2 sensing. These mechanisms, some of which are expected to rapidly respond to O_2 while others may provide a longer bias of H_2S concentration, are described in the following paragraphs and shown numerically in Fig. 2.2. While many of these pathways remain to be verified in the context of O_2 sensing, there is accumulating evidence, largely through studies on enzyme deficiencies, that their dysfunction will impact sulfide metabolism and either directly or indirectly increase H_2S concentration.

Rapid Effectors of H_2S Concentration

Electron transport: Disruption of electron flow down the respiratory chain by insufficient O_2 delivery to the mitochondrion at complex IV still remains a very likely and highly effective mechanism to regulate H_2S as it directly couples O_2 availability to H_2S inactivation. This will prevent any further oxidation of H_2S that was derived from transsulfurration. This pathway of H_2S oxidation has been well-established in the context of ATP synthesis (Goubern et al. 2007; Lagoutte et al. 2010; Modis et al. 2013).

ETHE1: The mitochondrial dioxygenase, ETHE1, uses molecular O_2 and water to oxidize the mobile persulfide from SQR to form sulfite. Inhibition of this pathway will prevent H_2S binding to SQR. Although the effects of O_2 on this pathway have not been examined in detail, they would presumably be similar to the well-characterized ETHE1 deficiencies in experimental animals (ETHE1$^{-/-}$mice) and humans, the pathology of which is characterized by greatly elevated tissue H_2S and thiosulfate (Di Meo et al. 2011; Drousiotou et al. 2011; Giordano et al. 2011; Tiranti et al. 2009).

Sulfite oxidase: Sulfite oxidase (SO) in the mitochondrial innermembrane space catalyzes the oxidation of sulfite to sulfate by transferring an atom of O_2 from water to sulfite and in the process the enzyme undergoes a 2-electron reduction (Rajapakshe et al. 2012). These electrons are then transferred from SO to cytochrome c and shuttled into the electron transport chain with molecular oxygen as the terminal acceptor. This couples sulfite concentration to O_2 availability and suggests that a hypoxia-induced increase in sulfite would increase thiosulfate concentration and ultimately H_2S production. SO deficiency in a human was first demonstrated in 1967 and, as might be expected, this patient presented with elevated urinary thiosulfate (Mudd et al. 1967).

Thiosulfate reduction: As described above, inhibition of either ETHE1 or SO will prevent further oxidation of thiosulfate and this thiosulfate can now directly produce H_2S under appropriate conditions by reduction. The sulfur atoms on thiosulfate have different oxidation states, the inner sulfur is +5 and the outer (persulfide) sulfur is −1. In the presence of a reducing reagent such as the endogenous mitochondrial reductant, dihydrolipoic acid (DHLA), 3-MST or thiosulfate reductase (TR; aka rhodanase, Rde) can catalyze the removal of the persulfide and generate H_2S (Mikami et al. 2011a; Villarejo and Westley 1963). H_2S generation from thiosulfate under these conditions has been demonstrated in a variety of mammalian and non-mammalian tissues and it is enhanced by hypoxia (Olson et al. 2013). Thiosulfate can also be generated from D-cysteine by D-amino acid oxidase through formation of 3-mercaptopyruvate (Huang et al. 1998) although the biological significance of this pathway is unknown. H_2S generation from thiosulfate will be greatest where thiosulfate concentrations are the highest and when the immediate environment becomes more reduced and the relevant reducing molecules become more available. Thiosulfate concentrations are probably highest in or near the inner mitochondrial membrane, the site of SQR. This is also likely to experience the greatest increase in hypoxia-induced reducing equivalents such as DHLA or thioredoxin because as cells become hypoxic, reactive oxygen species (ROS) increase in both the cytosol and mitochondrial intermembrane space, whereas ROS decrease in the mitochondrial matrix (Waypa et al. 2010). Thus, thiosulfate reduction could significantly contribute to the initial increase in H_2S concentration that activates the O_2 sensing cascade and by recycling sulfur it conserves biologically relevant thiols.

Long-Term Effectors of H_2S Concentration

Cysteine dioxygenase: Cysteine dioxygenase (CDO), a cytosolic enzyme, effectively eliminates sulfur from entering the H_2S pool by irreversibly catalyzing the oxidation of cysteine to cysteinesulfinate which can be further metabolized to hypotaurine, then taurine or sulfite and then sulfate for excretion (Stipanuk et al. 2009). One of the primary functions of CDO is believed to be the detoxification of excess dietary or metabolic cysteine as CDO activity is dynamically regulated by cysteine (as much as 450-fold) whereas cysteine desulfuration, the H_2S-forming transsulfuration pathway is not regulated (Stipanuk et al. 2009). CDO also contributes to degradation of methionine and homocysteine after their conversion to cysteine (Stipanuk and Ueki 2011). Impairment of this pathway, which has been demonstrated in CDO knockout ($CDO^{-/-}$) mice, redirects sulfur through the desulfuration pathway and increases thiosulfate and H_2S production (Ueki et al. 2011; Roman et al. 2013). As molecular O_2 is the only other substrate in CDO-mediated cysteine oxidation it is likely that hypoxia will impair cysteine oxidation and favor H_2S production. It is not likely that this would contribute to the rapid on/off signaling observed acute hypoxia, but it could place a long-term bias on chronic H_2S-mediated H_2S sensing. An inability of this pathway to handle a large transient cysteine load may partially explain how hypoxic responses are augmented by exogenous cysteine (see below). Ueki et al. (2011) also noted the striking similarities between $CDO^{-/-}$and $ETHE1^{-/-}$pathologies and suggested that sulfide/H_2S was the common factor.

CSE translocation to mitochondria: In vascular smooth muscle cells, hypoxia stimulates CSE translocation from the cytosol to the mitochondria where it utilizes the threefold increase in cysteine concentration to generate H_2S which can be subsequently used in ATP synthesis (Fu et al. 2012). It has been proposed that this H_2S improves mitochondrial ATP production and it is protective during hypoxia (Fu et al. 2012). There are, however, several problems with this hypothesis. First, vascular smooth muscle can derive sufficient energy from anaerobic metabolism and does not need oxidative phosphorylation to supply energy, even during hypoxia (Dromparis and Michelakis 2013). Second, because ATP generated from H_2S is ultimately dependent on O_2, it is unclear what the electron acceptor will be in the absence of O_2. However, the H_2S formed by CSE translocation could clearly contribute to O_2 sensing and hypoxic vasodilation.

Indirect O_2 Effects

3-MST and thioredoxin catalytic-site cysteines: Many enzymes contain cysteine in the catalytic site and because these cysteines generally have a low pKa, they are redox active (Nagahara 2011). The catalytic cysteine in 3-MST (Cys^{247}; rat) is one example. The exposed, Cys^{247} sulfur is readily oxidized to a sulfenyl (R-SO) by O_2, peroxide (H_2O_2) or other oxidants under mild oxidizing conditions; the sulfenyl is also reduced (to R-SH) by reduced thioredoxin (Trx) but not GSH (Nagahara 2012). Monomeric Rat 3-MST can also dimerize by mild oxidation of two other exposed cysteines, Cys^{154} and Cys^{263}; both oxidization of Cys^{247} and dimerization inactivate the enzyme. A defect in 3-MST activity, presented clinically as mercaptolactate-cysteine disulfidia, is believed to be associated with deficient H_2S production (Nagahara 2012). While the three external cysteines of 3-MST may well allow it to serve as an effective antioxidant (Nagahara 2012) it could also be a key component of the H2S-mediated O_2 sensing mechanism. Furthermore, as 3-MST is found in both the cytosol and mitochondrial matrix and during hypoxia ROS in the cytosol increase while ROS in the matrix decrease, in both pulmonary and systemic arterial smooth muscle cells (Waypa et al. 2010), this will favor mitochondrial H_2S production and inhibit it in the cytosol. Parenthetically, the now-oxidized Trx can be reduced by thioredoxin reductase (TRD) using NADPH; which may be an overlooked, but key explanation for why NADPH has been central in many O_2-sensing theories (Gupte and Wolin 2008).

2.3.3 Tissue H_2S Concentration Is Inversely Related to Po_2 at Physiologically Relevant Po_2s

If H_2S and O_2 coexist in the environment or in cells it is only transient. In fact, hypoxia, or more often anoxia, is generally (although rarely appreciated) a requisite for all measurements of tissue H_2S production, other than those using polarographic H_2S sensors (Olson 2012a). Using the polarographic sulfide sensor, originally developed by Jeroschewski et al. (1996), Kraus and Doeller (2004) observed that excised gills and gill mitochondria from sulfide-adapted mussel, *Geukensia demissa*, rapidly consumed H_2S in the presence of O_2 and that the rate of H_2S consumption was reduced 50-fold in anoxia and 75 % inhibited by cyanide. The authors estimated the P_{50} (the partial pressure of O_2 at half maximal rate) for H_2S oxidation in mitochondria of approximately 7.5 mmHg. Both Furne et al. (2008) and Doeller et al. (2005) measured the relationship between O_2 and H_2S production in rat tissues and observed that H_2S production was suppressed at normoxic Po_2 and Furne et al. (2008) observed a switch from H_2S production in hypoxia to H_2S consumption in normoxia in mouse liver and brain. Similar observations have been made in a variety of other tissues (Fig. 2.3; Dombkowski et al. 2011; Linden et al. 2011; Madden et al. 2012; Olson et al. 2008, 2010; Olson and Whitfield 2010; Whitfield et al. 2008). A compelling argument for H_2S-mediated O_2 sensing can be made by comparing bovine and sea lion lungs (Olson et al. 2010); while both tissues clearly demonstrate an identical and reciprocal relationship between H_2S production/consumption and O_2 (Fig. 2.3a, b), both hypoxia and H_2S constrict the former and dilate the latter (Fig. 2.1e).

To date, there has only been one study in which the rate of H_2S consumption by tissue has been measured at carefully controlled Po_2 (Olson et al. 2010). As shown in Fig. 2.3d, the efficiency of H_2S oxidation by bovine lung homogenate, bovine pulmonary arterial smooth muscle cells, or purified bovine heart mitochondria begins to fail at physiologically relevant Po_2s and at Po_2s routinely encountered during hypoxia H_2S metabolism becomes highly sensitive to O_2 availability. Further demonstration of the physiological relevancy of this process is the observation that the Po_2 at which the ability of pulmonary arterial smooth muscle cells to

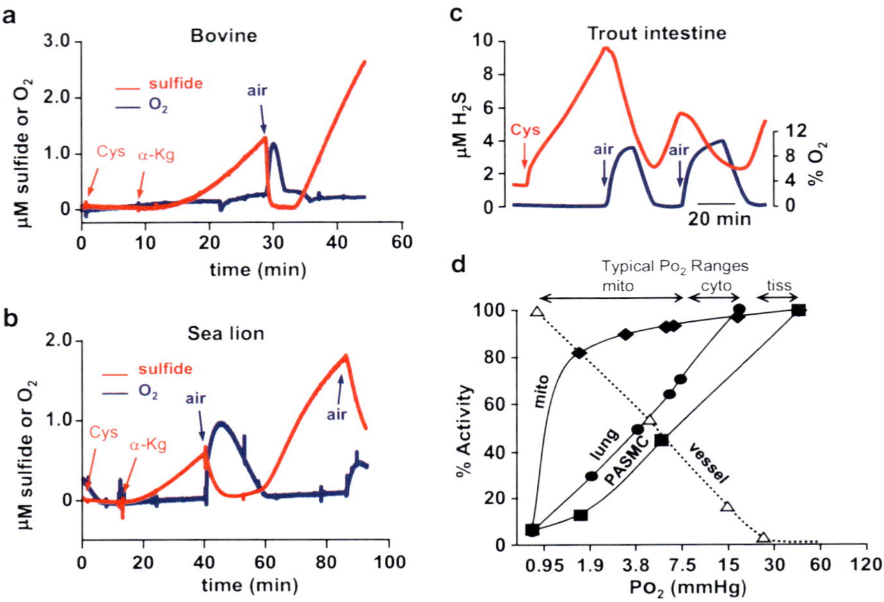

Fig. 2.3 Reciprocal relationship between O_2 and H_2S in tissues. (**a–c**) Realtime measurements of O_2 and H_2S in homogenized cow lung (**a**), homogenized sea lion lung (**b**) and homogenized trout intestine (**c**) using gas-specific polarographic electrodes. H_2S production increases after addition of 1 mM cysteine (Cys) and α-ketoglutarate (α-Kg) or Cys alone but only in the absence of O_2. Injection of a small air bubble (air) which increases O_2 concentration immediately reduces H_2S concentration all tissues. When the O_2 is consumed, H_2S concentration again increases. These results show that H_2S production is inversely related to O_2 in a variety of tissues, even though bovine pulmonary arteries are constricted by hypoxia whereas sea lion pulmonary arteries and trout intestines are relaxed. (**d**) H_2S consumption as a function of O_2 availability (Po_2) by bovine mitochondria (mito), homogenized bovine lung tissue (bovine lung) and bovine pulmonary artery smooth muscle cells (*PASMC*). H_2S is rapidly metabolized (100 % inactivation rate) until Po_2 falls to ~ 15 mmHg (1–2 mmHg for mito); H_2S metabolism rapidly fails as Po_2 continues to fall. The Po_2 values at which H_2S metabolism is impaired are at the low end of cytosolic and mitochondrial Po_2 and would be expected during hypoxia. The Po_2 at which H_2S metabolism is reduced to half (P_{50}) in lung tissue and PASMC is essentially the same Po_2 that produces half-maximal hypoxic vasoconstriction of bovine pulmonary arteries (vessel). (**a**, **b**, **d**, Modified from Olson et al. 2010, **c** modified from Dombkowski et al. 2011, with permission)

oxidize H_2S is halved (P_{50}) is identical to the P_{50} of hypoxic pulmonary vasoconstriction. As might be expected, mitochondria function at a Po_2 below the cytosolic Po_2 and it is evident in Fig. 2.3d that the H_2S oxidation curve is left-shifted accordingly. These studies clearly show that the metabolism of H_2S is O_2 dependent, that the ability of tissues to metabolize H_2S fails at physiologically relevant Po_2s, and this provides a sensitive and efficient mechanism for O_2 sensing.

2.3.4 Compounds that Augment or Inhibit H_2S Production Augment or Inhibit Hypoxic Responses

The ability of sulfur donors especially cysteine to augment hypoxic responses has been well documented (Fig. 2.4). Cysteine increases the magnitude of hypoxic vasoconstriction of isolated lamprey aortas (Olson et al. 2006), bovine pulmonary arteries (Olson et al. 2006, 2010) and it increases vascular resistance in the perfused rat lung (Madden et al. 2012). Both reduced and oxidized glutathione augment hypoxic vasoconstriction in pulmonary arteries and the perfused rat lung and cysteine plus α-ketoglutarate (presumably utilizing the CAT/3-MST pathway) increases hypoxic vasoconstriction in bovine pulmonary arteries (Madden et al. 2012; Olson et al. 2010). Continuous utilization of cysteine to sustain a hypoxic vasoconstriction is evident in Fig. 2.4d where it clearly sustains the hypoxic response. Exogenous cysteine also augments hypoxic relaxation of rat aortas (Bucci et al. 2010), the relaxation component of the perfused trout gill (Skovgaard and Olson 2012) as well as hypoxic relaxation of trout urinary bladder (Dombkowski et al. 2006) and salmon intestine (Dombkowski et al. 2011).

Hypoxic responses of lamprey aorta, bovine pulmonary arteries, rat aorta and the perfused trout gill and rat lung are also inhibited by inhibitors of H_2S synthesis (Fig. 2.5; Olson et al. 2006; Madden et al. 2012; Skovgaard and Olson 2012). Although inhibitors of H_2S are notoriously nonspecific and often poorly absorbed by tissues (Szabó 2007), their application can provide some information on the biosynthetic pathways that are being used to produce H_2S. Not surprisingly, CSE appears to be the major pathway for H_2S production by systemic vessels (Fig. 2.5c). However, in bovine pulmonary arteries inhibition of CBS, but not CSE, reduces the hypoxic response, whereas hydroxyl amine, which inhibits all pyridoxal phosphate dependent enzymes, including CBS, CSE and CAT, completely inhibits hypoxic vasoconstriction (Fig. 2.5b). This suggests that both the CBS and CAT/3-MST contribute to H_2S production in bovine pulmonary vessels. The CAT/3-MST pathway can also be utilized in the rat lung as the competitive inhibitor, aspartate, prevents the augmented hypoxic response produced by exogenous α-ketoglutarate (Fig. 2.5c). There appears to be some species variation in the enzymatic pathways employed as CSE may be a major component of H_2S production mediating hypoxic responses of the perfused rat lung (Madden et al. 2012). Inhibitors of H_2S biosynthesis have also been shown to inhibit hypoxic relaxation of rainbow trout urinary bladder (Dombkowski et al. 2006) and rainbow trout and Coho salmon intestine (Dombkowski et al. 2011).

Inhibiting CBS in the mouse carotid body decreases hypoxia-stimulated afferent nerve activity in vitro and blunts the hypoxic hyperventilation in vivo (Li et al. 2010). Conversely, Peng et al. (2010) observed an inverse Po_2-dependent increase in H_2S production by rat carotid bodies and both H_2S production and sinus nerve activity could be blocked by inhibiting CSE; in a mouse CSE knockout ($CSE^{-/-}$) hypoxic responses of glomus cells were significantly reduced. Hypoxia-evoked catecholamine

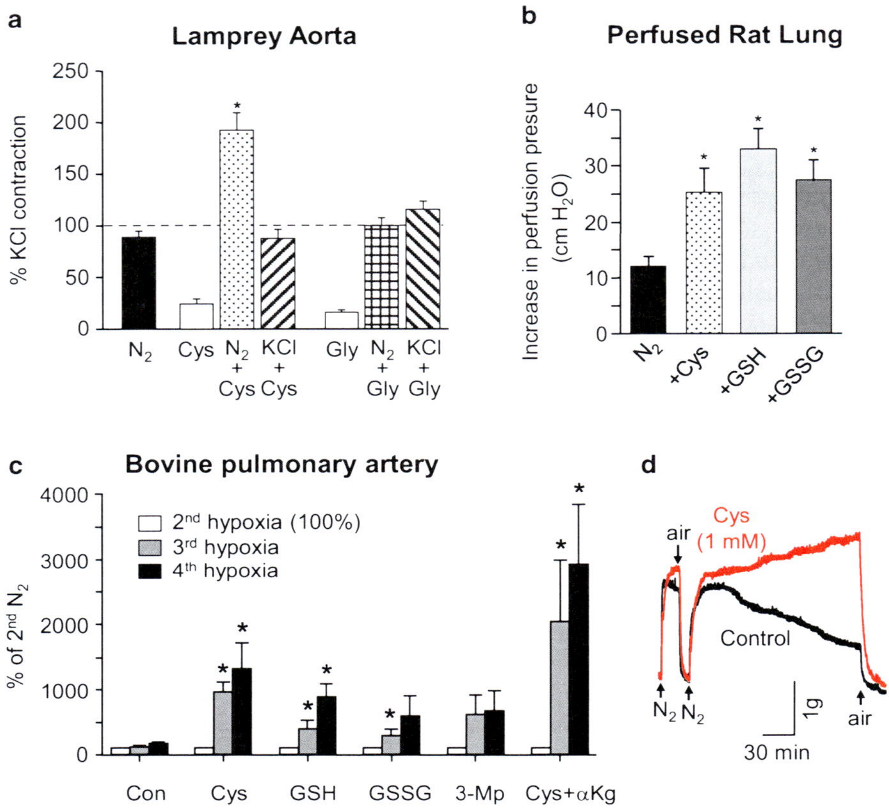

Fig. 2.4 Substrates for H_2S biosynthesis augment hypoxic responses. (**a**) In the lamprey aorta cysteine (Cys, 1 mM) nearly doubles a hypoxic contraction but does not affect a KCl (80 mM) contraction. Both Cys and glycine (Gly, 1 mM) had a slight effect but Gly does not affect the hypoxic response. (**b**) Cys (1 mM), reduced glutathione (GSH, 1 mM) and oxidized glutathione (GSSG, 1 mM) all significantly increase perfusion pressure in the perfused rat lung. (**c**) Cys (1 mM), GSH (1 mM), GSSG (1 mM) and Cys plus α-ketoglutarate (Cys + αKg, 1 mM) enhance consecutive hypoxic contractions of bovine pulmonary arteries. (**d**) Representative myograph traces illustrating the ability of Cys (1 mM) to prolong a hypoxic contraction in an isolated bovine pulmonary artery (Modified from Olson 2012b, with permission)

secretion from adrenal glands was also inhibited in $CSE^{-/-}$mice or by inhibiting CSE in rats Peng et al. (2010). Clearly additional studies are needed so sort out the specific metabolic pathways for H_2S production in the glomus cells, but nevertheless, a strong case can be made for the involvement of H_2S in hypoxic signal transduction.

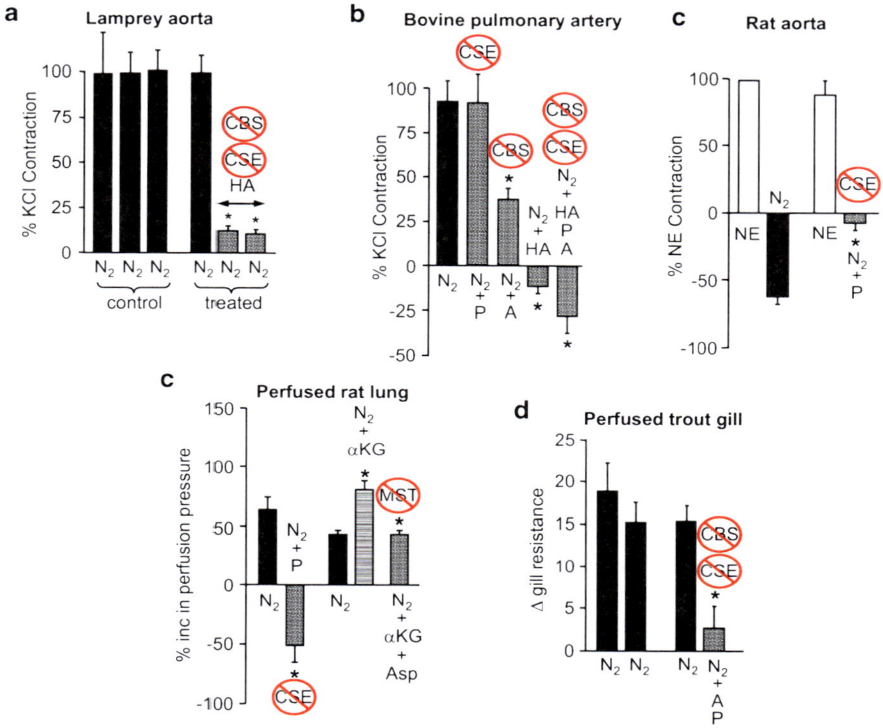

Fig. 2.5 Inhibitors of H_2S biosynthesis inhibit hypoxic vasoconstriction in the lamprey aorta (**a**), and bovine pulmonary artery (**b**) and hypoxic vasodilation of the norepinephrine (NE, 1 μM) precontracted rat aorta. Inhibiting H_2S biosynthesis also inhibits hypoxic vasoconstriction in the perfused rat lung (**c**) and perfused trout gill (**d**). *CBS* cystathionine β-synthase, *CSE* cystathionine γ-lyase, *A* amino-oxyacetate a CBS inhibitor (1 mM), *P* propargyl glycine a CSE inhibitor (10 mM), *HA*, *CBS* and *CSE* inhibitor hydroxylamine (1 mM), *α-Kg* α-ketoglutarate a substrate for mercaptopyruvate sulfur transferase (*MST*), *Asp* aspartic acid, an inhibitor of MST (10 mM) (From Olson 2012b, with permission)

2.3.5 H_2S Acts Upon Effector Mechanisms Known to Mediate Hypoxic Responses

The recent identification of H_2S signaling through sulfhydration of protein cysteine molecules (Mustafa et al. 2009) has not only contributed to our overall understanding of H_2S signaling pathways, but it has shed some light on the mechanisms of H_2S signaling in hypoxia. Because many proteins such as enzymes and structural proteins are regulated through active-site cysteines (Nagahara 2011) is also evident that H_2S signaling is most likely an autocrine activity and that even within the cell it must be highly spatially regulated. This is supported by the models predicting that hypoxic signaling proceeds in the immediate mitochondrial environment (Olson 2013).

As would be expected, the mechanisms with which inhibits or activates H_2S cells is commensurate with the intended outcome of H_2S signaling.

It is well-known that hypoxic vasodilation is mediated in part by K_{ATP} channels (Weir and Archer 1995) and these channels were one of the first targets identified for H_2S signaling (Zhao et al. 2001). Since then a variety of channels in vascular smooth muscle and endothelial cells have been shown to be affected by H_2S leading to vasodilation. These include K_{ATP}, intermediate conductance (IK_{Ca}) and K_v7 potassium channels (Jiang et al. 2010; Liang et al. 2011; Martelli et al. 2013; Mustafa et al. 2011). H_2S also relaxes newborn piglet cerebral arteries by increasing $[Ca^{2+}]$ in the sarcoplasmic reticulum. This stimulates Ca^{2+} sparks, increases current through K_{Ca} channels and hyperpolarizes the cells thereby lowering global intracellular $[Ca^{2+}]$ (Liang et al. 2012). H_2S did not directly affect K_{Ca} channels in these studies. H_2S activates the Kir 6.1 subunit of K_{ATP} channel through sulfhydration of specific cysteine residues, especially Cys-34, this decreases the inhibitory effect of ATP on these channels while increasing binding of the activator phosphatidylinositol (4,5)-bisphosphate (PIP2)28 to Kir 6.1; other channels such as the endothelial intermediate conductance (IK_{Ca}) channel and other cysteine residues such as Cys^6 and Cys^{26} may also be sulfhydrated and contribute to H_2S relaxation (Jiang et al. 2010; Mustafa et al. 2011). Intermittent hypoxia down regulates CSE and increases vascular tone via the loss of H_2S activation of BK_{Ca} channels (Jackson-Weaver et al. 2011) suggesting a longer time-scale of vascular regulation.

H_2S can also activate cells. We (Sudhahar et al. 2013) have recently shown that H_2S induces membrane trafficking of protein kinase Cε (PKCε) through specific sulfhydration of Cys-13 and Cys-74 in the C2 domain. This is consistent with the well known role of PKCε activation in hypoxic pulmonary vasoconstriction (Sylvester et al. 2012). H_2S also activates the carotid body through inhibition of large-conductance calcium activated potassium (BK_{Ca}) channels (Li et al. 2010; Telezhkin et al. 2009, 2010) and/or inhibition of background (TASK) potassium channels (Buckler 2012); both lead to membrane depolarization and voltage-gated Ca^{2+} entry. Although specific channels were not identified, inhibition of potassium channels is consistent with hypoxia- and H_2S-mediated depolarization of zebrafish neuroepithelial cells (Olson et al. 2008), bovine pulmonary arteries (Olson et al. 2006) and lamprey aortas (Madden and Olson, unpublished). H_2S stimulates catecholamine release from rat adrenal cells via H_2S inhibition of $I_{K(Ca)}$ current (Zhu et al. 2012). In other cells, H_2S directly increases BK channel activity in rat GH(3) pituitary tumor cells through its reducing action on sulfhydryl groups of the channel protein (Sitdikova et al. 2010). A direct link between H_2S signaling and hypoxia has been shown by Tao et al. (2012) where the Cys1045-Cys1024 disulfide bond of VEGFR2 is targeted by H_2S and serves as a specific molecular switch for hypoxia mediated migration of vascular endothelial cells.

It is not too surprising that the acute hypoxia signal, H_2S, interacts with the long-term hypoxia signaling hypoxia-inducible factors (HIFs). In general, H_2S decreases HIF-1α expression in a variety of mammalian tissues (Kai et al. 2012; Si et al. 2013; Wu et al. 2012). H_2S inhibits HIF-1α protein accumulation during hypoxia (1 % O_2) or hypoxia-mimetic conditions by enhancing HIF2α phosphorylation independent of protein synthesis or ubiquitin-proteasomal degradation (Kai et al. 2012; Wu et al. 2012). Hypoxic pre- (and post-)conditioning, which appears to involve H_2S signaling (Bian et al. 2006; Lavu et al. 2010; Pan et al. 2006; Yong et al. 2008) is also associated with a decrease in HIF1-α expression (Sims et al. 2012). Interestingly, H_2S has the opposite effect on the nematode, *Caenorhabditis elegans*, where it increases HIF-1 activity, although this effect appears to be independent of hypoxia-mediated HIF-1 expression (Budde and Roth 2010). Other effectors of H_2S-mediated cellular protection from hypoxia include heat shock protein 90 (Meng et al. 2011), inhibition of ROS-activated pathways such as NF-κB/COX-2 (Yang et al. 2011) and ERK1/2 and p38MAPK (Lan et al. 2011).

Direct competition between hypoxia and H_2S for the same effector response, seen as the inability of one stimulus to produce a response when the tissue is activated by the other, has been demonstrated in a variety of blood vessels (Olson et al. 2006; Skovgaard and Olson 2012). This provides additional, albeit indirect, evidence that H_2S is involved in hypoxic signaling.

2.3.6 H_2S-Mediated O_2 Sensing Has an Evolutionary Precedent and a Phylogenetic History

H_2S was likely an important energy source and structural entity in the origin of life and for the first 500 million years after the origin of eukaryotes, the latter occurring in sulfidic and anoxic environments (reviewed in Olson 2012c). Thus, there is a long evolutionary history of H_2S in intracellular energy trafficking and signaling. Some of these facets are retained today as SQR, the initial enzyme in H_2S oxidation, is an integral component of the mitochondrial electron transport chain, as is SO (Hildebrandt 2011). It is evident in the present day, however, that H_2S and O_2 are not mutually compatible either in the environment, or in cells. Thus, as the Earth's O_2 levels began to rise some 800 million years ago, environmental H_2S fell and H_2S was no longer available as a substrate for energy production. However, it is clear that cells retained much of their metabolic capabilities and with an ironic, but well established metabolic twist, they now use H_2S as a reporter of O_2 availability.

Conclusions

There is now considerable evidence that the reciprocal relationship between H_2S and O_2 provides cells with an effective and accurate mechanism with which to couple O_2 availability to a variety of effector responses in O_2 sensing and perhaps all cells. This hypothesis is supported by the ubiquitous similarity between the effects of hypoxia and H_2S on a variety of tissues, that net tissue H_2S is exquisitely controlled by O_2 availability at physiologically relevant O_2 levels, that factors that augment or inhibit H_2S production have a similar effect on hypoxic responses and that the downstream effectors for hypoxia and H_2S appear to be identical. This mechanism appears to have originated early on in evolution and it is likely widespread in the animal kingdom. The next step in evaluating this signaling mechanism will be to determine how, and where, this mechanism operates at the sub-cellular level.

Acknowledgments The author wishes to acknowledge the numerous colleagues that contributed to this research. The author's work has been supported by National Science Foundation Grants, IBN 0235223, IOS 0641436 and IOS 1051627.

References

Ackermann M, Kubitza M, Maier K, Brawanski A, Hauska G, Pina AL (2011) The vertebrate homolog of sulfide-quinone reductase is expressed in mitochondria of neuronal tissues. Neuroscience 199:1–12

Banerjee R, Zou CG (2005) Redox regulation and reaction mechanism of human cystathionine-beta-synthase: a PLP-dependent hemesensor protein. Arch Biochem Biophys 433:144–156

Baragatti B, Ciofini E, Sodini D, Luin S, Scebba F, Coceani F (2013) Hydrogen sulfide in the mouse ductus arteriosus: a naturally occurring relaxant with potential EDHF function. Am J Physiol Heart Circ Physiol 304:H927–H934

Bearden SE, Beard RS Jr, Pfau JC (2010) Extracellular transsulfuration generates hydrogen sulfide from homocysteine and protects endothelium from redox stress. Am J Physiol Heart Circ Physiol 299:H1568–H1576

Beauchamp RO Jr, Bus JS, Popp JA, Boreiko CJ, Andjelkovich DA (1984) A critical review of the literature on hydrogen sulfide toxicity. Crit Rev Toxicol 13:25–97

Bian JS, Yong QC, Pan TT, Feng ZN, Ali MY, Zhou S, Moore PK (2006) Role of hydrogen sulfide in the cardioprotection caused by ischemic preconditioning in the rat heart and cardiac myocytes. J Pharmacol Exp Ther 316:670–678

Bickler PE, Buck LT (2007) Hypoxia tolerance in reptiles, amphibians, and fishes: life with variable oxygen availability. Annu Rev Physiol 69:145–170

Bouillaud F, Blachier F (2011) Mitochondria and sulfide: a very old story of poisoning, feeding and signaling? Antioxid Redox Signal 15:379–391

Bucci M, Papapetropoulos A, Vellecco V, Zhou Z, Pyriochou A, Roussos C, Roviezzo F, Brancaleone V, Cirino G (2010) Hydrogen sulfide is an endogenous inhibitor of phosphodiesterase activity. Arterioscler Thromb Vasc Biol 30:1998–2004

Buckler KJ (2012) Effects of exogenous hydrogen sulphide on calcium signalling, background (TASK) K channel activity and mitochondrial function in chemoreceptor cells. Pflugers Arch 463:743–754

Budde MW, Roth MB (2010) Hydrogen sulfide increases hypoxia-inducible factor-1 activity independently of von Hippel-Lindau tumor suppressor-1 in C. elegans. Mol Biol Cell 21:212–217

Cai WJ, Wang MJ, Moore PK, Jin HM, Yao T, Zhu YC (2007) The novel proangiogenic effect of hydrogen sulfide is dependent on Akt phosphorylation. Cardiovasc Res 76:29–40

Calvert JW, Elston M, Nicholson CK, Gundewar S, Jha S, Elrod JW, Ramachandran A, Lefer DJ (2010) Genetic and pharmacologic hydrogen sulfide therapy attenuates ischemia-induced heart failure in mice. Circulation 122:11–19

Chen L, Zhang J, Ding Y, Li H, Nie L, Zhou H, Tang Y, Zheng Y (2013) Site-specific hydrogen sulfide-mediated central regulation of respiratory rhythm in medullary slices of neonatal rats. Neuroscience 233:118–126

Cheng Y, Ndisang JF, Tang G, Cao K, Wang R (2004) Hydrogen sulfide-induced relaxation of resistance mesenteric artery beds of rats. Am J Physiol Heart Circ Physiol 287:H2316–H2323

Derwall M, Francis RC, Kida K, Bougaki M, Crimi E, Adrie C, Zapol WM, Ichinose F (2011) Administration of hydrogen sulfide via extracorporeal membrane lung ventilation in sheep with partial cardiopulmonary bypass perfusion: a proof of concept study on metabolic and vasomotor effects. Crit Care 15:R51

di d'Emmanuele Villa Bianca, Sorrentino R, Maffia P, Mirone V, Imbimbo C, Fusco F, De PR, Ignarro LJ, Cirino G (2009) Hydrogen sulfide as a mediator of human corpus cavernosum smooth-muscle relaxation. Proc Natl Acad Sci USA 106:4513–4518

Di Meo I, Fagiolari G, Prelle A, Viscomi C, Zeviani M, Tiranti V (2011) Chronic exposure to sulfide causes accelerated degradation of cytochrome C oxidase in ethylmalonic encephalopathy. Antioxid Redox Signal 15:353–362

Doeller JE, Isbell TS, Benavides G, Koenitzer J, Patel H, Patel RP, Lancaster JR Jr, Drley-Usmar VM, Kraus DW (2005) Polarographic measurement of hydrogen sulfide production and consumption by mammalian tissues. Anal Biochem 341:40–51

Dombkowski RA, Russell MJ, Schulman AA, Doellman MM, Olson KR (2005) Vertebrate phylogeny of hydrogen sulfide vasoactivity. Am J Physiol Regul Integr Comp Physiol 288: R243–R252

Dombkowski RA, Doellman MM, Head SK, Olson KR (2006) Hydrogen sulfide mediates hypoxia-induced relaxation of trout urinary bladder smooth muscle. J Exp Biol 209:3234–3240

Dombkowski RA, Naylor MG, Shoemaker E, Smith M, DeLeon ER, Stoy GF, Gao Y, Olson KR (2011) Hydrogen sulfide (HS) and hypoxia inhibit salmonid gastrointestinal motility: evidence for HS as an oxygen sensor. J Exp Biol 214:4030–4040

Dromparis P, Michelakis ED (2013) Mitochondria in vascular health and disease. Annu Rev Physiol 75:95–126

Drousiotou A, Di Meo I, Mineri R, Georgiou T, Stylianidou G, Tiranti V (2011) Ethylmalonic encephalopathy: application of improved biochemical and molecular diagnostic approaches. Clin Genet 79:385–390

Evans AM, Hardie DG, Peers C, Mahmoud A (2011) Hypoxic pulmonary vasoconstriction: mechanisms of oxygen-sensing. Curr Opin Anaesthesiol 24:13–20

Fu M, Zhang W, Wu L, Yang G, Li H, Wang R (2012) Hydrogen sulfide (H_2S) metabolism in mitochondria and its regulatory role in energy production. Proc Natl Acad Sci USA 109:2943–2948

Furne J, Saeed A, Levitt MD (2008) Whole tissue hydrogen sulfide concentrations are orders of magnitude lower than presently accepted values. Am J Physiol Regul Integr Comp Physiol 295:R1479–R1485

Fusco F, di d'Emmanuele V, Mitidieri E, Cirino G, Sorrentino R, Mirone V (2012) Sildenafil effect on the human bladder involves the L-cysteine/hydrogen sulfide pathway: a novel mechanism of action of phosphodiesterase type 5 inhibitors. Eur Urol 62:1174–1180

Giordano C, Viscomi C, Orlandi M, Papoff P, Spalice A, Burlina A, Di Meo I, Tiranti V, Leuzzi V, di Amati G, Zeviani M (2011) Morphologic evidence of diffuse vascular damage in human and in the experimental model of ethylmalonic encephalopathy. J Inherit Metab Dis 35(3):451–458

Gonzalez C, Agapito MT, Rocher A, Gomez-Niño A, Rigual R, Castañeda J, Conde SV, Obeso A (2010) A revisit to O_2 sensing and transduction in the carotid body chemoreceptors in the context of reactive oxygen species biology. Respir Physiol Neurobiol 174:317–330

Goubern M, Andriamihaja M, Nubel T, Blachier F, Bouillaud F (2007) Sulfide, the first inorganic substrate for human cells. FASEB J 21:1699–1706

Gupte SA, Wolin MS (2008) Oxidant and redox signaling in vascular oxygen sensing: implications for systemic and pulmonary hypertension. Antioxid Redox Signal 10:1137–1152

Haggard HW, Henderson Y (1922) The influence of hydrogensulphide upon respiration. Am J Physiol 61:289–297

Haldar SM, Stamler JS (2013) S-nitrosylation: integrator of cardiovascular performance and oxygen delivery. J Clin Invest 123:101–110

Haouzi P (2012) Ventilatory and metabolic effects of exogenous hydrogen sulfide. Respir Physiol Neurobiol 184:170–177

Haouzi P, Bell HJ, Notet V, Bihain B (2009) Comparison of the metabolic and ventilatory response to hypoxia and H_2S in unsedated mice and rats. Respir Physiol Neurobiol 167:316–322

Haouzi P, Bell H, Philmon M (2011) Hydrogen sulfide oxidation and the arterial chemoreflex: effect of methemoglobin. Respir Physiol Neurobiol 177:273–283

Hildebrandt TM (2011) Modulation of sulfide oxidation and toxicity in rat mitochondria by dehydroascorbic acid. Biochim Biophys Acta 1807:1206–1213

Hildebrandt TM, Grieshaber MK (2008) Three enzymatic activities catalyze the oxidation of sulfide to thiosulfate in mammalian and invertebrate mitochondria. FEBS J 275:3352–3361

Hosoki R, Matsuki N, Kimura H (1997) The possible role of hydrogen sulfide as an endogenous smooth muscle relaxant in synergy with nitric oxide. Biochem Biophys Res Commun 237:527–531

Hu H, Shi Y, Chen Q, Yang W, Zhou H, Chen L, Tang Y, Zheng Y (2008) Endogenous hydrogen sulfide is involved in regulation of respiration in medullary slice of neonatal rats. Neuroscience 156:1074–1082

Huang J, Khan S, O'Brien PJ (1998) The glutathione dependence of inorganic sulfate formation from L- or D-cysteine in isolated rat hepatocytes. Chem Biol Interact 110:189–202

Jackson MR, Melideo SL, Jorns MS (2012) Human sulfide: quinone oxidoreductase catalyzes the first step in hydrogen sulfide metabolism and produces a sulfane sulfur metabolite. Biochemistry 51:6804–6815

Jackson-Weaver O, Paredes DA, Gonzalez Bosc LV, Walker BR, Kanagy NL (2011) Intermittent hypoxia in rats increases myogenic tone through loss of hydrogen sulfide activation of large-conductance Ca(2+)-activated potassium channels. Circ Res 108:1439–1447

Jeroschewski P, Steuckart C, Kuhl M (1996) An amperometric microsensor for the determination of H_2S in aquatic environments. Anal Chem 68:4351–4357

Jiang B, Tang G, Cao K, Wu L, Wang R (2010) Molecular mechanism for H(2)S-induced activation of K(ATP) channels. Antioxid Redox Signal 12:1167–1178

Julian D, Statile JL, Wohlgemuth SE, Arp AJ (2002) Enzymatic hydrogen sulfide production in marine invertebrate tissues. Comp Biochem Physiol A Mol Integr Physiol 133:105–115

Kabil O, Banerjee R (2010) Redox biochemistry of hydrogen sulfide. J Biol Chem 285:21903–21907

Kai S, Tanaka T, Daijo H, Harada H, Kishimoto S, Suzuki K, Takabuchi S, Takenaga K, Fukuda K, Hirota K (2012) Hydrogen sulfide inhibits hypoxia- but not anoxia-induced hypoxia-inducible factor 1 activation in a von hippel-lindau- and mitochondria-dependent manner. Antioxid Redox Signal 16:203–216

Kamoun P (2004) Endogenous production of hydrogen sulfide in mammals. Amino Acids 26:243–254

Kemp PJ, Lewis A, Hartness ME, Searle GJ, Miller P, O'Kelly I, Peers C (2002) Airway chemotransduction: from oxygen sensor to cellular effector. Am J Respir Crit Care Med 166:S17–S24

Kimura H (2010) Hydrogen sulfide: its production, release and functions. Amino Acids 41:113–121

Klentz RD, Fedde MR (1978) Hydrogen sulfide: effects on avian respiratory control and intrapulmonary CO_2 receptors. Respir Physiol 32:355–367

Koj A, Frendo J, Wojtczak L (1975) Subcellular distribution and intramitochondrial localization of three sulfurtransferases in rat liver. FEBS Lett 57:42–46

Kraus DW, Doeller JE (2004) Sulfide consumption by mussel gill mitochondria is not strictly tied to oxygen reduction: measurements using a novel polargraphic sulfide sensor. J Exp Biol 207:3667–3679

Kwiatkoski M, Soriano RN, Francescato HD, Batalhao ME, Coimbra TM, Carnio EC, Branco LG (2012) Hydrogen sulfide as a cryogenic mediator of hypoxia-induced anapyrexia. Neuroscience 201:146–156

Lagoutte E, Mimoun S, Andriamihaja M, Chaumontet C, Blachier F, Bouillaud F (2010) Oxidation of hydrogen sulfide remains a priority in mammalian cells and causes reverse electron transfer in colonocytes. Biochim Biophys Acta 1797:1500–1511

Lan A, Liao X, Mo L, Yang C, Yang Z, Wang X, Hu F, Chen P, Feng J, Zheng D, Xiao L (2011) Hydrogen sulfide protects against chemical hypoxia-induced injury by inhibiting ROS-activated ERK1/2 and p38MAPK signaling pathways in PC12 cells. PLoS One 6:e25921

Lavu M, Bhushan S, Lefer DJ (2010) Hydrogen sulfide-mediated cardioprotection: mechanisms and therapeutic potential. Clin Sci (Lond) 120:219–229

Leffler CW, Parfenova H, Basuroy S, Jaggar JH, Umstot ES, Fedinec AL (2010) Hydrogen sulfide and cerebral microvascular tone in newborn pigs. Am J Physiol Heart Circ Physiol 300: H440–H447

Li Q, Sun B, Wang X, Jin Z, Zhou Y, Dong L, Jiang LH, Rong W (2010) A crucial role for hydrogen sulfide in oxygen sensing via modulating large conductance calcium-activated potassium channels. Antioxid Redox Signal 12:1179–1189

Liang GH, Adebiyi A, Leo MD, McNally EM, Leffler CW, Jaggar JH (2011) Hydrogen sulfide dilates cerebral arterioles by activating smooth muscle cell plasma membrane KATP channels. Am J Physiol Heart Circ Physiol 300:H2088–H2095

Liang GH, Xi Q, Leffler CW, Jaggar JH (2012) Hydrogen sulfide activates Ca(2)(+) sparks to induce cerebral arteriole dilatation. J Physiol 590:2709–2720

Linden DR, Furne J, Stoltz GJ, Ddel-Rehim MS, Levitt MD, Szurszewski JH (2011) Sulfide quinone reductase contributes to hydrogen sulfide metabolism in murine peripheral tissues but not in the central nervous system. Br J Pharmacol 165:2178–2190

Liu X, Pan L, Zhuo Y, Gong Q, Rose P, Zhu Y (2010) Hypoxia-inducible factor-1 alpha is involved in the pro-angiogenic effect of hydrogen sulfide under hypoxic stress. Biol Pharm Bull 33:1550–1554

Liu WQ, Chai C, Li XY, Yuan WJ, Wang WZ, Lu Y (2011) The cardiovascular effects of central hydrogen sulphide are related to K(ATP) channels activation. Physiol Res 60:729–738

Madden JA, Vadula MS, Kurup VP (1992) Effects of hypoxia and other vasoactive agents on pulmonary and cerebral artery smooth muscle cells. Am J Physiol Lung Cell Mol Physiol 263: L384–L393

Madden JA, Ahlf SB, Dantuma MW, Olson KR, Roerig DL (2012) Precursors and inhibitors of hydrogen sulfide synthesis affect acute hypoxic pulmonary vasoconstriction in the intact lung. J Appl Physiol 112:411–418

Makarenko VV, Nanduri J, Raghuraman G, Fox AP, Gadalla MM, Kumar GK, Snyder SH, Prabhakar NR (2012) Endogenous H_2S is required for hypoxic sensing by carotid body glomus cells. Am J Physiol Cell Physiol 303:C916–C923

Marcia M, Ermler U, Peng G, Michel H (2010) A new structure-based classification of sulfide: quinone oxidoreductases. Proteins 78:1073–1083

Martelli A, Testai L, Breschi MC, Lawson K, McKay NG, Miceli F, Taglialatela M, Calderone V (2013) Vasorelaxation by hydrogen sulphide involves activation of Kv7 potassium channels. Pharmacol Res 70:27–34

Meng JL, Mei WY, Dong YF, Wang JH, Zhao CM, Lan AP, Yang CT, Chen PX, Feng JQ, Hu CH (2011) Heat shock protein 90 mediates cytoprotection by H_2S against chemical hypoxia-induced injury in PC12 cells. Clin Exp Pharmacol Physiol 38:42–49

Mikami Y, Shibuya N, Kimura Y, Nagahara N, Ogasawara Y, Kimura H (2011a) Thioredoxin and dihydrolipoic acid are required for 3-mercaptopyruvate sulfurtransferase to produce hydrogen sulfide. Biochem J 439:479–485

Mikami Y, Shibuya N, Kimura Y, Nagahara N, Yamada M, Kimura H (2011b) Hydrogen sulfide protects the retina from light-induced degeneration by the modulation of Ca2+ influx. J Biol Chem 286:39379–39386

Mikami Y, Shibuya N, Ogasawara Y, Kimura H (2013) Hydrogen sulfide is produced by cystathionine gamma-lyase at the steady-state low intracellular Ca(2+) concentrations. Biochem Biophys Res Commun 431:131–135

Milsom WK, Burleson ML (2007) Peripheral arterial chemoreceptors and the evolution of the carotid body. Respir Physiol Neurobiol 157:4–11

Modis K, Coletta C, Erdelyi K, Papapetropoulos A, Szabo C (2013) Intramitochondrial hydrogen sulfide production by 3-mercaptopyruvate sulfurtransferase maintains mitochondrial electron flow and supports cellular bioenergetics. FASEB J 27:601–611

Mudd SH, Irreverre F, Laster L (1967) Sulfite oxidase deficiency in man: demonstration of the enzymatic defect. Science 156:1599–1602

Mustafa AK, Gadalla MM, Sen N, Kim S, Mu W, Gazi SK, Barrow RK, Yang G, Wang R, Snyder SH (2009) H_2S signals through protein S-sulfhydration. Sci Signal 2:ra72

Mustafa AK, Sikka G, Gazi SK, Steppan J, Jung SM, Bhunia AK, Barodka VM, Gazi FK, Barrow RK, Wang R, Amzel LM, Berkowitz DE, Snyder SH (2011) Hydrogen sulfide as endothelium-derived hyperpolarizing factor sulfhydrates potassium channels. Circ Res 109:1259–1268

Nagahara N (2011) Catalytic site cysteines of thiol enzyme: sulfurtransferases. J Amino Acids 2011:1–7

Nagahara N (2012) Regulation of mercaptopyruvate sulfurtransferase activity via intrasubunit and intersubunit redox-sensing switches. Antioxid Redox Signal

Nurse CA, Buttigieg J, Thompson R, Zhang M, Cutz E (2006) Oxygen sensing in neuroepithelial and adrenal chromaffin cells. Novartis Found Symp 272:106–114

Olson KR (2011) Hydrogen sulfide is an oxygen sensor in the carotid body. Respir Physiol Neurobiol 179:103–110

Olson KR (2012a) A practical look at the chemistry and biology of hydrogen sulfide. Antioxid Redox Signal 17:32–44

Olson KR (2012b) Hydrogen sulfide as an oxygen sensor. Clin Chem Lab Med 51:623–632

Olson KR (2012c) Mitochondrial adaptations to utilize hydrogen sulfide for energy and signaling. J Comp Physiol B 182:881–897

Olson KR (2013) A theoretical examination of hydrogen sulfide metabolism and its potential in autocrine/paracrine oxygen sensing. Respir Physiol Neurobiol 186:173–179

Olson KR, Whitfield NL (2010) Hydrogen sulfide and oxygen sensing in the cardiovascular system. Antioxid Redox Signal 12:1219–1234

Olson KR, Dombkowski RA, Russell MJ, Doellman MM, Head SK, Whitfield NL, Madden JA (2006) Hydrogen sulfide as an oxygen sensor/transducer in vertebrate hypoxic vasoconstriction and hypoxic vasodilation. J Exp Biol 209:4011–4023

Olson KR, Healy M, Qin J, Skovgaard N, Vulesevic B, Duff DW, Whitfield NL, Yang G, Wang R, Perry SF (2008) Hydrogen sulfide as an oxygen sensor in trout gill chemoreceptors. Am J Physiol Regul Integr Comp Physiol 295:R669–R680

Olson KR, Whitfield NL, Bearden SE, St. Leger J, Nilson E, Gao Y, Madden JA (2010) Hypoxic pulmonary vasodilation: a paradigm shift with a hydrogen sulfide mechanism. Am J Physiol Regul Integr Comp Physiol 298:R51–R60

Olson KR, DeLeon ER, Gao Y, Hurley K, Sadauskas V, Batz C (2013) Thiosulfate: a readily accessible source of hydrogen sulfide in oxygen sensing. Am J Physiol Regul Integr Comp Physiol (in press)

Pan TT, Feng ZN, Lee SW, Moore PK, Bian JS (2006) Endogenous hydrogen sulfide contributes to the cardioprotection by metabolic inhibition preconditioning in the rat ventricular myocytes. J Mol Cell Cardiol 40:119–130

Papapetropoulos A, Pyriochou A, Altaany Z, Yang G, Marazioti A, Zhou Z, Jeschke MG, Branski LK, Herndon DN, Wang R, Szabo C (2009) Hydrogen sulfide is an endogenous stimulator of angiogenesis. Proc Natl Acad Sci USA 106:21972–21977

Peng YJ, Nanduri J, Raghuraman G, Souvannakitti D, Gadalla MM, Kumar GK, Snyder SH, Prabhakar NR (2010) H_2S mediates O_2 sensing in the carotid body. Proc Natl Acad Sci USA 107:10719–10724

Perry SF, Montpetit CJ, Borowska M (2000) The effects of acute hypoxia on chemically or neuronally induced catecholamine secretion in rainbow trout (Oncorhynchus mykiss) in situ and in vivo. J Exp Biol 203:1487–1495

Perry SF, McNeill B, Elia E, Nagpal A, Vulesevic B (2009) Hydrogen sulfide stimulates catecholamine secretion in rainbow trout (Oncorhynchus mykiss). Am J Physiol Regul Integr Comp Physiol 296:R133–R140

Rajapakshe A, Tollin G, Enemark JH (2012) Kinetic and thermodynamic effects of mutations of human sulfite oxidase. Chem Biodivers 9:1621–1634

Reiffenstein RJ, Hulbert WC, Roth SH (1992) Toxicology of hydrogen sulfide. Annu Rev Pharmacol Toxicol 32:109–134

Roman HB, Hirschberger LL, Krijt J, Valli A, Kozich V, Stipanuk MH (2013) The cysteine dioxygenase knockout mouse: altered cysteine metabolism in nonhepatic tissues leads to excess H2S/HS production and evidence of pancreatic and lung toxicity. Antioxid Redox Signal

Russell MJ, Dombkowski RA, Olson KR (2008) Effects of hypoxia on vertebrate blood vessels. J Exp Zool A Ecol Genet Physiol 309:55–63

Schultz HD, Del RR, Ding Y, Marcus NJ (2012) Role of neurotransmitter gases in the control of the carotid body in heart failure. Respir Physiol Neurobiol 184:197–203

Shibuya N, Koike S Tanaka M, Ishigami-Yuasa M, Kimura Y, Ogasawara Y, Fukui K, Nagahara N, Kimura H (2013) A novel pathway for the production of hydrogen sulfide from D-cysteine in mammalian cells. Nat Commun 4:1366

Si YF, Wang J, Guan J, Zhou L, Sheng Y, Zhao J (2013) Treatment with hydrogen sulfide alleviates streptozotocin-induced diabetic retinopathy in rats. Br J Pharmacol 169:619–631

Sims B, Clarke M, Francillion L, Kindred E, Hopkins ES, Sontheimer H (2012) Hypoxic preconditioning involves system Xc- regulation in mouse neural stem cells. Stem Cell Res 8:285–291

Sitdikova GF, Weiger TM, Hermann A (2010) Hydrogen sulfide increases calcium-activated potassium (BK) channel activity of rat pituitary tumor cells. Pflugers Arch 459:389–397

Skovgaard N, Olson KR (2012) Hydrogen sulfide mediates hypoxic vasoconstriction through a production of mitochondrial ROS in trout gills. Am J Physiol Regul Integr Comp Physiol 303: R487–R494

Sowmya S, Swathi Y, Yeo AL, Shoon ML, Moore PK, Bhatia M (2010) Hydrogen sulfide: regulatory role on blood pressure in hyperhomocysteinemia. Vascul Pharmacol 53:138–143

Stipanuk MH (2004) Sulfur amino acid metabolism: pathways for production and removal of homocysteine and cysteine. Annu Rev Nutr 24:539–577

Stipanuk MH, Ueki I (2011) Dealing with methionine/homocysteine sulfur: cysteine metabolism to taurine and inorganic sulfur. J Inherit Metab Dis 34:17–32

Stipanuk MH, Ueki I, Dominy JE Jr, Simmons CR, Hirschberger LL (2009) Cysteine dioxygenase: a robust system for regulation of cellular cysteine levels. Amino Acids 37:55–63

Sudhahar CG, Lee J, Buck LE, Olson KR, Stahelin RV (2013) Hydrogen sulfide induces the membrane translocation of protein kinase Cε through sulfhydration of the C2 domain. Antioxid Redox Signal (in press)

Sylvester JT, Shimoda LA, Aaronson PI, Ward JP (2012) Hypoxic pulmonary vasoconstriction. Physiol Rev 92:367–520

Szabó C (2007) Hydrogen sulphide and its therapeutic potential. Nat Rev Drug Discov 6:917–935

Szabo C, Papapetropoulos A (2011) Hydrogen sulfide and angiogenesis: mechanisms and applications. Br J Pharmacol 164:853–865

Tao BB, Liu SY, Zhang CC, Fu W, Cai WJ, Wang Y, Shen Q, Wang MJ, Chen Y, Zhu Y, Zhu YC (2012) VEGFR2 functions as an H_2S-targeting receptor protein kinase with its novel Cys1045-Cys1024 disulfide bond serving as a specific molecular switch for hydrogen sulfide actions in vascular endothelial cells. Antioxid Redox Signal. doi:10.1089/ars.2012.4565

Telezhkin V, Brazier SP, Cayzac S, Muller CT, Riccardi D, Kemp PJ (2009) Hydrogen sulfide inhibits human BK(Ca) channels. Adv Exp Med Biol 648:65–72

Telezhkin V, Brazier SP, Cayzac SH, Wilkinson WJ, Riccardi D, Kemp PJ (2010) Mechanism of inhibition by hydrogen sulfide of native and recombinant BKCa channels. Respir Physiol Neurobiol 172:169–178

Theissen U, Martin W (2008) Sulfide: quinone oxidoreductase (SQR) from the lugworm Arenicola marina shows cyanide- and thioredoxin-dependent activity. FEBS J 275:1131–1139

Tiranti V, Viscomi C, Hildebrandt T, Di Meo I, Mineri R, Tiveron C, Levitt MD, Prelle A, Fagiolari G, Rimoldi M, Zeviani M (2009) Loss of ETHE1, a mitochondrial dioxygenase, causes fatal sulfide toxicity in ethylmalonic encephalopathy. Nat Med 15:200–205

Ueki I, Roman HB, Valli A, Fieselmann K, Lam J, Peters R, Hirschberger LL, Stipanuk MH (2011) Knockout of the murine cysteine dioxygenase gene results in severe impairment in ability to synthesize taurine and an increased catabolism of cysteine to hydrogen sulfide. Am J Physiol Endocrinol Metab 301:E668–E684

Van de Louw A, Haouzi P (2012) Inhibitory effects of hyperoxia and methemoglobinemia on H(2) S induced ventilatory stimulation in the rat. Respir Physiol Neurobiol 181:326–334

Villarejo M, Westley J (1963) Mechanism of rhodanese catalysis of thiosulfate-lipoate oxidation-reduction. J Biol Chem 238:4016–4020

Wang MJ, Cai WJ, Li N, Ding YJ, Chen Y, Zhu YC (2010a) The hydrogen sulfide donor NaHS promotes angiogenesis in a rat model of hind limb ischemia. Antioxid Redox Signal 12:1065–1077

Wang MJ, Cai WJ, Zhu YC (2010b) Mechanisms of angiogenesis: role of hydrogen sulphide. Clin Exp Pharmacol Physiol 37:764–771

Waypa GB, Schumacker PT (2010) Hypoxia-induced changes in pulmonary and systemic vascular resistance: where is the O_2 sensor? Respir Physiol Neurobiol 174:201–211

Waypa GB, Marks JD, Guzy R, Mungai PT, Schriewer J, Dokic D, Schumacker PT (2010) Hypoxia triggers subcellular compartmental redox signaling in vascular smooth muscle cells. Circ Res 106:526–535

Weir EK, Archer SL (1995) The mechanism of acute hypoxic pulmonary vasoconstriction: the tale of two channels. FASEB J 9:183–189

Whitfield NL, Kreimier EL, Verdial FC, Skovgaard N, Olson KR (2008) Reappraisal of H_2S/sulfide concentration in vertebrate blood and its potential significance in ischemic preconditioning and vascular signaling. Am J Physiol Regul Integr Comp Physiol 294:R1930–R1937

Wolin MS, Gupte SA, Neo BH, Gao Q, Ahmad M (2010) Oxidant-redox regulation of pulmonary vascular responses to hypoxia and nitric oxide-cGMP signaling. Cardiol Rev 18:89–93

Wu B, Teng H, Yang G, Wu L, Wang R (2012) Hydrogen sulfide inhibits the translational expression of hypoxia-inducible factor-1alpha. Br J Pharmacol 167:1492–1505

Yan H, Do J, Tang C (2004) The possible role of hydrogen sulfide on the pathogenesis of spontaneous hypertension in rats. Biochem Biophys Res Commun 313:22–27

Yang G, Wu L, Jiang B, Yang W, Qi J, Cao K, Meng Q, Mustafa AK, Mu W, Zhang S, Snyder SH, Wang R (2008) H_2S as a physiologic vasorelaxant: hypertension in mice with deletion of cystathionine gamma-lyase. Science 322:587–590

Yang C, Yang Z, Zhang M, Dong Q, Wang X, Lan A, Zeng F, Chen P, Wang C, Feng J (2011) Hydrogen Sulfide Protects against Chemical Hypoxia-Induced Cytotoxicity and Inflammation in HaCaT Cells through Inhibition of ROS/NF-kappaB/COX-2 Pathway. PLoS One 6:e21971

Yong QC, Lee SW, Foo CS, Neo KL, Chen X, Bian JS (2008) Endogenous hydrogen sulphide mediates the cardioprotection induced by ischemic postconditioning. Am J Physiol Heart Circ Physiol 295:H1330–H1340

Zhang J, Xie Y, Xu Y, Pan Y, Shao C (2011) Hydrogen sulfide contributes to hypoxia-induced radioresistance on hepatoma cells. J Radiat Res 52:622–628

Zhao W, Zhang J, Lu Y, Wang R (2001) The vasorelaxant effect of H_2S as a novel endogenous KATP channel opener. EMBO J 20:6008–6016

Zhu D, Yu X, Sun J, Li J, Ma X, Yao W (2012) H_2S induces catecholamine secretion in rat adrenal chromaffin cells. Toxicology 302:40–43

Multiple Roles of H_2S in Inflammation: A New Class of Therapeutics?

3

Ling Li, Mohamed Shirhan Bin Mohamed, and Philip K. Moore

Abstract

H_2S exhibits complex roles in inflammation. Work over the last decade has shown that this gas can be pro-inflammatory, anti-inflammatory and/or can promote the resolution of an ongoing inflammatory response. The precise effect will depend on a range of, as yet, not completely understood factors, such as stage of the disease and the concentration and cellular localization of the gas. H_2S is synthesized by a wide range of inflammatory cells and affects multiple biological processes involved in initiating, sustaining and resolving inflammation. These include blood vessel dilatation, adhesion, migration and survival of leukocytes, oedema and pain and hyperalgesia. In addition, H_2S helps to regulate the release and effects of numerous other inflammatory mediators. Drugs which target the H_2S system can provide a new approach to the treatment of inflammation.

Keywords

Hydrogen sulfide • Inflammation • Pain • NF-κB

L. Li
p53 Laboratory, A*Star 8A Biomedical Grove, Immunos, Singapore 138648, Singapore

M.S.B. Mohamed • P.K. Moore (✉)
Department of Pharmacology, Yong Loo Lin School of Medicine,
National University of Singapore, Singapore 117597, Singapore
e-mail: dprmpk@nus.edu.sg

H. Kimura (ed.), *Hydrogen Sulfide and its Therapeutic Applications*,
DOI 10.1007/978-3-7091-1550-3_3, © Springer-Verlag Wien 2013

Abbreviations

3-MST	3-Mercaptopyruvate transferase
ADT-OH	5-(*p*-hydroxyphenyl)-1,2-dithione-3-thione
CBS	Cystathionine β synthetase
CO	Carbon monoxide
COPD	Chronic obstructive pulmonary disease
COX-2	Cyclooxygenase-2
CSE	Cystathionine γ lyase
EDHF	Endothelium derived hyperpolarising factor
GYY4137	Morpholin-4-ium 4 methoxyphenyl(morpholino) phosphinodithioate
H_2S	Hydrogen sulfide
ICAM-1	Intercellular adhesion molecule 1
IFN-γ	Interferon γ
IKKB	Inhibitor of nuclear factor kappa-B kinase subunit beta
IL	Interleukin
iNOS	Inducible nitric oxide synthase
LPS	Lipopolysaccharide
MAPK	Mitogen-activated protein kinase
MPO	Myeloperoxidase
Na_2S	Sodium sulfide
NaHS	Sodium hydrosulfide
NF-κB	Nuclear factor kappa-light-chain-enhancer of activated B cells
NO	Nitric oxide
NSAID	Nonsteroidal antiinflammatory drug
oxLDL	Oxidised low density lipoprotein
PAG	DL-propargylglycine
PKB	Protein kinase B
PPP	Pyridoxal 5′phosphate
SAC	S-allylcysteine
SPRC	S-propargylcysteine
STAT	Signal transducer and activator of transcription
TRPA	Transient receptor potential ion channels
VCAM	Vascular cell adhesion protein

3.1 Introduction

Hydrogen sulfide (H_2S), nitric oxide (NO) and carbon monoxide (CO) together form a group of biologically active gases that have been termed gasotransmitters. H_2S is the most recently discovered of these and, in the last decade, appears to have attracted interest and controversy in equal measure. One of the most controversial areas is the part played by H_2S in inflammation. Early research suggested that H_2S exhibited both pro- and anti-inflammatory effects, some times in the same animal model, whilst more recent work has concentrated on the ability of this gas to augment the resolution of an inflammatory response. Either way, the possibility that manipulating the H_2S system, using inhibitors of its biosynthesis or chemical donors, might have therapeutic potential to treat acute and/or chronic inflammation has been 'centre stage'.

3.2 Overview of Inflammation: Where Does H_2S Fit in?

Inflammation is a protective response of the body, triggered by injurious stimuli or pathogens and designed to result in tissue healing, repair and a rapid return to full functionality. Inflammation is often classified as being either 'acute' or 'chronic' although these sub-divisions can sometimes be arbitrary. An acute inflammatory response is the immediate response of the host to injury or infection and usually takes place over a period of hours. It can be as simple as the swelling and pain associated with an insect bite or as complex and life-threatening as septic shock and the accompanying failure of major organs. Chronic inflammation is of prolonged duration and examples include such debilitating conditions as rheumatoid arthritis, asthma, atherosclerosis and colitis.

Multiple cell types and multiple biological mediators have been implicated in inflammation. The weight of evidence, most of it gathered in the last decade, strongly suggest that H_2S now fully warrants its classification as an inflammatory mediator. Whilst few would argue with this statement, the precise role of H_2S in inflammation is not clear and a number of questions remain to be answered. Of these, probably the most pertinent questions at this time are: (i) is H_2S a pro- or an anti-inflammatory mediator or does it have a mixed effect depending on the prevailing conditions? (ii) what are the molecular targets for H_2S in terms of inflammation? (iii) what is the role, if any, of H_2S in inflammation in man? (iv) might drugs which affect the H_2S system be useful to treat patients with inflammatory disease?

This chapter provides an overview of the topic and seeks to address these questions. However, with the volume of information now available on this topic, it is not possible herein to be exhaustive, either in scope or in breadth, and interested readers are therefore referred to other, more specialised reviews for further information (Coletta and Szabo 2013; Li et al. 2011a; Rivers et al. 2012; Wallace et al. 2012; Whiteman and Winyard 2011).

3.3 Biosynthesis of H_2S in Inflammatory Cells

H_2S is synthesised by a number of different enzymes including cystathionine γ lyase (CSE), cystathione β synthetase (CBS) and 3-mercaptopyruvate transferase (3-MST) (reviewed in, Li et al. 2011a; Wang 2012). CSE and CBS are pyridoxal $5'$ phosphate (PPP)-dependent enzymes and use L-cysteine as substrate whilst

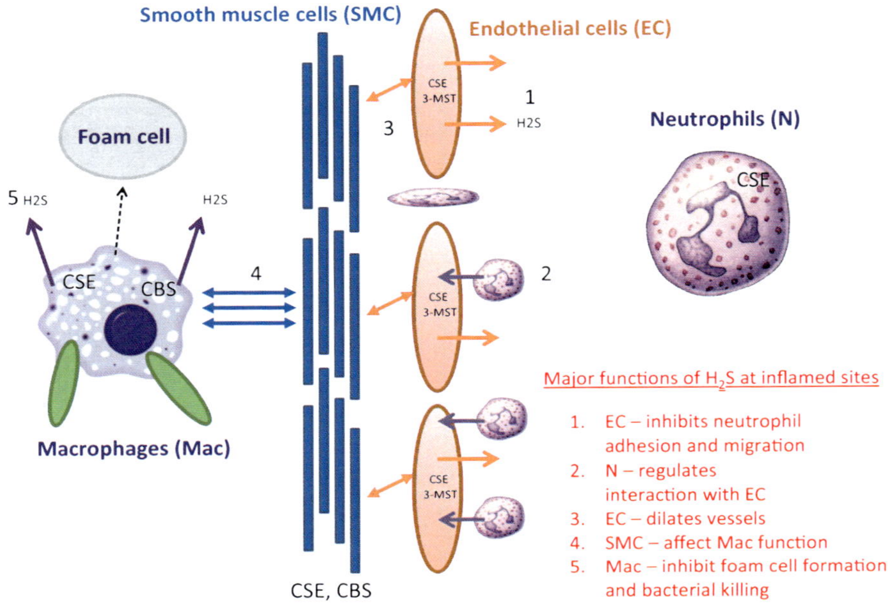

Fig. 3.1 Schematic showing H_2S biosynthesis and effects in the major inflammatory cell types. Key: *EC* endothelial cell, *N* neutrophil, *SMC* smooth muscle cell and *Mac* macrophage

3-MST is not PPP-dependent and uses 3-mercaptopyruvate as substrate. Each of these enzymes is evolutionally conserved across the phyla suggesting that they likely serve a very fundamental biochemical role in life. As an example, the nematode, *C. elegans*, has two CSE orthologues, three CBS orthologues and as many as seven 3-MST orthologues (Mathew et al. 2011). The distribution of CSE and CBS in mammalian cells and tissues is widespread and likely ubiquitous although CSE is often stated to predominate in the vasculature whilst CBS occurs in relatively larger amounts in nerves. The distribution of 3-MST in mammals has not been so well studied although large amounts have been found in liver and kidney (Ubuka et al. 1985).

It has been known for a number of years that many cell types involved in acute and/or chronic inflammation contain one or more of the enzymes needed to synthesise H_2S along with plentiful supplies of substrates and cofactors. The list includes vascular endothelial (Baragatti et al. 2013; Shibuya et al. 2009; Yang et al. 2008) and smooth muscle (Yang et al. 2004) cells as well as neutrophils (Li et al. 2009a), macrophages (Oh et al. 2006; Dufton et al. 2012; Whiteman et al. 2010a; Zhu et al. 2010), lymphocytes (Barathi et al. 2007) and dorsal root ganglion (DRG) cells (Takahashi et al. 2010). An overview of the generation of H_2S by different inflammatory cells and some of their major biological effects is shown schematically in Fig. 3.1.

Vascular endothelial cells play a crucial role in inflammation not only by regulating blood flow through the inflamed area but also by affecting leukocyte rolling, adhesion and migration into the tissues. Endothelial cells contain CSE (Yang et al. 2008) and synthesise H_2S in cell culture when challenged with agents such as the calcium ionophore, A23187 (Yang et al. 2008). However, the precise role of calcium in H_2S biosynthesis is likely not clear cut since, steady state calcium concentration is needed for CSE activity in vitro whilst, an increase in intracellular calcium reduces CSE activity (Mikami et al. 2013). Interestingly, retinal H_2S biosynthesis via 3-MST was also inhibited by calcium (Mikami et al. 2011). The H_2S which is generated by endothelial cells dilates blood vessels mostly likely by opening vascular smooth muscle K_{ATP} (Zhao et al. 2001) or voltage-operated Kv7.4 channels (Martelli et al. 2013) and in this way H_2S can act as an endothelium-derived hyperpolarizing factor (EDHF, Tang et al. 2013). At the molecular level, H_2S sulfhydrates K_{ATP} channels which may trigger their opening (Mustafa et al. 2011).

Most studies of the cellular sites of H_2S biosynthesis in inflammation have centred on leukocytes as important sources of this gas during an inflammatory response. One of the earliest reports in this regard highlighted a cytotoxic response to high concentrations of H_2S by rat pulmonary macrophages as evidenced by dramatically reduced zymosan-evoked respiration (Khan et al. 1991). Macrophage viability was unchanged in this particular study. In later work, exposure to H_2S was found to promote the survival of neutrophils (but not lymphocytes or eosinophils) by inhibiting caspase-3 cleavage and p38 MAP kinase phosphorylation and thereby impair apoptosis (Rinaldi et al. 2006). That H_2S can be cytoprotective in this way is counterintuitive bearing in mind its effect on cell respiration but seems to be a property of this gas conserved across different cells types.

Numerous researchers have identified CSE and CBS in macrophages and have further shown that CSE is upregulated by inflammatory stimuli such as *E. coli* lipopolysaccharide (LPS) or cytokines (e.g. TNF-α) (Whiteman et al. 2010a; Zhu et al. 2010). LPS also upregulates CSE expression in cultured primary rat neutrophils – an effect reversed by the anti-inflammatory steroid, dexamethasone (Li et al. 2009b). Similar data has been reported using anti-inflammatory steroids in primary mouse macrophages (Zhu et al. 2010) Whether anti-inflammatory steroids also alter H_2S formation in man is not yet known. Upregulation of H_2S synthesising enzymes is also a feature of a number of other animal models of inflammation including endotoxic (Li et al. 2005) and hemorrhagic (Mok et al. 2004) shock, pancreatitis (Bhatia et al. 2005a), myocarditis (Hua et al. 2010), peritonitis (Dufton et al. 2012) and carrageenan-evoked hindpaw oedema (Bhatia et al. 2005b; Ekundi-Valentim et al. 2010; Li et al. 2007; Sidhapuriwala et al. 2007). Interestingly, increased ovalbumin-induced airways hyperresponsivenes and lung inflammation has recently been noted in CSE knockout mice suggesting that allergic inflammation of this type might be associated with lack of pulmonary H_2S formation although the precise cellular origin is not clear (Zhang et al. 2013).

3.4 Effects of H_2S in Inflammation: Cells, Tissues and Animals

As noted above, many of the cells involved in inflammation including endothelial cells, neutrophils and macrophages are equipped with the enzyme(s) and substrates needed to synthesise H_2S. Since H_2S is rapidly broken down it seems logical that the major biological effects of this gas will take place at or close to the site of its generation. Thus, to understand better the role of H_2S in inflammation we should concentrate on the effect of this gas on the functioning of its synthesising cells and their close neighbours.

Considerable evidence now shows that H_2S, generated from endothelial cells, and likely monocytes/macrophages and neutrophils, affects vascular homeostasis. As noted above, H_2S dilates both large capacitance blood vessels (e.g. aorta) and smaller resistance blood vessels and promotes blood flow through the microcirculation (reviewed in Whiteman and Moore 2009). In this respect, H_2S works alongside many other vasodilator inflammatory mediators including other gases such as nitric oxide (NO) and carbon monoxide (CO). The interaction between H_2S and NO in the vasculature is a complex one with reports describing either a synergistic (Hosoki et al. 1997) or an antagonist (Yusuf et al. 2006) relationship between the two gases. In addition, H_2S can react with NO to form a novel nitrosothiol (HSNO) signalling molecule (Filipovic et al. 2012) the role of which in vascular homeostasis and inflammation has yet to be explored. Both H_2S and NO are made by endothelial cells and likely work in concert to help regulate blood flow in the microcirculation.

In addition to dilating the vasculature, H_2S also plays a key role in regulating the leukocyte-endothelium interface. Several years ago, sodium hydrosulfide (NaHS, fast-releasing H_2S donor) was shown to inhibit aspirin-induced leukocyte adhesion to the vascular endothelium as well as the resulting extravasation of leukocytes by opening K_{ATP} channels (Zanardo et al. 2006). Subsequently, intraperitoneal injection of either LPS (able to increase endogenous H_2S production by increasing CSE expression) or NaHS was shown to increase myeloperoxidase (MPO) activity (indicative of neutrophil infiltration) in several mouse tissues (Li et al. 2005). This would suggest that H_2S promotes neutrophil access from the blood into the tissues. Other researchers subsequently used the CSE inhibitor, DL-propargylglycine (PAG), as an indirect means to probe the role of endogenous H_2S in the neutrophil/endothelial cell interaction. In these experiments, treating mice with PAG reduced, whilst NaHS enhanced, LPS-evoked peritoneal leukocyte rolling, adhesion and migration (Dal-Secco et al. 2008). At the molecular level, H_2S increased leukocyte adhesion by upregulating expression of adhesion molecules such as P-selectin and ICAM-1. Somewhat different conclusions were obtained in a later study in which NaHS concentration-dependently suppressed TNF-α-evoked adhesion of U937 monocytes to endothelial cells which effect was associated with diminished expression of ICAM-1, VCAM-1, P-selectin and E-selectin (Pan et al. 2011a). Recently, Ball and colleagues (2013) have similarly reported that H_2S downregulates L-selectin expression using human neutrophils. All in all, it would appear most likely that H_2S tonically downregulates leukocyte adherence to the endothelium which agrees well with the ability of CSE inhibitors (e.g. PAG) to increase mucosal inflammation and injury (Fiorucci et al. 2005; Wallace et al. 2007, 2012).

Macrophages are not only excellent sources of H_2S but also prime targets for this gas. This is especially the case in atherosclerosis in which this gas inhibits oxidised LDL (oxLDL)-induced macrophage foam cell formation (Zhao et al. 2011). That

PAG promotes foam cell production in the same conditions strongly suggests a role, not only for exogenous, but also for endogenous, H_2S in this process. H_2S also inhibited upregulation of CX3CR1 in macrophages present in atherosclerotic plaques of high fat-fed $ApoE^{-/-}$ mice by a peroxisome proliferator-activated receptor (PPAR)-γ dependent mechanism (Zhang et al. 2012a). These data add to the growing body of evidence that the inflammatory component of atherosclerosis is, at least partly, an H_2S deficiency disease (Lynn and Austin 2011). Apart from a role in atherosclerosis, H_2S also affects other basic functions of macrophages, for example, promoting their ability to phagocytose *E. coli* (Dufton et al. 2012) and inhibiting LPS-induced NO production (Oh et al. 2006).

Early work suggested that H_2S, usually in the form of injected NaHS or Na_2S, promoted oedema formation in various tissues when injected into the mouse (Li et al. 2005). Both NaHS and L-cysteine challenge rapidly caused oedema in the mouse hindpaw by a mechanism dependent on phospholipase A_2 and prostaglandins (di Villa Bianca et al. 2010). PAG also reduced hindpaw swelling following intraplantar injection of carrageenan (Bhatia et al. 2005a) implying that endogenous H_2S exhibits pro-oedema activity. Other published work comes to the opposite conclusion i.e. that H_2S has anti- and not pro-oedema activity. For example, NaHS reduces carrageenan-induced hindpaw oedema (Zanardo et al. 2006) whilst intra-articular injection of Na_2S (another fast-releasing H_2S donor) reduced synovial blood flow in the rat keen joint (Andruski et al. 2008). Clearly, the precise effect of H_2S on inflammatory oedema is variable.

In addition, H_2S also plays a part in neurogenic inflammation. For example, NaHS activates primary sensory nerves, most likely by an effect on TRPA1 (Streng et al. 2008) or TRPV1 (Trevisani et al. 2005) ion channels to release neuropeptides such as substance P and calcitonin gene-related peptide (CGRP). These neuropeptides can elicit hyperalgesia in their own right as well as contracting bladder and airways smooth muscle (Patacchini et al. 2004; Pozsgai et al. 2012). Indeed, NaHS injection in the mouse hindpaw elicited acute pain by activating TRPA1 channels (Ogawa et al. 2012). It is now becoming increasingly clear that H_2S increases signalling from visceral pain organs, such as the pancreas and colon, by activating T-type calcium channels on primary sensory nerves (Matsunami et al. 2009). A similar mechanism may underlie the ability of endogenous H_2S to cause bladder pain and referred hyperalgesia in mice with cyclophosphamide-evoked cystitis (Matsunami et al. 2012). Finally, in inflamed colon, due to local instillation of trinitrobenzenesulfonic acid, H_2S modulates the responsiveness of colonic K_{ATP} channels by sulfhydrating cysteine residues of the SUR2B subunit (Gade et al. 2013). The precise relationship between H_2S, sensory neuropeptides and TRP ion channels in triggering and sustaining hyperalgesia during inflammation remains to be determined. However, sufficient experimental evidence has already been gathered to suggest that H_2S donors can promote hyperalgesia at sites of inflammation. Whether this proves a drawback in the clinical setting remains to be seen.

Much attention has been paid to the possible roles of H_2S as a mediator of the acute and/or chronic inflammatory response. For example, NaHS ameliorated tobacco smoke-induced emphysema in mice (Han et al. 2011), protected the lung against the combined insult of burn and smoke inhalation (Esechie et al. 2008) and reduced airways inflammation in a rat model of allergic asthma (Chen et al. 2009a) whilst inhaled H_2S reduced LPS-evoked septic shock (Tokuda et al. 2012). Recently, chronic (12 week) treatment of ovariectomised rats with estrogen has been shown not only to upregulate tissue CSE expression and H_2S production but also to limit myocardial inflammation as evidence by reduced levels of IL-6 and TNF-α implying that endogenous, as well as exogenous, can also be anti-inflammatory (Zhu et al. 2013). However, whether H_2S plays a part in resolving ongoing inflammation has only recently been suggested (Dufton et al. 2012; Wallace et al. 2009, 2012). Many of the biological effects of H_2S in inflammation already outlined in this review will no doubt contribute to such 'pro-resolution' activity. The best examples of such behaviour can be seen in the gastrointestinal tract. Thus, H_2S promotes healing of experimentally-induced stomach ulcers in rats whilst treatment with PAG has the opposite effect (Wallace et al. 2007, 2010). Furthermore, H_2S-releasing naproxen (ABT-346) caused markedly less or no gastrointestinal toxicity (c.f. naproxen) in healthy rats (Wallace et al. 2010) and remarkably, this protection was also apparent in arthritic, obese or aged animals, in which the damaging effect of naproxen is more pronounced (Blackler et al. 2012). These experiments not only increase our understanding of the complex role of H_2S in inflammation but open up further opportunities for the treatment of existing inflammatory disease.

3.5 Effects of H_2S in Inflammation: Molecular Mechanisms?

The molecular basis of this effect of H_2S in inflammation has received considerable attention in recent years with one particular molecular target, the IκB/NF-κB pathway, consistently being implicated. Oh and colleagues (2006) first reported that H_2S inhibited LPS-induced NF-κB activation in cultured mouse macrophages. Since that time, many other studies have confirmed that H_2S impairs transcription via NF-κB in a number of cells leading to reduced expression of a plethora of pro-inflammatory mediators including inducible nitric oxide synthase (iNOS), cyclooxygenase-2 (COX-2), TNF- α, IL-1β, IL-6, IL-8, IL-18 and a range of adhesion molecules. NaHS acts by inhibiting IκB-α degradation and thereby reducing translocation of NF-κB into the nucleus in both endothelial (Wang et al. 2009) and acinar (Tamizhselvi et al. 2009) cells challenged with inflammatory stimuli such as LPS. In intact tissues, a similar mechanism of action of H_2S was noted following ischemia-reperfusion in the kidney (Tripatara et al. 2009) and heart (Sivarajah et al. 2009) and in the heart after NaHS injection in rats (Gao et al. 2012). Other H_2S donors such as GYY4137 (morpholin-4-ium 4 methoxyphenyl (morpholino) phosphinodithioate, Li et al. 2008) have parallel effects in cultured mouse macrophages (Whiteman et al. 2010a) and in the liver and other tissues removed from LPS-injected rats (Li et al. 2009a). S-diclofenac, a slow-releasing H_2S adduct drug, also reduced liver NF-κB activation in rats with endotoxic shock (Li et al. 2007) whilst S-propargyl-cysteine (SPRC), an alternative H_2S donor, had a similar effect in the hypothalamus after LPS-triggered neuroinflammation (Gong et al. 2011). However, all is not straightforward in that NaHS has also been reported to augment (not inhibit) IκB-α degradation and thence increase (not reduce) NF-κB activation. This occurs, for example, in an interferon-γ (IFN-γ) primed human monocytic cell line (U937) (Zhi et al. 2007). To make thing even more complicated, the IKKβ/NF-κB pathway can also be anti-inflammatory by antagonising STAT-1 (Fong et al. 2008). Clearly, there are variations in the way in which the NF-κB system responds to H_2S. Whether this reflects differences between cell type, incubation conditions, the concentration of H_2S achieved or other, as yet unidentified, experimental conditions remains to be determined.

Whilst NF-κB is a key target for H_2S in inflammatory cells this does not, of course, exclude the possibility that other transduction mechanisms are involved in the response to this gas. NaHS is known to affect AKT/PKB and MAPK pathways and these pathways may be involved in H_2S-mediated cell survival (Shao et al. 2011). There is also evidence that H_2S targets other molecules inside the cell such as STAT-3 (Li et al. 2009a) and Nrf-2 (Calvert et al. 2009; Kalayarasan et al. 2009). The effect of H_2S on cell survival is highly relevant to its overall role in inflammation. For instance, sulfhydration of the p65 subunit of NF-κB at cysteine-38 has been suggested to mediate the anti-apoptotic effect of NF-κB in response to pro-inflammatory agents such as TNF-α and LPS in macrophages (Sen et al. 2012). Such an action may underlie, at least some, of the beneficial effects of H_2S in, for example, atherosclerosis where apoptosis of macrophages leads to plaque necrosis. However, H_2S generally appears to inhibit MAPK pathways which are required for the survival of several cell types including macrophages and neutrophils. Interestingly, a pro-apoptotic effect of H_2S in neutrophils has been described (Rinaldi et al. 2006) and this may have implications for the pro-resolution effect of H_2S, since neutrophil apoptosis is central to the successful resolution of an inflammatory response.

That H_2S is a gas, which allows it to access different intracellular components rapidly, coupled with the possibility that many of the biological effects of this gas occur by protein sulfydration, suggest that other intracellular proteins may also be targets. Since H_2S is a powerful anti-oxidant it seems logical that its effect within the cell will be promiscuous affecting a variety of redox-sensitive intracellular targets rather than just the one. This perspective is particularly relevant as many immune cells including neutrophils and macrophages maintain a highly oxidative intracellular environment thereby facilitating their pathogen killing function. If this is indeed the case then it does raise the obvious question of how inflammatory cells are able to 'coordinate' their response to such a pervasive influence.

3.6 Effects of H_2S in Inflammation: Studies in Man, Does it 'Translate'?

Whilst much has been much written in recent years about the possible physiological and pathophysiological roles of H_2S in inflammation the vast majority of these reports are based on data using isolated cells or animal models of disease. To date, there has been relatively little published on the possible roles of H_2S in inflammation in man and what has been reported relies largely on the measurement of H_2S in plasma or serum from patients with one or other inflammatory condition. Results of this type are very difficult to interpret appropriately for a number of reasons. For example, plasma or serum concentration of H_2S is, (i) very unlikely to be indicative of the 'real' concentration at the inflammatory site which will vary with the particular type of inflammatory disease being considered, (ii) very likely to vary with the severity of the disease (which may change over time) and other factors such as age and sex of the patient and other underlying pathologies (e.g. diabetes, obesity), (iii) be subject to alteration, in a largely unknown way, by administration of therapeutic drugs (particularly anti-inflammatory steroids which are known to affect H_2S synthesis, e.g. Li et al. 2009a; Zhu et al. 2010) and (iv) inherently difficult to measure accurately and reliably with the currently available technology (reviewed by Tangerman 2009).

With these caveats in mind, several studies have indeed shown changes in the H_2S system in human inflammatory disease. For example, serum H_2S concentration, determined by zinc trapping spectrophotometry, is reduced by about 25 % in adult patients with stable asthma compared with healthy subjects (Wu et al. 2008). A similar conclusion was reached when measuring serum H_2S levels in 64 children with asthma compared with 60 healthy children. In this case, the decline in serum H_2S was positively associated with diminished lung function indices suggesting a relationship between the two parameters (Tian et al. 2012). Interestingly, serum H_2S concentration was also reduced (by 36 %) in patients with pneumonia but not in those with chronic obstructive pulmonary disease (COPD) (Chen et al. 2009b). Overall, it might therefore be reasoned that the respiratory tract responds to a disease challenge by reducing synthesis of H_2S. Which particular cells show diminished H_2S formation and are hence responsible for the decline in plasma/serum levels and whether the seeming H_2S deficiency contributes to, or opposes, the ongoing inflammation is not known at this stage. However, a recently published epidemiological study may be able to shed some light. In a survey of 1,637 adults exposed to high ambient levels of geothermal H_2S in Rotorua (New Zealand) in the period 2008–2010 there was no overall evidence of increased asthma risk from H_2S exposure in this population. However, there was a suggestion from the study of a reduced risk of diagnosed asthma as well as asthma symptoms (e.g. wheezing) in higher H_2S exposure areas, perhaps implying that inhaled H_2S can be beneficial in this condition in the general population (Bates et al. 2013). More studies of this type in man are clearly needed.

Plasma/serum H_2S has also been measured in other forms of inflammatory disease in man. For example, plasma H_2S, measured using a sulfide-selective electrode, was lower in 35 children with Kawasaki disease (autoimmune blood vessel inflammation) than in 32 healthy children, suggesting that H_2S may be involved in the vascular inflammation associated with this childhood disorder (Li et al. 2011b). Moreover, serum H_2S was reduced by 51 % in human psoriasis sufferers – an effect which was also correlated with the severity of the clinical symptoms (Alshorafa et al. 2012).

Whilst inflammation in man seems to be tied to a lack of H_2S this is not always the case. For example, Park and colleagues (2013) have recently reported increased (not decreased) plasma levels of H_2S in patients with mild to severe allergic rhinitis which change was correlated with upregulated CSE expression in the nasal mucosal endothelium, vascular smooth muscle and submucosal glands. Moreover, both plasma and synovial fluid H_2S concentration is increased in rheumatoid arthritis patients and this was positively correlated with plasma white cell count (Whiteman et al. 2010b).

Notwithstanding the obvious technical and interpretational issues with studies of this type it seems increasingly likely that H_2S metabolism in inflammation is indeed disordered in humans. However, no clear pattern of the nature of this disorder has emerged so far and which cells are involved also remains to be determined. Clearly, more work is needed firstly to improve the sensitivity and specificity of H_2S assays and also to standardise patient selection and clinical history in studies of this type.

3.7 H_2S Donors: An Approach to the Treatment of Inflammation?

Traditionally, researchers have used sulfide salts such as NaHS and Na_2S as H_2S donors (sometimes called 'classical' donors). These compounds release H_2S instantaneously in aqueous solution and have proved to be very useful tools to study the biology of H_2S. A number of 'alternative' H_2S donors, many of them releasing their H_2S 'load' more slowly, are now available and these have also been scrutinised biologically.

The general chemistry and biology of these donors has been the subject of a number of excellent recent reviews (Gu and Zhu 2011; Kashfi and Olson 2013; Sparatore et al. 2011) and will not be discussed further here. Rather, we will concentrate here on the possible therapeutic applications of some of these agents in inflammatory disease. To date, H_2S donors (excluding the classical sulfide salts) which have been assessed in animal models of inflammation range from cysteine analogues, particularly S-allylcysteine (SAC) and S-SPRC, thiones such as 5-(*p*-hydroxyphenyl)-1,2-dithione-3-thione (ADT-OH), GYY4137 and H_2S-releasing derivatives of non-steroidal anti-inflammatory drugs (NSAID) such as S-diclofenac, S-naproxen and S-mesalamine. The latter compounds (chimeras) are thione-based derivatives, linked to the parent compound by an ester bond, which is readily split in the body by esterase action thereby releasing H_2S plus the parent compound. To this list we should also add the recently reported NO/H_2S releasing aspirin hybrid (NOSH-aspirin, Kodela et al. 2012). The chemical structures of a selection of these compounds are shown in Fig. 3.2.

Of these, attention has been focussed on the S-substituted cysteine compounds. Before they were even described as H_2S donors, SAC and SPRC were reported to alleviate inflammation by reducing plasma levels of pro-inflammatory C-reactive protein, IL-6, IL-10 and TNF-α in ethanol treated mice (Yan and Yin 2007).

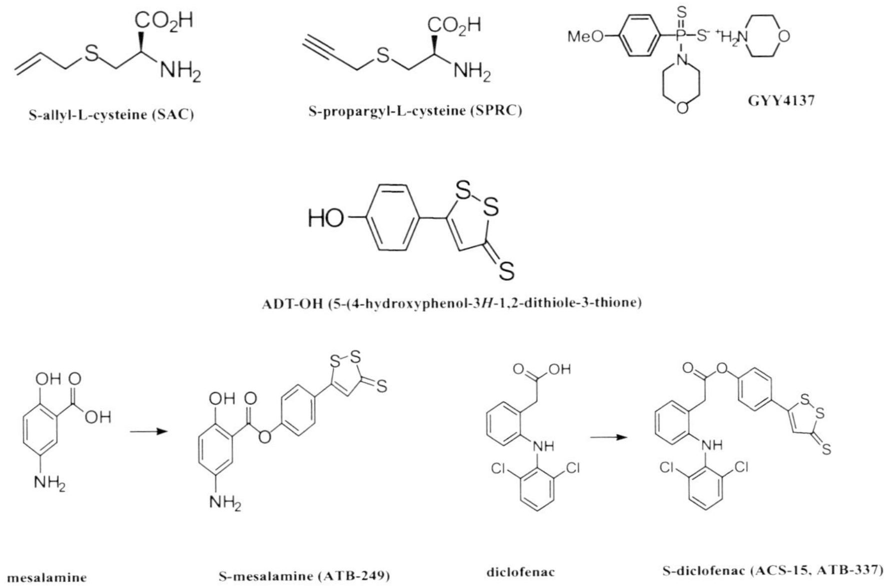

Fig. 3.2 Chemical structures of some commonly used H_2S donors

More recent work has confirmed that SAC reduces TNF- α and iNOS expression in the striatum of MPTP-injected mice (García et al. 2010) suggesting a possible role for H_2S in the inflammation associated with neurodegenerative disease whilst SPRC inhibited LPS-induced inflammation in cultured cardiac myocytes (Pan et al. 2011b) and protected mice against experimentally induced pancreatitis (Sidhapuriwala et al. 2012) in both cases mediated by release of H_2S. However, some degree of caution is advised in extrapolating these data to a beneficial effect of H_2S in inflammation since both SAC and SPRC exhibit anti-oxidant activity in their own right and this might also contribute to their effects.

There have also been a number of studies reporting the anti-inflammatory effect of H_2S-releasing derivatives of NSAIDs. For example, H_2S-releasing diclofenac (ATB-337 or ACS-15) is a more potent inhibitor of carrageenan-induced hindpaw than is the parent compound, diclofenac (Wallace et al. 2007; Sidhapuriwala et al. 2007) presumably due to the additional release of anti-inflammatory H_2S. As noted above, chronic treatment of rats with ATB337 was associated with markedly less gastrointestinal inflammation and damage than was diclofenac (Wallace et al. 2007) suggesting that inclusion of H_2S into the molecule had the added benefit of diminishing gastrointestinal complications which is a characteristic, and major drawback of, NSAID use. In addition, H_2S-releasing mesalamine (ATB-429) reduced colitis-associated leukocyte infiltration and expression of several pro-inflammatory cytokines (Fiorucci et al. 2007) and decreased the hyperalgesia associated with colorectal distension in the same species (Distrutti et al. 2006). Finally, in a recent report, H_2S-releasing aspirin (ACS-014) reduced IFN- γ and

LPS-induced upregulation of CX3CR1 in mouse macrophages by a PPAR-γ mechanism suggesting a possible use in the treatment of atherosclerotic disease (Zhang et al. 2012b).

The slow-releasing H_2S donor, GYY4137, also reduces LPS-evoked septic shock (Li et al. 2009a) and knee joint oedema in response to an intra-articular injection of Freund's adjuvant (Li et al. 2013) and inhibits generation of inflammatory mediators such as IL-1β, IL-6 and TNF-α in LPS-challenged macrophages in culture (Whiteman et al. 2010a). Interestingly, in vivo, the effect of GYY4137 against inflammation was determined to some extent by the timing of its administration. Thus, negligible anti-inflammatory activity was seen against LPS-evoked septic shock in the rat when GYY4137 was given 1 h before but inflammatory signs were reduced when the drug was given 2 h after LPS injection (Li et al. 2009a). In a mouse knee joint model of arthritis, GYY4137 potentiated knee joint oedema when given 1 h beforehand but was anti-inflammatory when administered 18 h afterwards (Li et al. 2013). One interpretation of these results is that H_2S can be either pro- or anti-inflammatory in these models depending on the stage of inflammation, local inflammatory environment and perhaps the concentration of H_2S achieved at the inflamed site. That GYY4137 is anti-inflammatory when administered at a time when knee joint swelling was higher might be said to auger well for a possible clinical use in this condition.

Whilst a number of H_2S donors have been assessed for anti-inflammatory activity both in vitro and in vivo none can yet be said to have been characterised in depth in experimental animals or even tested in man. Hence, definitive conclusions about H_2S donation as a possible approach to the treatment of inflammation are premature at this stage. Clearly, there are difficulties with using (or indeed testing) H_2S donors in man. One potential problem is the possibility of H_2S-induced organ toxicity which may be a factor especially at higher doses. Another is our current inability to be able to accurately 'titrate' a therapeutic (i.e. anti-inflammatory) effect with the plasma/serum concentration of H_2S. This is due both to technical difficulties with the available assays as noted previously and also because the downstream metabolites of H_2S have yet to be identified and hence a full pharmacokinetic analysis is not feasible at present. Nevertheless, different chemical classes of H_2S donor, which presumably all release H_2S, albeit at different rates, exhibit largely anti-inflammatory effects in different animal models. Moreover, some of the effects observed (e.g. ulcer healing) are dramatic. For these reasons, further efforts are required to establish whether H_2S donors can be used in man to treat inflammatory disease.

Conclusions

Research over the last decade has established a role for H_2S in inflammation. However, perhaps remarkably given the huge growth in interest in this area, the precise nature of that role remains elusive. H_2S has a very wide range of often diametrically opposite effects in inflammation. Thus, H_2S can be pro-inflammatory, anti-inflammatory or contribute to the resolution of inflammation.

Whilst some of the effects of H_2S reported in the literature are possibly artefactual, it is nevertheless becoming increasingly clear, that this gas does indeed exert complex, biological functions in inflammation. One important distinction which is often lost is the need to draw a distinction between endogenous vis-a-via exogenous H_2S. Thus, the effects of endogenous H_2S in inflammation are not always mirrored in the response to exogenous (i.e. donor-derived) H_2S probably because the concentrations of the gas achieved using H_2S donors and the timing of their action in an ongoing inflammatory response are very different to naturally generated H_2S. It is also important to consider the interaction between endogenous and exogenous H_2S. Put simply, injection of H_2S donors to naïve animals tends to promote inflammation whereas an anti-inflammatory effect is often the norm when these same donors are injected into animals with ongoing inflammatory disease. Perhaps endogenous H_2S, generated locally during inflammation, primes the anti-inflammatory response to a subsequent injection of an H_2S donor? This concept is not new since H_2S has been known for some time to contribute to ischemic preconditioning in heart, liver, kidney and other organs (Nicholson and Calvert 2010). Whether H_2S, generated locally at the inflamed site, also 'preconditions' the surrounding cells to a subsequent exposure of the gas is intriguing and warrants further study.

Acknowledgements We wish to thank the Biomedical Research Council (BMRC) and the Agency for Science, Technology and Research (A*STAR), Singapore for support.

References

Alshorafa AK, Guo Q, Zeng F, Chen M, Tan G, Tang Z, Yin R (2012) Psoriasis is associated with low serum levels of hydrogen sulfide, a potential anti-inflammatory molecule. Tohoku J Exp Med 228:325–332

Andruski B, McCafferty DM, Ignacy T, Millen B, McDougall JJ (2008) Leukocyte trafficking and pain behavioral responses to a hydrogen sulfide donor in acute monoarthritis. Am J Physiol 295:R814–R820

Ball CJ, Reiffel AJ, Chintalapani S, Kim M, Spector JA, King MR (2013) Hydrogen sulfide reduces neutrophil recruitment in hind-limb ischemia-reperfusion injury in an L-selectin and ADAM-17-dependent manner. Plast Reconstr Surg 131:487–497

Baragatti B, Ciofini E, Sodini D, Luin S, Scebba F, Coceani F (2013) Hydrogen sulfide in the mouse ductus arteriosus: a naturally occurring relaxant with potential EDHF function. Am J Physiol Heart Circ Physiol 304:H927–H934

Barathi S, Vadhana P, Angayarkanni N, Ramakrishnan S (2007) Estimation of hydrogen sulphide in the human lymphocytes. Indian J Biochem Biophys 44:179–182

Bates MN, Garrett N, Crane J, Balmes JR (2013) Associations of ambient hydrogen sulfide exposure with self-reported asthma and asthma symptoms. Environ Res 122:81–87

Bhatia M, Wong FL, Fu D, Lau HY, Moochhala SM, Moore PK (2005a) Role of hydrogen sulfide in acute pancreatitis and associated lung injury. FASEB J 19:623–625

Bhatia M, Sidhapuriwala J, Moochhala SM, Moore PK (2005b) Hydrogen sulphide is a mediator of carrageenan-induced hindpaw oedema in the rat. Br J Pharmacol 145:141–144

Blackler R, Syer S, Bolla M, Ongini E, Wallace JL (2012) Gastrointestinal-sparing effects of novel NSAIDs in rats with compromised mucosal defence. PLoS One 7:1–8e35196

Calvert JW, Jha S, Gundewar S, Elrod JW, Ramachandran A, Patillo CB, Kevil CG, Lefer DJ (2009) Hydrogen sulfide mediates cardioprotection through Nrf2 signaling. Circ Res 105: 365–367

Chen YH, Wu R, Geng B, Qi YF, Wang PP, Yao WZ, Tang CS (2009a) Endogenous hydrogen sulfide reduces airway inflammation and remodeling in a rat model of asthma. Cytokine 45:117–123

Chen YH, Yao WZ, Cao JZ, Geng B, Wang PP, Tang CS (2009b) Serum hydrogen sulfide as a novel marker predicting bacterial involvement in patients with community-acquired lower respiratory tract infections. Respirology 14:746–752

Coletta C, Szabo C (2013) Potential role of hydrogen sulfide in the pathogenesis of vascular dysfunction in septic shock. Curr Vasc Pharmacol 11:208–221

Dal-Secco D, Cunha TM, Freitas A, Alves-Filho JC, Souto FO, Fukada SY, Grespan R, Alencar NM, Neto AF, Rossi MA, Ferreira SH, Hothersall JS, Cunha FQ (2008) Hydrogen sulfide augments neutrophil migration through enhancement of adhesion molecule expression and prevention of CXCR2 internalization: role of ATP-sensitive potassium channels. J Immunol 181:4287–4298

di Villa Bianca RD, Coletta C, Mitidieri E, De Dominicis G, Rossi A, Sautebin L, Cirino G, Bucci M, Sorrentino R (2010) Hydrogen sulphide induces mouse paw oedema through activation of phospholipase A2. Br J Pharmacol 161:1835–1842

Distrutti E, Sediari L, Mencarelli A, Renga B, Orlandi S, Russo G, Caliendo G, Santagada V, Cirino G, Wallace JL, Fiorucci S (2006) 5-Amino-2-hydroxybenzoic acid 4-(5-thioxo-5H-[1,2] dithiol-3yl)-phenyl ester (ATB-429), a hydrogen sulfide-releasing derivative of mesalamine, exerts antinociceptive effects in a model of postinflammatory hypersensitivity. J Pharmacol Exp Ther 319:447–458

Dufton N, Natividad J, Verdu EF, Wallace JL (2012) Hydrogen sulphide and resolution of inflammation: a comparative study utilizing a novel fluorescent probe. Sci Reps 2:1–11

Ekundi-Valentim E, Santos KT, Camargo EA, Denadai-Souza A, Teixeira SA, Zanoni CI, Grant AD, Wallace J, Muscará MN, Costa SK (2010) Differing effects of exogenous and endogenous hydrogen sulphide in carrageenan-induced knee joint synovitis in the rat. Br J Pharmacol 159:1463–1474

Esechie A, Kiss L, Olah G, Horváth EM, Hawkins H, Szabo C, Traber DL (2008) Protective effect of hydrogen sulfide in a murine model of acute lung injury induced by combined burn and smoke inhalation. Clin Sci 115:91–97

Filipovic MR, Miljkovic JLJ, Nauser T, Royzen M, Klos K, Shubina T, Koppenol WH, Lippard SJ, Ivanović-Burmazović I (2012) Chemical characterization of the smallest S-nitrosothiol, HSNO; cellular cross-talk of H_2S and S-nitrosothiols. J Am Chem Soc 134:12016–12027

Fiorucci S, Antonelli E, Distrutti E, Rizzo G, Mencarelli A, Orlandi S, Zanardo R, Renga B, Di Sante M, Morelli A, Cirino G, Wallace JL (2005) Inhibition of hydrogen sulfide generation contributes to gastric injury caused by anti-inflammatory nonsteroidal drugs. Gastroenterology 129:1210–1224

Fiorucci S, Orlandi S, Mencarelli A, Caliendo G, Santagada V, Distrutti E, Santucci L, Cirino G, Wallace JL (2007) Enhanced activity of a hydrogen sulphide-releasing derivative of mesalamine (ATB-429) in a mouse model of colitis. Br J Pharmacol 150:996–1002

Fong CH, Bebien M, Didierlaurent A, Nebauer R, Hussell T, Broide D, Karin M, Lawrence T (2008) An antiinflammatory role for IKKbeta through the inhibition of "classical" macrophage activation. J Exp Med 205:1269–1276

Gade AR, Kang M, Akbarali HI (2013) Hydrogen sulfide as an allosteric modulator of ATP-sensitive potassium channels in colonic inflammation. Mol Pharmacol 83:294–306

Gao C, Xu DQ, Gao CJ, Ding Q, Yao LN, Li ZC, Chai W (2012) An exogenous hydrogen sulphide donor, NaHS, inhibits the nuclear factor κB inhibitor kinase/nuclear factor κb inhibitor/nuclear factor-κB signaling pathway and exerts cardioprotective effects in a rat hemorrhagic shock model. Biol Pharm Bull 35:1029–1034

García E, Villeda-Hernández J, Pedraza-Chaverrí J, Maldonado PD, Santamaría A (2010) S-allylcysteine reduces the MPTP-induced striatal cell damage via inhibition of pro-inflammatory cytokine tumor necrosis factor-α and inducible nitric oxide synthase expressions in mice. Phytomedicine 18:65–73

Gong QH, Wang Q, Pan LL, Liu XH, Xin H, Zhu YZ (2011) S-propargyl-cysteine, a novel hydrogen sulfide-modulated agent, attenuates lipopolysaccharide-induced spatial learning and memory impairment: involvement of TNF signaling and NF-κB pathway in rats. Brain Behav Immun 25:110–119

Gu X, Zhu YZ (2011) Therapeutic applications of organosulfur compounds as novel hydrogen sulfide donors and/or mediators. Expert Rev Clin Pharmacol 4:123–133

Han W, Dong Z, Dimitropoulou C, Su Y (2011) Hydrogen sulfide ameliorates tobacco smoke-induced oxidative stress and emphysema in mice. Antioxid Redox Signal 15:2121–2134

Hosoki R, Matsuki N, Kimura H (1997) The possible role of hydrogen sulfide as an endogenous smooth muscle relaxant in synergy with nitric oxide. Biochem Biophys Res Commun 237:527–531

Hua W, Jiang JB, Rong X, Wu RZ, Qiu HX, Zhang YH, Chen Q (2010) Expression of cystathionine-γ-lyase/hydrogen sulfide pathway in CVB3-induced myocarditis in mice. Zhongguo Dang Dai Er Ke Za Zhi 12:744–748

Kalayarasan S, Prabhu PN, Sriram N, Manikandan R, Arumugam M, Sudhandiran G (2009) Diallyl sulfide enhances antioxidants and inhibits inflammation through the activation of Nrf2 against gentamicin-induced nephrotoxicity in Wistar rats. Eur J Pharmacol 606:162–171

Kashfi K, Olson KR (2013) Biology and therapeutic potential of hydrogen sulfide and hydrogen sulfide-releasing chimeras. Biochem Pharmacol 8:689–703

Khan AA, Yong S, Prior MG, Lillie LE (1991) Cytotoxic effects of hydrogen sulfide on pulmonary alveolar macrophages in rats. J Toxicol Environ Health 33:57–64

Kodela R, Chattopadhyay M, Kashfi K (2012) NOSH-aspirin: a novel nitric oxide-hydrogen sulfide-releasing hybrid: a new class of anti-inflammatory pharmaceuticals. ACS Med Chem Lett 3:257–262

Li L, Bhatia M, Zhu YZ, Zhu YC, Ramnath RD, Wang ZJ, Anaur F, Whiteman M, Salto-Tellez M, Moore PK (2005) Hydrogen sulfide is a novel mediator of endotoxic shock. FASEB J 19:1196–1198

Li L, Rossoni G, Sparatore A, Lee LC, Del Soldato P, Moore PK (2007) Anti-inflammatory and gastrointestinal effects of a novel diclofenac derivative. Free Radic Biol Med 42:706–719

Li L, Whiteman M, Guan YY, Neo KL, Cheng Y, Lee SW, Zhao Y, Baskar R, Tan CH, Moore PK (2008) Characterisation of a novel, water soluble hydrogen sulfide releasing molecule (GYY4137): new insights into the biology of hydrogen sulphide. Circulation 117:2351–2356

Li L, Salto-Tellez M, Tan CH, Whiteman M, Moore PK (2009a) GYY4137, a novel hydrogen sulfide-releasing molecule, protects against endotoxic shock in the rat. Free Radic Biol Med 47:103–113

Li L, Whiteman M, Moore PK (2009b) Dexamethasone inhibits lipopolysaccharide-induced hydrogen sulphide biosynthesis in intact cells and in an animal model of endotoxic shock. J Cell Mol Med 13:2684–2692

Li L, Rose P, Moore PK (2011a) The biological significance of hydrogen sulfide. Ann Rev Pharmacol 51:169–187

Li XH, Zhang CY, Wu JX, Zhang T (2011b) Changes in plasma hydrogen sulfide and nitric oxide levels and their clinical significance in children with Kawasaki disease. Chin Med J (Engl) 124:3445–3449

Li L, Fox B, Keeble J, Salto-Tellez M, Winyard PG, Wood ME, Moore PK, Whiteman M (2013) The complex effects of the slow-releasing hydrogen sulfide donor GYY4137 in a model of acute joint inflammation and in human cartilage cells. J Cell Mol Med 17:365–376

Lynn EG, Austin RC (2011) Hydrogen sulfide in the pathogenesis of atherosclerosis and its therapeutic potential. Expert Rev Clin Pharmacol 4:97–108

Martelli A, Testai L, Breschi MC, Lawson K, McKay NG, Miceli F, Taglialatela M, Calderone V (2013) Vasorelaxation by hydrogen sulphide involves activation of Kv7 potassium channels. Pharmacol Res 70:27–34

Mathew ND, Schlipalius DI, Ebert PR (2011) Sulfurous gases as biological messengers and toxins: comparative genetics of their metabolism in model organisms. J Toxicol 394970:1–14

Matsunami M, Tarui T, Mitani K, Nagasawa K, Fukushima O, Okubo K, Yoshida S, Takemura M, Kawabata A (2009) Luminal hydrogen sulfide plays a pronociceptive role in mouse colon. Gut 58:751–761

Matsunami M, Miki T, Nishiura K, Hayashi Y, Okawa Y, Nishikawa H, Sekiguchi F, Kubo L, Ozaki T, Tsujiuchi T, Kawabata A (2012) Involvement of the endogenous hydrogen sulfide/Ca (v) 3.2 T-type Ca2+ channel pathway in cystitis-related bladder pain in mice. Br J Pharmacol 167:917–928

Mikami Y, Shibuya N, Kimura Y, Nagahara N, Yamada M, Kimura H (2011) Hydrogen sulfide protects the retina from light-induced degeneration by the modulation of Ca2+ influx. J Biol Chem 286:39379–39386

Mikami Y, Shibuya N, Ogasawara Y, Kimura H (2013) Hydrogen sulfide is produced by cystathionine γ-lyase at the steady-state low intracellular Ca(2+) concentrations. Biochem Biophys Res Commun 431:131–135

Mok YYP, Atan MS, Cheung YP, Wang ZJ, Bhatia M, Moochhala SM, Moore PK (2004) Role of hydrogen sulphide in haemorragic shock in the rat: protective effect of inhibitors of hydrogen sulphide biosynthesis. Br J Pharmacol 143:881–889

Mustafa AK, Sikka G, Gazi SK, Steppan J, Jung SM, Bhunia AK, Barodka VM, Gazi FK, Barrow RK, Wang R, Amzel LM, Berkowitz DE, Snyder SH (2011) Hydrogen sulfide as endothelium-derived hyperpolarizing factor sulfhydrates potassium channels. Circ Res 109:1259–1268

Nicholson CK, Calvert JW (2010) Hydrogen sulfide and ischemia-reperfusion injury. Pharmacol Res 62:289–297

Ogawa H, Takahashi K, Miura S, Imagawa T, Saito S, Tominaga M, Ohta T (2012) H(2)S functions as a nociceptive messenger through transient receptor potential ankyrin 1 (TRPA1) activation. Neuroscience 218:335–343

Oh GS, Pae HO, Lee BS, Kim BN, Kim JM, Kim HR, Jeon SB, Jeon WK, Chae HJ, Chung HT (2006) Hydrogen sulfide inhibits nitric oxide production and nuclear factor-kappaB via heme oxygenase-1 expression in RAW264.7 macrophages stimulated with lipopolysaccharide. Free Radic Biol Med 41:106–119

Pan LL, Liu XH, Gong QH, Wu D, Zhu YZ (2011a) Hydrogen sulfide attenuated tumor necrosis factor-α-induced inflammatory signaling and dysfunction in vascular endothelial cells. PLoS One 6:1-13e19766

Pan LL, Liu XH, Gong QH, Zhu YZ (2011b) S-Propargyl-cysteine (SPRC) attenuated lipopolysaccharide-induced inflammatory response in H9c2 cells involved in a hydrogen sulfide-dependent mechanism. Amino Acids 41:205–215

Park SJ, Kim TH, Lee SH, Ryu HY, Hong KH, Jung JY, Hwang GH, Lee SH (2013) Expression levels of endogenous hydrogen sulfide are altered in patients with allergic rhinitis. Laryngoscope 123:557–563

Patacchini R, Santicioli P, Giuliani S, Maggi CA (2004) Hydrogen sulfide (H_2S) stimulates capsaicin-sensitive primary afferent neurons in the rat urinary bladder. Br J Pharmacol 142:31–34

Pozsgai G, Hajna Z, Bagoly T, Boros M, Kemény Á, Materazzi S, Nassini R, Helyes Z, Szolcsányi J, Pintér E (2012) The role of transient receptor potential ankyrin 1 (TRPA1) receptor activation in hydrogen-sulphide-induced CGRP-release and vasodilation. Eur J Pharmacol 689:56–64

Rinaldi L, Gobbi G, Pambianco M, Micheloni C, Mirandola P, Vitale M (2006) Hydrogen sulfide prevents apoptosis of human PMN via inhibition of p38 and caspase 3. Lab Invest 86:391–397

Rivers JR, Badiei A, Bhatia M (2012) Hydrogen sulfide as a therapeutic target for inflammation. Expert Opin Ther Targets 16:439–449

Sen N, Paul BD, Gadalla MM, Mustafa AK, Sen T, Xu R, Kim S, Snyder SH (2012) Hydrogen sulfide-linked sulfhydration of NF-κB mediates its antiapoptotic actions. Mol Cell 45:13–24

Shao JL, Wan XH, Chen Y, Bi C, Chen HM, Zhong Y, Heng XH, Qian JQ (2011) H2S protects hippocampal neurons from anoxia-reoxygenation through cAMP-mediated PI3K/Akt/p70S6K cell-survival signaling pathways. J Mol Neurosci 43:453–460

Shibuya N, Mikami Y, Kimura Y, Nagahara N, Kimura H (2009) Vascular endothelium expresses 3-mercaptopyruvate sulfurtransferase and produces hydrogen sulfide. J Biochem 146:623–626

Sidhapuriwala J, Li L, Sparatore A, Bhatia M, Moore PK (2007) Effect of S-diclofenac, a novel hydrogen sulfide releasing derivative, on carrageenan-induced hindpaw oedema formation in the rat. Eur J Pharmacol 569:149–154

Sidhapuriwala JN, Hegde A, Ang AD, Zhu YZ, Bhatia M (2012) Effects of S-propargyl-cysteine (SPRC) in caerulein-induced acute pancreatitis in mice. PLoS One 7:1–8e32574

Sivarajah A, Collino M, Yasin M, Benetti E, Gallicchio M, Mazzon E, Cuzzocrea S, Fantozzi R, Thiemermann C (2009) Anti-apoptotic and anti-inflammatory effects of hydrogen sulfide in a rat model of regional myocardial I/R. Shock 31:267–274

Sparatore A, Santus G, Giustarini D, Rossi R, Del Soldato P (2011) Therapeutic potential of new hydrogen sulfide-releasing hybrids. Expert Rev Clin Pharmacol 4:109–121

Streng T, Axelsson HE, Hedlund P, Andersson DA, Jordt SE, Bevan S, Andersson KE, Högestätt ED, Zygmunt PM (2008) Distribution and function of the hydrogen sulfide-sensitive TRPA1 ion channel in rat urinary bladder. Eur Urol 53:391–399

Takahashi T, Aoki Y, Okubo K, Maeda Y, Sekiguchi F, Mitani K, Nishikawa H, Kawabata A (2010) Upregulation of Ca(v)3.2 T-type calcium channels targeted by endogenous hydrogen sulfide contributes to maintenance of neuropathic pain. Pain 150:183–191

Tamizhselvi R, Sun J, Koh YH, Bhatia M (2009) Effect of hydrogen sulfide on the phosphatidylinositol 3-kinase-protein kinase B pathway and on caerulein-induced cytokine production in isolated mouse pancreatic acinar cells. J Pharmacol Exp Ther 329:1166–1177

Tang G, Yang G, Jiang B, Ju Y, Wu L, Wang R (2013) H_2S is an endothelium-derived hyperpolarizing factor. Antioxid Redox Signal (Epub ahead of print)

Tangerman A (2009) Measurement and biological significance of the volatile sulfur compounds hydrogen sulfide, methanethiol and dimethyl sulfide in various biological matrices. J Chromatogr B Analyt Technol Biomed Life Sci 877:3366–3377

Tian M, Wang Y, Lu YQ, Yan M, Jiang YH, Zhao DY (2012) Correlation between serum H_2S and pulmonary function in children with bronchial asthma. Mol Med Rep 6:335–338

Tokuda K, Kida K, Marutani E, Crimi E, Bougaki M, Khatri A, Kimura H, Ichinose F (2012) Inhaled hydrogen sulfide prevents endotoxin-induced systemic inflammation and improves survival by altering sulfide metabolism in mice. Antioxid Redox Signal 17:11–21

Trevisani M, Patacchini R, Nicoletti P, Gatti R, Gazzieri D, Lissi N, Zagli G, Creminon C, Geppetti P, Harrison S (2005) Hydrogen sulfide causes vanilloid receptor 1-mediated neurogenic inflammation in the airways. Br J Pharmacol 145:1123–1131

Tripatara P, Patel NS, Brancaleone V, Renshaw D, Rocha J, Sepodes B, Mota-Filipe H, Perretti M, Thiemermann C (2009) Characterisation of cystathionine gamma-lyase/hydrogen sulphide pathway in ischaemia/reperfusion injury of the mouse kidney: an in vivo study. Eur J Pharmacol 606:205–209

Ubuka T, Hosaki Y, Nishina H, Ikeda T (1985) 3-Mercaptopyruvate sulfurtransferase activity in guinea pig and rat tissues. Physiol Chem Phys Med NMR 17:41–43

Wallace JL, Caliendo G, Santagada V, Cirino G, Fiorucci S (2007) Gastrointestinal safety and anti-inflammatory effects of a hydrogen sulfide-releasing diclofenac derivative in the rat. Gastroenterology 132:261–271

Wallace JL, Vong L, McKnight W, Dicay M, Martin GR (2009) Endogenous and exogenous hydrogen sulfide promotes resolution of colitis in rats. Gastroenterology 129:569–578

Wallace JL, Caliendo G, Santagada V, Cirino G (2010) Markedly reduced toxicity of a hydrogen sulphide-releasing derivative of naproxen (ATB-346). Br J Pharmacol 159:1236–1246

Wallace JL, Ferraz JG, Muscara MN (2012) Hydrogen sulfide: an endogenous mediator of resolution of inflammation and injury. Antioxid Redox Signal 17:58–67

Wang R (2012) Physiological implications of hydrogen sulfide: a whiff exploration that blossomed. Physiol Rev 92:791–896

Wang Y, Zhao X, Jin H, Wei H, Li W, Bu D, Tang X, Ren Y, Tang C, Du J (2009) Role of hydrogen sulfide in the development of atherosclerotic lesions in apolipoprotein E knockout mice. Arterioscler Thromb Vasc Biol 29:173–179

Whiteman M, Moore PK (2009) Hydrogen sulfide and the vasculature: a novel vasculoprotective entity and regulator of nitric oxide bioavailability? J Cell Mol Med 13:488–507

Whiteman M, Winyard PG (2011) Hydrogen sulfide and inflammation: the good, the bad, the ugly and the promising. Expert Rev Clin Pharmacol 4:13–32

Whiteman M, Haigh R, Tarr JM, Gooding KM, Shore AC, Winyard PG (2010a) Detection of hydrogen sulfide in plasma and knee-joint synovial fluid from rheumatoid arthritis patients: relation to clinical and laboratory measures of inflammation. Ann N Y Acad Sci 1203:146–150

Whiteman M, Li L, Rose P, Tan CH, Parkinson DB, Moore PK (2010b) The effect of hydrogen sulfide donors on lipopolysaccharide-induced formation of inflammatory mediators in macrophages. Antioxid Redox Signal 12:1147–1154

Wu R, Yao WZ, Chen YH, Geng B, Tang CS (2008) Plasma level of endogenous hydrogen sulfide in patients with acute asthma. Beijing Da Xue Xue Bao 40:505–508

Yan SL, Yin MC (2007) Protective and alleviative effects from 4 cysteine-containing compounds on ethanol-induced acute liver injury through suppression of oxidation and inflammation. J Food Sci 72:S511–S515

Yang G, Sun X, Wang R (2004) Hydrogen sulfide-induced apoptosis of human aorta smooth muscle cells via the activation of mitogen-activated protein kinases and caspase-3. FASEB J 18:1782–1784

Yang G, Wu L, Jiang B, Yang W, Qi J, Cao K, Meng Q, Mustafa AK, Mu W, Zhang S, Snyder SH, Wang R (2008) H_2S as a physiologic vasorelaxant: hypertension in mice with deletion of cystathionine gamma-lyase. Science 322:587–590

Yusuf M, Whiteman M, Mok YYP, Li L, Kostetski I, Chu SH, Siau JL, Cheong YP, Bhatia M, Moore PK (2006) Effect of hydrogen sulfide and nitric oxide alone and together on rat aortic contractility and blood pressure. Br J Pharmacol 149:625–634

Zanardo RC, Brancaleone V, Distrutti E, Fiorucci S, Cirino G, Wallace JL (2006) Hydrogen sulfide is an endogenous modulator of leukocyte-mediated inflammation. FASEB J 20:2118–2120

Zhang H, Guo C, Wu D, Zhang A, Gu T, Wang L, Wang C (2012a) Hydrogen sulfide inhibits the development of atherosclerosis with suppressing CX3CR1 and CX3CL1 expression. PLoS One 7:1-18e41147

Zhang H, Guo C, Zhang A, Fan Y, Gu T, Wu D, Sparatore A, Wang C (2012b) Effect of S-aspirin, a novel hydrogen-sulfide-releasing aspirin (ACS14), on atherosclerosis in apoE-deficient mice. Eur J Pharmacol 697:106–116

Zhang G, Wang P, Yang G, Cao Q, Wang R (2013) The inhibitory role of hydrogen sulfide in airway hyperresponsiveness and inflammation in a mouse model of asthma. Am J Pathol 182:1188–1195

Zhao W, Zhang J, Lu Y, Wang R (2001) The vasorelaxant effect of H_2S as a novel endogenous gaseous K_{ATP} channel opener. EMBO J 20:6008–60016

Zhao ZZ, Wang Z, Li GH, Wang R, Tan JM, Cao X, Suo R, Jiang ZS (2011) Hydrogen sulfide inhibits macrophage-derived foam cell formation. Exp Biol Med (Maywood) 236:169–176

Zhi L, Ang AD, Zhang H, Moore PK, Bhatia M (2007) Hydrogen sulfide induces the synthesis of proinflammatory cytokines in human monocyte cell line U937 via the ERK-NF-kappaB pathway. J Leukoc Biol 81:1322–1332

Zhu XY, Liu SJ, Liu YJ, Wang S, Ni X (2010) Glucocorticoids suppress cystathionine gamma-lyase expression and H_2S production in lipopolysaccharide-treated macrophages. Cell Mol Life Sci 67:1119–1132

Zhu X, Tang Z, Cong B, Du J, Wang C, Wang L, Ni X, Lu J (2013) Estrogens increase cystathionine-γ-lyase expression and decrease inflammation and oxidative stress in the myocardium of ovariectomized rats. Menopause epub ahead of print

Signaling Mechanisms Underlying the Hydrogen Sulfide Effects: Identification of Hydrogen Sulfide "Receptors"

4

Yi-Chun Zhu

Abstract

After extensive research efforts worldwide in the last decades, it is no doubt now that hydrogen sulfide (H_2S) has important physiological roles in many organs/tissues and is also involved in the pathogenesis of some important diseases. However, the mechanisms underlying the H_2S effects are largely unknown. One of the most challenging questions in the fielf of H_2S biology would be identification of the "receptors" for H_2S. Recent studies indicate that H_2S may react with its "receptors" with some mechanisms beyond the typical ligand-receptor mechanisms. Some atomic mechanisms including nucleophilic attack is involved in H_2S-induced regulation of its "receptor" by certain molecular switches such as a disulfide bond. A certain protein containing a disulfide bond molecular switch labile to H_2S regulation may serve as an H_2S "receptors". The receptor tyrosine kinase family may contain an H_2S switch in their common intracellular kinase core. More future works are required to identify potential H_2S switches in another kind of protein, i.e., the ion channels.

Keywords

Hydrogen sulfide • Receptor • Molecular switch • Disulfide bond • Receptor tyrosine kinase • Ion channels

Y.-C. Zhu (✉)
Department of Physiology and Pathophysiology, Fudan University Shanghai Medical College, Shanghai, China
e-mail: yczhu@shmu.edu.cn

H. Kimura (ed.), *Hydrogen Sulfide and its Therapeutic Applications*,
DOI 10.1007/978-3-7091-1550-3_4, © Springer-Verlag Wien 2013

Abbreviations

3MST	3-Mercaptopyruvate sulfurtransferase
$[Ca^{2+}]i$	Intracellular calcium transient
$[Ca^{2+}]SR$	SR Ca^{2+} load
CAT	Catalase
cGMP	Cyclic guanosine monophosphate
CSE	Cystathionine γ-lyase
Cys	Cysteine
Cys-Gly	Cysteinylglycine
EDHF	Endothelium derived hyperpolarization factor
eNOS	Endothelial nitric oxide synthase
ESI-MS	Electrospray ionization mass spectrometry
GAPDH	Glyceraldehyde-3-phosphate dehydrogenase
GK	Goto-Kakizaki
GSH	Glutathione
H_2S	Hydrogen sulfide
HOMO	The highest occupied molecular orbital
$I_{Ca\ L}$	The L-type calcium channel current
K_{ATP} channels	ATP-sensitive potassium channels
K_{Ca}	Transient Ca^{2+}-activated K^+
K_{ir}	Inward rectifier potassium channel
K_V	Voltage dependent potassium channel
LUMO	The lowest unoccupied molecular orbital
NaHS	Sodium hydrosulfide
NMDA	N-methyl-D-aspartate
NO	Nitric oxide
PI3K	Phosphoinositide 3-kinase
PPG	DL-propargylglycine
ROS	Reactive oxygen species
SHR	Spontaneously hypertensive rats
SR	Sarcoplasmic reticulum
TRPA1	Transient receptor potential ankyrin-1
VEGF	Vascular endothelial growth factor
VEGFR2	Vascular endothelial growth factor receptor 2
WKY	Wistar-Kyoto

4.1 Introduction

Despite numerous studies reporting various biological effects of hydrogen sulfide (H_2S) in physiology and diseases in different organs/tissues, little is known about the signaling mechanisms underlying the H_2S effects (Li et al. 2011; Kimura et al. 2012). One of the most challenging questions in the field is identification of the "receptor" for H_2S to mediate numerous biological effects. It is sure that H_2S must have some "receptors" to mediate effects. The term receptor means the targeting molecule, usually a protein on the membrane of a cell, which specifically binds with a ligand. The chemical nature of the ligand is multifarious – peptides, some amino acids, or some endogenous chemicals such as steroids or monoamines. The conformation of the ligands matches with the pocket structure contained in their receptors. Upon binding with the ligands, the conformation and function of the receptors are changed. Conformational matching is the key rule of a typical ligand-receptor binding. However, gaseous molecules such as H_2S are too small to have a conformation being specific enough to recognize and bind with their potential "receptors". This gives rise to a question that either the various biological effects of H_2S reported to date are non-specific effects, or H_2S acts with a new mechanism beyond typical ligand-receptor binding. In this article, recent works aiming to answer this question are reviewed.

4.2 H_2S Is a Promoter of Angiogenesis

We have first reported a proangiogenic effect of H_2S in an in vitro model of migration of vascular endothelial cells and in an in vivo model of angiogenesis in 2007 (Cai et al. 2007). H_2S promotes migration of vascular endothelial cells in both transwell migration assay and scratch would healing assay and these effects are blocked by dominant-negative mutant of Akt. In vascular endothelial cells cultured in three dimensional medium, H_2S treatment promotes microvessel tube formation and this effect is again blocked by dominant-negative mutant of Akt. The data suggest that the proangiogenic effect of H_2S is dependent on Akt. We also observe this proangiogenic effect of H_2S in an in vivo model of Matrigel plug assay (Fig. 4.1).

Two years later in 2009, the proangiogenic effect of H_2S is confirmed by Papapetropoulos et al. in the same in vitro models of cell migration (Papapetropoulos et al. 2009). They further show the proangiogenic role of endogenous H_2S using aortic rings isolated from cystathionine γ-lyase (CSE, one of the H_2S generating enzymes) knockout mice where microvessel formation in response to vascular endothelial growth factor (VEGF) is reduced as compared to wild-type littermates (Papapetropoulos et al. 2009). Moreover, skin wound healing is delayed in CSE knockout mice and H_2S treatment enhances wound healing in a rat model of skin wound (Papapetropoulos et al. 2009). At this stage, the next challenge is to clarify whether H_2S would promote angiogenesis in chronic ischemia. To address this question, we show a proangiogenic role of H_2S in a model of hind limb

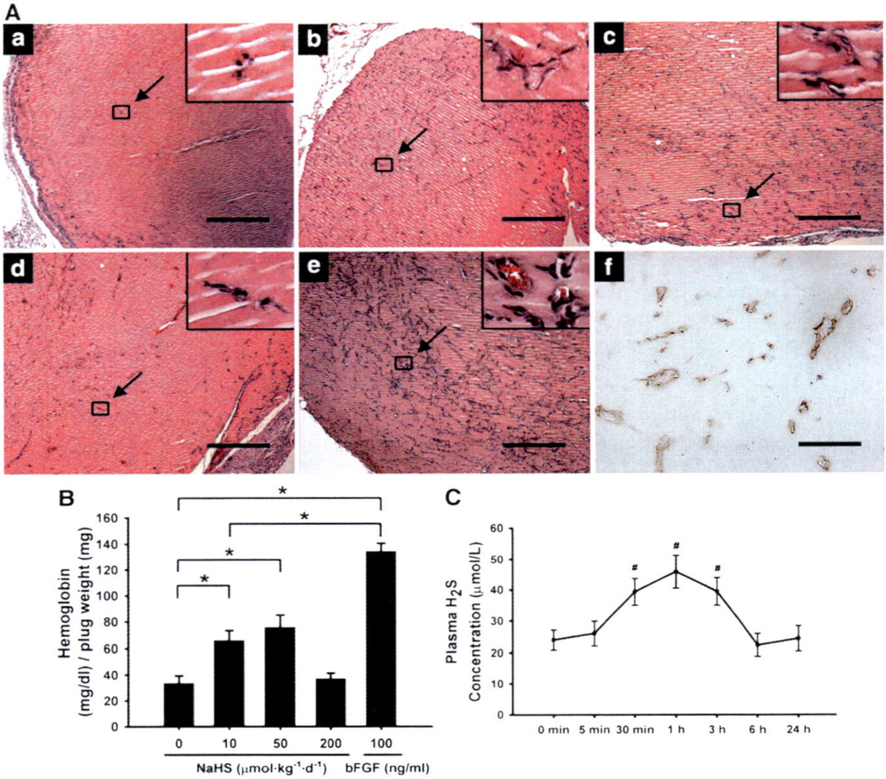

Fig. 4.1 H_2S treatment promotes angiogenesis in vivo. The effects of H_2S on in vivo angiogenesis were assessed using Matrigel plug assay in mice. (**A**), Representative photomicrographs of hematoxylin–eosin stained Matrigel sections of mice treated with vehicle (*a*), various doses of NaHS (*b*, *c* and *d* for 10, 50 and 200 μmol/kg/day NaHS, respectively) and bFGF (*e*, 100 ng/ml in Matrigel). *Insets* are higher magnifications of the areas marked in *squares* (*arrow*). Capillaries were defined as tubular structures (*brown signals*) stained with rabbit polyclonal anti-CD31 antibodies in Matrigel sections from the mice treated with 50 μmol/kg/day NaHS (*f*). (**B**), Neovascularization in the Matrigel plugs was quantified by measuring hemoglobin content using the tetramethylbenzidine method. NaHS treatment (10 and 50 μmol/kg/day) significantly promoted neovascularization in the Matrigel plugs in mice. This effect of NaHS was less potent than that of bFGF. (**C**), Time course of plasma H_2S concentrations in mice after an intraperitoneal injection of NaHS (100 μmol/kg). Data represent the mean ± SEM of five mice in each group. Bar = 400 μm in (**A**) *a–e* (for higher magnifications of the areas marked in *squares* (*arrow*), bar = 50 μm). Bar = 50 μm in Af. $^{*}p < 0.05$ in (**B**); $^{\#}p < 0.05$ vs. 0 min in (**C**) (Reproduced from Cai et al. 2007)

ischemia induced by femoral artery occlusion in rat. We find that chronic H_2S treatment for 4 weeks increases collateral vessel growth, regional tissue blood flow (Fig. 4.2), and capillary density in ischemic hind limb muscles as compared with the controls. These effects are associated with an increase in VEGF expression in the skeletal muscles and vascular endothelial growth factor receptor 2 (VEGFR2)

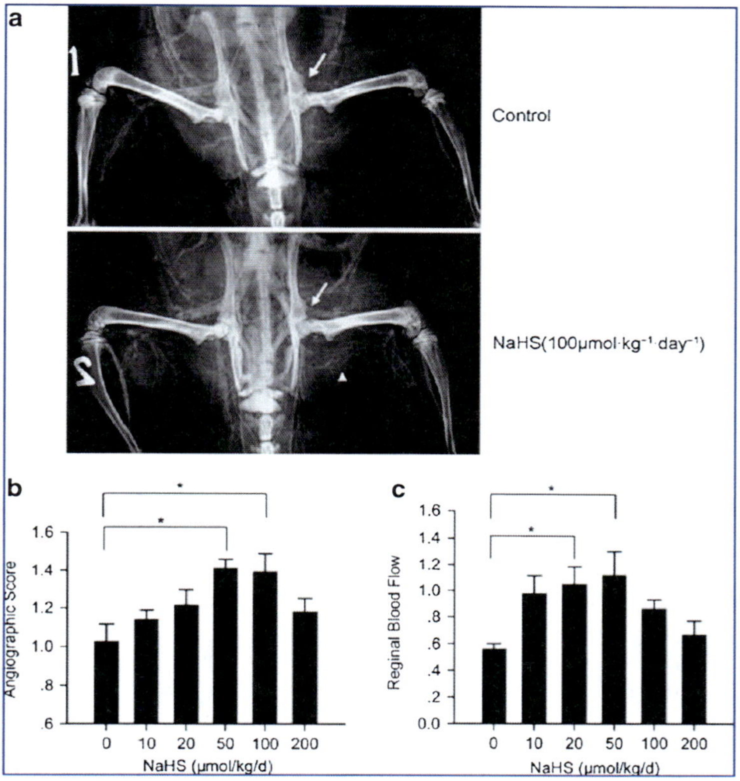

Fig. 4.2 NaHS treatment promoted collateral vessel formation and regional blood flow after femoral artery occlusion in the rat hind limb ischemia model. (**a**) Representative postmortem angiograms obtained 4 weeks after surgery. There was more collateral vessel formation in the ischemic left hind limb of the rats treated with NaHS at a dose of 100 μmol/kg/day. *Arrow* denotes the site of ligation at the femoral artery. *Arrow head* indicates the typical "corkscrew" appearance of collateral vessels. (**b**) Quantitative analysis of collateral vessel development was performed by measuring the total length of the contrast-opacified vessels. The angiographic score was significantly greater in the rats receiving NaHS (50 and 100 μmol/kg/day) than in the control animals ($^*p < 0.05$ vs. vehicle by ANOVA). Data represent the mean ± SEM of three (10 and 20 μmol/kg/day NaHS) or five (the rest) experiments in each group. (**c**) Blood flow measured with microsphere assay. The regional blood flow in ischemic limb was standardized with tissue weight and represented as the ratio of fluorescence intensity in the ischemic hind limb to that of the contralateral nonischemic hind limb in each animal. NaHS treatment (20 and 50 μmol/kg/day) significantly improved regional blood flow in the ischemic limb ($^*p < 0.05$ vs. vehicle by ANOVA). Data represent the mean ± SEM of nine (the control) or seven (the NaHS groups) experiments in each group (Reproduced from Wang et al. 2010)

phosphorylation in the neighboring vascular endothelial cells, suggesting a role of VEGF in mediating the H_2S effects in a cell–cell interaction pattern (Wang et al. 2010). These data not only suggest a possibility that chronic organ ischemia may be treated with a new H_2S-based pathway, but also reveal some more complicated cell–cell interaction mechanisms underlying the in vivo proangiogenic effect of H_2S. More recently in 2012, Bir et al. duplicate the proangiogenic role of H_2S in a mouse model of hind limb ischemia and further show the role of endothelial nitric oxide (NO) synthase in mediating the proangiogenic effect of H_2S. They find that the proangiogenic effect of H_2S treatment was blunted in endothelial nitric oxide synthase (eNOS) knockout mice, suggesting a role of NO in mediating the proangiogenic role of H_2S (Bir et al. 2012). However, it is unknown if NO is the only pathway to mediate the H_2S effects in angiogenesis. Coletta et al. recently show that inhibition of eNOS abolishes H_2S-stimulated angiogenesis. On the other hand, knockdown of the H_2S-generating enzyme cystathionine-γ-lyase abolishes NO-induced cyclic guanosine monophosphate (cGMP) increase and angiogenesis. Therefore, the authors conclude that H_2S and NO are mutually dependent in the regulation of angiogenesis. The two pathways cross talk at cGMP (Coletta et al. 2012). These studies do not support the idea that the H_2S effects are solely mediated by the NO signals. Though the mechanisms underlying the interaction between the H_2S signals and the NO signals are largely unknown, accumulating data suggest that H_2S has its independent signaling pathways.

In addition to above-mentioned proangiogenic effect of H_2S in hind limb ischemia models (Wang et al. 2010; Bir et al. 2012), H_2S has also been shown to promote angiogenesis and tissue blood flow in a mouse model of myocardial infarction. These effects are associated with an increase in the expression of VEGF, flk-1 and flt-1, and a decrease in the level of endostatin, angiostatin and parstatin in the myocardium of the mice treated with the H_2S donor, sodium hydrosulfide (NaHS) (Qipshidze et al. 2012). Though the proangiogenic role of H_2S is associated with an inhibition of anti-angiogenic proteins and stimulation of angiogenic factors, the mechanisms underlying H_2S-induced angiogenesis in the myocardium remain unknown.

The key to uncover the signaling mechanisms underlying the proangiogenic effects of H_2S and other numerous biological effects of H_2S is to identify the "receptors" for H_2S and the molecular and atomic mechanisms of H_2S-induced regulation in structure and functions of the H_2S "receptors".

4.3 H_2S Signals Mediated via Ion Channels

4.3.1 The Role of H_2S in the Regulation of Calcium Channels

The first piece of direct evidence illustrating the inhibitory role of H_2S on calcium channels emerges in 2008 (Sun et al. 2008). The effects of H_2S on L-type calcium channels are examined using patch-clamp techniques in acutely isolated cardiomyocytes (Fig. 4.3). H_2S causes an immediate inhibition on the L-type calcium channels in the membrane of cardiomyocytes in ~2 min after administration of exogenous H_2S. This effect is also observed in the presence of isoprenaline. The H_2S effects are dose-dependent and can be washed away. H_2S shows dose-dependent inhibition on the L-type calcium channel current ($I_{Ca,\,L}$) at concentrations

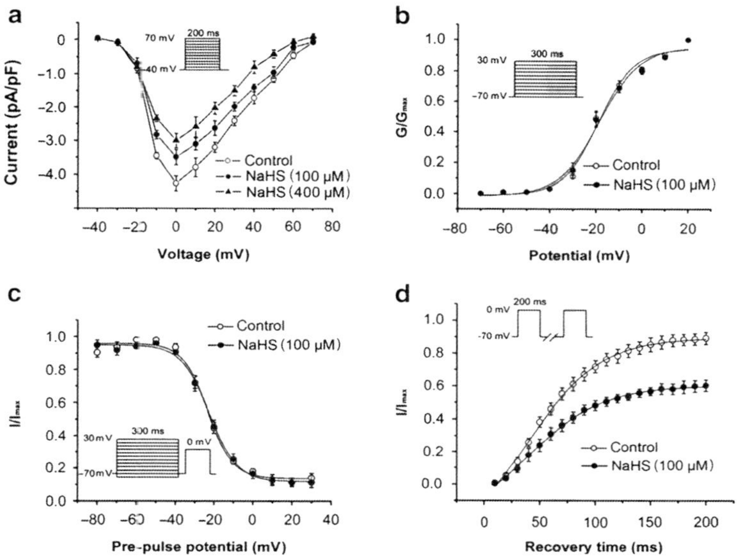

Fig. 4.3 The effects of H_2S on (**a**) the I–V relationship of $I_{Ca,L}$; (**b**) the steady-state activation of $I_{Ca,L}$; (**c**) the steady-state inactivation of $I_{Ca,L}$; (**d**) the kinetics of the recovery of $I_{Ca,L}$ following inactivation ($n = 6$ for each group studied) (Reproduced from Sun et al. 2008)

of 25, 50, 100, 200 and 400 μmol/L in cardiomyocytes isolated from spontaneously hypertensive rats (SHR) and Wistar-Kyoto (WKY) rats. The H_2S donor, NaHS, shows similar effects as compared with the H_2S gas solution. Except the concentration of 200 and 400 μmol/L NaHS which may induce toxic levels of H_2S, the rest concentrations (25, 50 and 100 μmol/L) yields physiologically relevant H_2S levels. Therefore, the inhibition of H_2S on L-type channels is not just a pharmacological phenomenon, but is physiologically relevant. It is note worthy that there is no shift in the steady state activation curve of $I_{Ca,\,L}$ nor in the steady state inactivation of $I_{Ca,\,L}$ (Fig. 4.3) in cardiomyocytes treated with H_2S. While H_2S causes a shift in the recovery curve of $I_{Ca,\,L}$ (Fig. 4.3). Though the molecular mechanisms underlying H_2S-induced inhibition on L-type calcium channels have not yet been clarified in this study, the data suggest that H_2S acts on the motifs related to channel opening but not on the motifs related to the voltage-dependency character of L-type calcium channels (Sun et al. 2008). More future works are required to testify this hypothesis. In particular, it remains to be clarified how H_2S can regulate the structure and function of L-type calcium channels and if there are some motifs contained in the channel molecule that serves as molecular switch for H_2S regulation.

On the other hand, the physiological relevance of H_2S-induced L-type calcium channel inhibition is also investigated. Electric field induced intracellular calcium transient ($[Ca^{2+}]$i) and contraction of single cardiomyocytes and isolated papillary muscles are also reduced by H_2S treatment. In contrast, caffeine induces an increase in $[Ca^{2+}]$i that is not altered by H_2S (Sun et al. 2008). These data indicate that

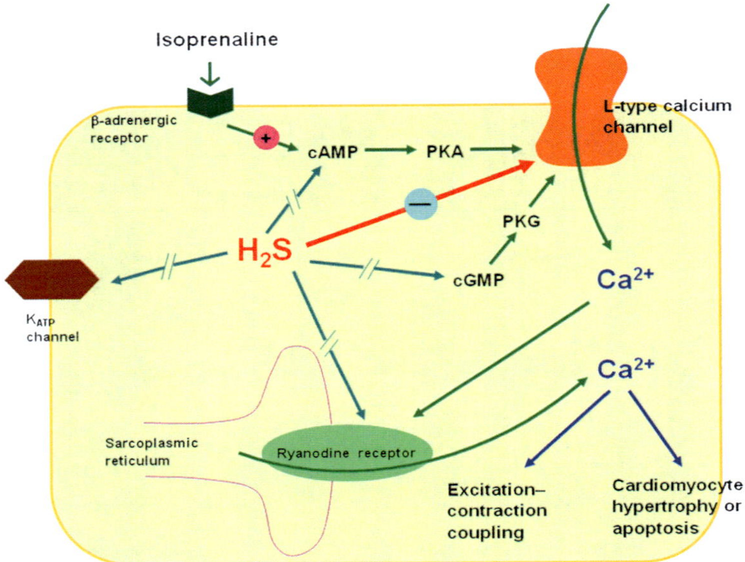

Fig. 4.4 The inhibitory effects of H_2S on the L-type calcium channel in the cardiomyocytes

the inhibition on $[Ca^{2+}]i$ and contraction of cardiomyocytes is an effect secondary to H_2S-induced inhibition on L-type calcium channels (Fig. 4.4). H_2S may provide protection for the myocardium by inhibition of $[Ca^{2+}]i$ transients.

In line with our results, Tian et al. report a similar inhibition of H_2S on L-type calcium channels using patch clamp recordings in single myocytes isolated from the cerebral artery of rat. $[Ca^{2+}]i$ and contraction of the cerebral artery are also reduced by H_2S treatment (Tian et al. 2012). On the other hand, the decay of both electric- and caffeine-induced $[Ca^{2+}]i$ transients is significantly accelerated in the cardiomyocytes preconditioned with H_2S for 30 min (Pan et al. 2008).

However, H_2S does not exert any effect on ATP-sensitive potassium channels (K_{ATP} channels) in the cardiomyocytes at concentrations of 50 and 100 μmol/L NaHS. This is different from the results of Zhao et al. who show an opening effect of H_2S on the K_{ATP} channels in the vascular smooth muscle cells (Zhao et al. 2001). The reason underlying this controversy remains to be investigated. It is possible that the K_{ATP} channels are regulated in the cardiomyocytes via some signaling pathways different from that in the vascular smooth muscle cells. Moreover, the subtypes of the K_{ATP} channels in the cardiomyocytes are different from that in the vascular smooth muscle cells (Burke et al. 2008). This may account for the differential effects of H_2S in the cardiomyocytes and vascular smooth muscle cells.

There are also a series of reports implicating a role of T-type calcium channels in mediating H_2S effects (Okubo et al. 2011, 2012; Matsunami et al. 2012). For example, in mice repeatedly treated with antisense oligodeoxynucleotides to silence Cav3.2 T-type calcium or transient receptor potential ankyrin-1 (TRPA1) channels in the sensory neurons, H_2S-induced hyperalgesia and allodynia are prevented in mice (Okubo et al. 2012). In a rat model of colitis induced by 2,4,6-trinitrobenzenesulfonic acid, H_2S treatment significantly suppresses this colitis and the H_2S effect is prevented by T-type Ca^{2+} channel blockers (Matsunami et al. 2012). However, all the data suggesting a role of H_2S in the regulation of T-type calcium channels reported to date are indirect evidences. The conclusions are all based on experiments using T-type calcium channel blockers or knockdown of the T-type calcium channels. Direct evidence such as recording of the T-type calcium currents using patch clamp technique has not been provided yet.

Moreover, H_2S has been shown to regulate recycle of $[Ca^{2+}]i$ into the sarcoplasmic reticulum (SR). Using high-speed confocal imaging, Liang et al. find that H_2S increases SR Ca^{2+} load ($[Ca^{2+}]SR$), leading to an increase in the frequency of Ca^{2+} spark-induced transient Ca^{2+}-activated K^+ (K_{Ca}) currents, but does not alter the amplitude of these events. This results in consequent hyperpolarization of vascular smooth muscle cells and vasodilatation of piglet cerebral arterioles (Liang et al. 2012).

4.3.2 The Role of H_2S on Potassium Channels

The first report describing a role of H_2S in the regulation of K_{ATP} channels is provided by Wang's group (Zhao et al. 2001). H_2S has been shown to open the K_{ATP} channels using patch clamp recording in vascular smooth muscle cells. The cell membrane is consequently hyperpolarized and blood vessels are therefore dilated. It is note worthy that the K_{ATP} channel opening effect of H_2S is observed at very high concentrations of NaHS (an H_2S donor) at 300 and 600 μmol/L (Zhao et al. 2001) which yield H_2S concentrations at about 10- to 20-folds higher than physiologically relevant concentrations of H_2S since plasma H_2S levels have been reported to be ~50 μmol/L in rats (Zhao et al. 2001), ~34 μmol/L in mice (Li et al. 2005) and ~44 μmol/L in human (Li et al. 2005). On the other hand, the K_{ATP} channels do not play a major role in determining membrane potential in vascular smooth muscles under physiological conditions where the K_{ATP} channel current is very small. The K_{ATP} channels only have a significant role in membrane potential regulation with low ATP levels since it is highly sensitive to a decrease in ATP levels (Noma 1983). In this context, the vasodilatation effect of H_2S which has been first reported early in 1997 (Hosoki et al. 1997) may not be ascribed to H_2S-induced opening of the K_{ATP} channels. Some mechanisms beyond the K_{ATP} channels may be involved in H_2S-induced vasodilatation. In fact, there are some contradictory evidences that do not support the role of H_2S-induced opening of the K_{ATP} channels. In both small mesenteric arteries and cerebral arteries stimulated with intermittent hypoxia, exogenous H_2S dilated and hyperpolarized both sham and intermittent hypoxia arteries, however, this dilation was not blocked by the K_{ATP} channel inhibitor, glibenclamide (Jackson-Weaver et al. 2011). This suggests that some potassium channels other than the K_{ATP} channels may mediate H_2S-induced vasodilatation and hyperpolarization of the vascular smooth muscle cells.

Streeter et al. also exclude the role of the possible K_{ATP} channel opening effect of H_2S in mediating H_2S-induced vasodilatation. They find that selective blockers of K_{ATP}, calcium sensitive (K_{Ca}), voltage dependent potassium channel (K_V), or inward rectifier potassium channel (K_{ir}) channels do not prevent H_2S-induced relaxation of middle cerebral arteries of rats. However, blockade of L-type calcium channels significantly decreases the maximum relaxation induced by H_2S. In addition inhibition of K^+ conductance with 50 mmol/L K^+ significantly attenuates H_2S-mediated vasodilatation. Based on these results, the authors conclude that H_2S-induced relaxation of middle cerebral arteries is partly mediated by inhibition of L-type calcium channels and some potassium channels but not the K_{ATP}, K_{Ca}, K_V, or K_{ir} subtypes (Streeter et al. 2012). More future works are required to identify why H_2S can cause hyperpolarization of vascular smooth muscle cells. In particular, specific potassium channels which may mediate H_2S-induced hyperpolarization remain to be clarified. Is it possible that some unknown mechanisms are involved in this H_2S action in vascular smooth muscle cells? In fact, Liang et al. reports an H_2S-induced increase in the frequency but not the amplitude of calcium transient, leading to an increase in transient K_{Ca} currents and consequent hyperpolarization of vascular smooth muscle cells and vasodilatation of piglet cerebral arterioles. However, H_2S does not change the activity of single K_{Ca} channels recorded in the absence of Ca^{2+} sparks, remaining the potassium channels that may directly mediate H_2S-induced hyperpolarization unknown (Liang et al. 2012).

Moreover, it has been reported that H_2S activates the ATP-sensitive, intermediate conductance and small conductance potassium channels through cysteine S-sulfhydration and therefore results in hyperpolarization and vasorelaxation of vascular endothelial and smooth muscle cells. Based on these data, H_2S has been considered as a major endothelium derived hyperpolarization factor (EDHF) (Mustafa et al. 2011). However, Streeter et al. show that L-cysteine-induced vasorelaxation is attenuated by the CSE inhibitor DL-propargylglycine and this relaxation is independent of endothelium. The data also suggest that H_2S is produced in the vascular smooth muscle (Streeter et al. 2012). This contradicts with the idea that H_2S is an EDHF. Indeed, H_2S may be produced in the endothelium by 3MST (3-mercaptopyruvate sulfurtransferase) and CAT (catalase) (Shibuya et al. 2009). To date, it is unknown about the half life of endogenous H_2S within a cell and in the interstitial between the cells. H_2S generated in the vascular endothelial cells may be degraded before it may reach the vascular endothelial cells. The current evidence is not sufficient to conclude that H_2S is an EDHF.

Despite of the debate about the mechanisms underlying H_2S-induced hyperpolarization, the effect of H_2S in inducing hyperpolarization in vascular smooth muscle cells is accordant in all of the above-mentioned studies. Moreover, H_2S also cause hyperpolarization in cell types other than the vascular endothelial cells. Early in 1993, Kombian et al. show that H_2S causes hyperpolarization in dorsal serotonergic cells using intracellular sharp microelectrode and whole-cell recording techniques (Kombian et al. 1993). However, in the type 1 cells of the chemoreceptors, H_2S seems to cause depolarization. In perforated patch clamp recordings, H_2S inhibits background potassium channels leading to membrane depolarization and subsequent voltage-gated Ca^{2+} entry and sensitization of the type 1 cells of the chemoreceptors (Buckler 2012).

On the other hand, despite numerous reports about the role of H_2S in regulating ion channels, little is known about how H_2S can regulate channel function. The first question is if H_2S directly targets at some ion channels or acts via some signaling molecules. If H_2S directly targets at the ion channels, the second question would be how H_2S targets at the channel molecules? Does the "S-sulfhydration theory" or the "disulfide bond breaking theory" apply to H_2S-mediated regulation of the channels? The further question would be which theory is right? Or both theories are right?

4.4 H_2S Regulates Insulin Signals: The Role of H_2S in Diabetes

The role of H_2S in diabetes is of significant scientific importance and also has potential values in translational medicine. However, extensive research efforts in this area result in contradictory conclusions. An acute injection with NaHS at a dose of 2 mg/kg causes an increase in glucose levels and a decrease in insulin levels in Wistar Albino rats (a Wistar strain without diabetes). Whereas, chronic treatment with NaHS at a dose of 2 mg/kg (a very high dosage) for 30 days has no effect on the levels of glucose and insulin, nor on insulin sensitivity (Patel and Shah 2010). An acute pretreatment of NaHS (39 μmol/kg body weight) significantly deteriorates glucose tolerance of wild-type mice, whereas NaHS alone has no effect on blood glucose levels of these mice (Yang et al. 2011). In streptozotocin-diabetic rats, supplementation of NaHS at a dose of 16 μg/kg per min (equals 411.3792 μmol/kg/day) for 28 days via osmotic pumps improves the endothelium-dependent relaxant function of vascular rings, however, has no effect on blood glucose (Suzuki et al. 2011). Similarly in streptozotocin-diabetic rats, 1 week treatment with NaHS at dose of 50 μmol/kg/day exerts no effect on blood glucose (Yuan et al. 2011). Streptozotocin injection caused delayed onset of diabetic status in CSE knockout mice as compared with wild-type mice, suggesting a deleterious role of H_2S in streptozotocin-diabetic rats (Yang et al. 2011). All these studies suggest that H_2S either deteriorates diabetic status or has no effect on glucose metabolism. In contrast, clinical investigations suggest an association between plasma H_2S levels and the occurrence of diabetes. For example, plasma H_2S levels are significantly lower in patients with type 2 diabetes compared with lean (Whiteman et al. 2010). Type 2 diabetic patients with sleep apnea have significantly lower H_2S levels compared with patients with normal sleep patterns (Jain et al. 2012). In addition, experiments in rats show a down-regulation of CSE gene expression induced by glucose (Zhang et al. 2011). Though these clinical and experimental evidences are not yet sufficient to clarify the role of H_2S in diabetes, however, a decrease in H_2S levels in diabetic patients suggest that H_2S supplementation may be beneficial for diabetes. This leads to the main hypothesis we tried to testify in our recent study (Xue et al. 2013). We find that H_2S promotes glucose uptake in myotubes and adipocytes cultured in medium containing low (5.5 mmol/L) or high (25 mmol/L) levels of glucose. These H_2S effects are ascribed to a sensitization of the insulin

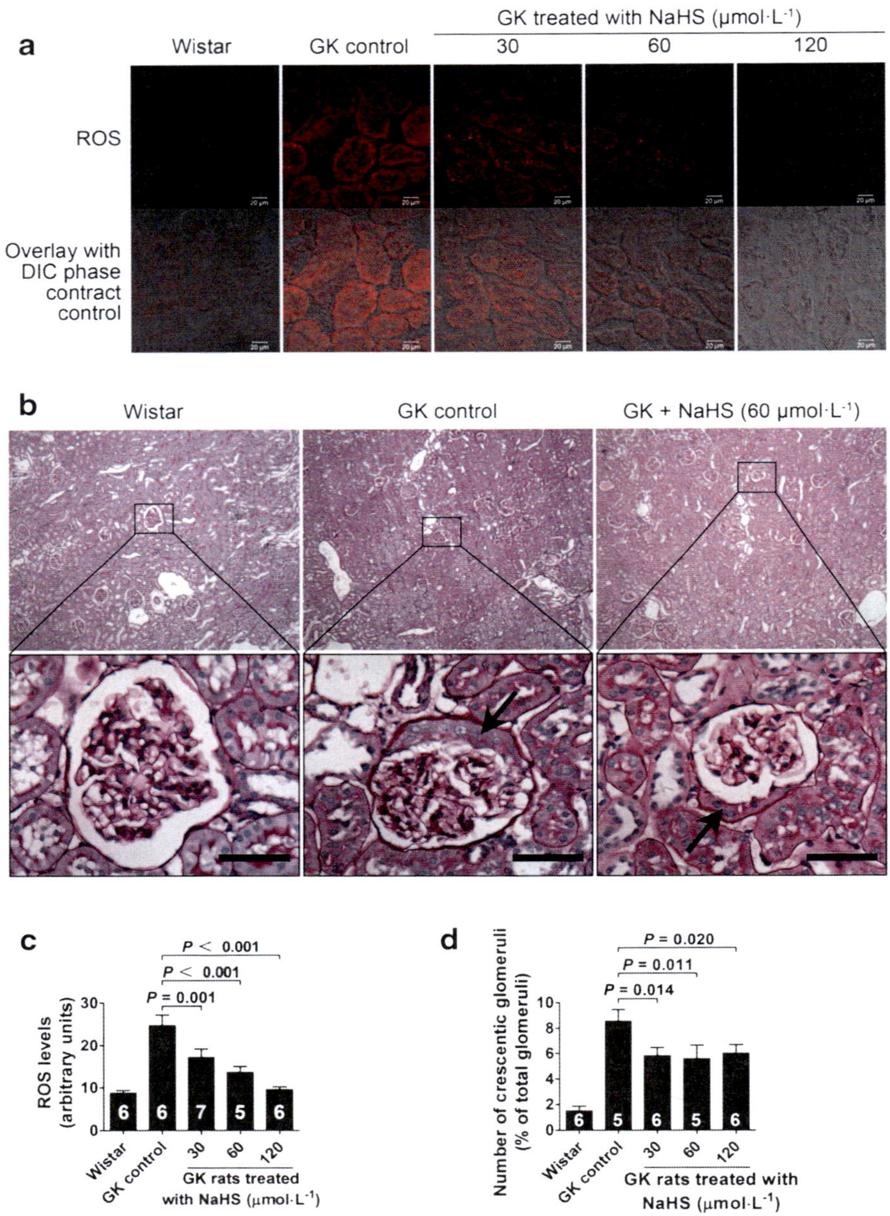

Fig. 4.5 Kidney morphology of the rats treated with NaHS. (**a**) Micrographs showing ROS expression in the kidney of GK rats treated with NaHS for 10 weeks. (**b**) Morphological change of the number of glomerular crescent observed in the kidney of GK rats. The number of crescentic glomeruli was decreased in the kidney of GK rats chronically treated with NaHS at doses of 30, 60 and 120 μmol/kg/day compared with the GK control group. Scale bar, 250 μm. (**c**) Graphs showing

receptors. Phosphorylation of the insulin receptor, phosphoinositide 3-kinase (PI3K) and Akt is increased in response to H_2S treatment. Moreover, the H_2S effect in promoting glucose uptake is attenuated either by inhibitors of PI3K and the insulin receptor or by siRNA-mediated knockdown of the insulin receptor. In addition, H_2S directly increases the kinase activity of the insulin receptors. Knockdown of CSE shows some effect in reducing glucose uptake and this effect is more pronounced in the presence of exogenous H_2S, suggesting a role of endogenous H_2S. In long-term treatment experiments in Goto-Kakizaki (GK) diabetic rats and Wistar control rats, we find that chronic H_2S treatment increases insulin sensitivity and glucose tolerance and decreases fasting blood glucose. More interestingly, chronic H_2S treatment seems to prevent diabetic end organ damage. H_2S reduces reactive oxygen species (ROS) levels and the number of crescentic glomeruli in the kidney of GK rats (Fig. 4.5). Our data is in line with a study about the effect of garlic on glucose metabolism in fructose fed rats where raw garlic homogenate (250 mg/kg/day) improves insulin sensitivity (Padiya et al. 2011). Fresh raw garlic homogenate contains water 70 %, protein 4.4 %, fat 0.2 %, carbohydrate 23.0 %, fiber 0.7 %, Vitamin A, Vitamin B, Vitamin C, Fe etc. In addition, raw garlic contains allitride which is composed of 16 kinds of sulfur compounds, including diallyl trisulfide (~60 %), allyl disulfide (23–39 %), allylpropyl disulfide (13–19 %), propane disulfide (4–5 %) etc. Since allitride releases H_2S, this study supports our hypothesis about a beneficial role of H_2S in diabetes though at the stage of this study the beneficial role of H_2S remains to be validated since raw garlic contains numerous components including some active metabolites of allitride. Another novel finding of our current study is an amelioration of diabetic end organ damage as demonstrated with improvement of kidney morphology. This is in line with an in vitro experiment where H_2S inhibits high glucose-induced matrix protein synthesis in renal epithelial cells (Lee et al. 2012).

The reason of why do we but not other investigators find a beneficial role of H_2S in improving glucose metabolism and morphology may be ascribed to the long-term treatment protocol (10 weeks of chronic treatment with a physiologically relevant low dose) applied in our current study, while other investigators never used a treatment protocol longer than 4 weeks. Wu et al. have not investigated the effect of chronic NaHS treatment on fasting glucose, insulin resistance and kidney morphology in an in vivo model of type 2 diabetes (Wu et al. 2009). Wu et al. treat Zucker diabetic rats with a CSE inhibitor, DL-propargylglycine (PPG) (33.9 mg/kg for 4 weeks), and observe a decrease in blood glucose. Based on these results Wu et al. conclude that endogenous H_2S increases blood glucose in diabetic rats.

Fig. 4.5 (Continued) statistical analysis of ROS levels in the kidney of GK rats treated with NaHS at doses of 30, 60 and 120 μmol/kg/day. (**d**) Graphs showing statistical analysis of the number of crescentic glomeruli in the kidney of GK rats. Data represent means ± SEM. A *P* value < 0.05 represents statistical significance. *ROS* reduced oxygen species. The *arrows* indicate the crescents observed in the glomeruli of GK rats (Reproduced from Xue et al. 2013)

However, these data may not be sufficient to make such a conclusion since PPG is not a high specific inhibitor for CSE. PPG has been shown to have some additional effects such as an inhibition on other vitamin B6-dependent enzymes. PPG is a lead-containing compound and is considered to be toxic after long-term administration. Most recently, there is a piece of new evidence about protection of H_2S for diabetic retinopathy in streptozotocin-induced diabetes in rats (Si et al. 2013). This supports our idea that H_2S provides protection in diabetes.

4.5 Identification of the "Receptor" for H_2S

To date, H_2S has emerged as a new gas transducer for numerous important biological processes. For example, H_2S causes hyperpolarization in neurons (Kombian et al. 1993), increases N-methyl-D-aspartate (NMDA) receptor-mediated responses and facilitate the induction of hippocampal long-term potentiation (Abe and Kimura 1996). H_2S also causes vasodilatation (Hosoki et al. 1997), angiogenesis (Cai et al. 2007; Wang et al. 2010), myocardial protection (Bian et al. 2006; Shi et al. 2007; Yao et al. 2010), inhibition of cardiac L-type calcium channels (Sun et al. 2008), regulation of inflammation (Li et al. 2005), sensitization of insulin signals (Xue et al. 2013), hyperalgesia and allodynia (Okubo et al. 2012) and suppression of colitis (Matsunami et al. 2012). An internet search tells us that there is an explosion of information about the biological role of H_2S which is mostly published in recent years. However, most of the reports are about the biological effects of H_2S. It is unknown about how H_2S can exert numerous biological effects in various organs/tissues. Though there have been some studies intend to clarify the signaling mechanisms underlying numerous H_2S effects, the most important question, i.e., what is the direct target molecule for H_2S, remains largely unknown.

We have recently proposed a theory that the disulfide bonds serve as targeting motif for H_2S to function as molecular switch for H_2S-induced regulation of the structure and function of the H_2S "receptors" (Tao et al. 2013). This theory is based on identification of an H_2S "receptor", VEGFR2, which mediates the proangiogenic effect of H_2S in vascular endothelial cells. H_2S specifically targets and breaks the new found Cys1045-Cys1024 disulfide bond (Fig. 4.6). The Cys1045-Cys1024 disulfide bond is an intrinsic inhibitory motif which resides in the intracellular kinase core of VEGFR2. This is evidenced by significant increase in the kinase activity of a VEGFR2 mutant, C1045A, which prevents the formation of the Cys1045-Cys1024 disulfide bond. Moreover, transfection and expression of C1045A to vascular endothelial cells cause an increase in migration of the cells and exogenous H_2S treatment does not promote cell migration any more. Molecular dynamics simulations show that the Cys1045-Cys1024 disulfide bond disrupts the integrity of the intracellular kinase core for ATP binding by turning the kinase core into the inactive "Phe-out" conformation. Upon breaking of the Cys1045-Cys1024 disulfide bond by H_2S, the integrity of the kinase core is restored with its structure being shifted into the active "Phe-in" conformation which is suitable for ATP binding (Fig. 4.7).

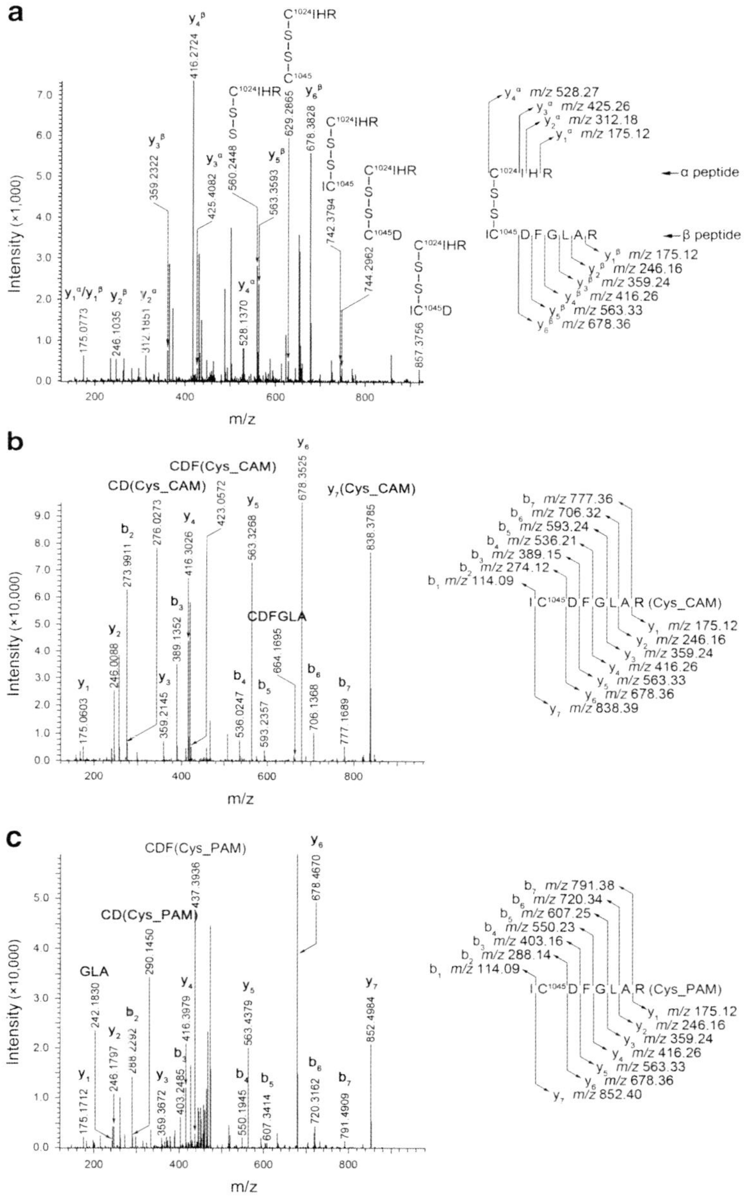

Fig. 4.6 ESI-CID-MS-MS spectra of VEGFR2. (**a**) CID spectra of $[M + 3H]^{3+}$ *m/z* 473.90 from a tryptic digest of VEGFR2 in the absence of NaHS showing an S–S bond between Cys1045 and Cys1024. (**b**) CID spectra of $[M + 2H]^{2+}$ *m/z* 476.24 from a tryptic digest of VEGFR2 in the presence of dithiothreitol showing the β peptide containing Cys1045. (**c**) CID spectra of $[M + 2H]^{2+}$ *m/z* 483.25 from a tryptic digest of VEGFR2 in the presence of NaHS showing the β peptide containing Cys1045. *CID* collision-induced dissociation, *Cys_CAM* carboxyamidomethyl cysteine, *Cys_PAM* propionamide cysteine (Reproduced from Tao et al. 2013)

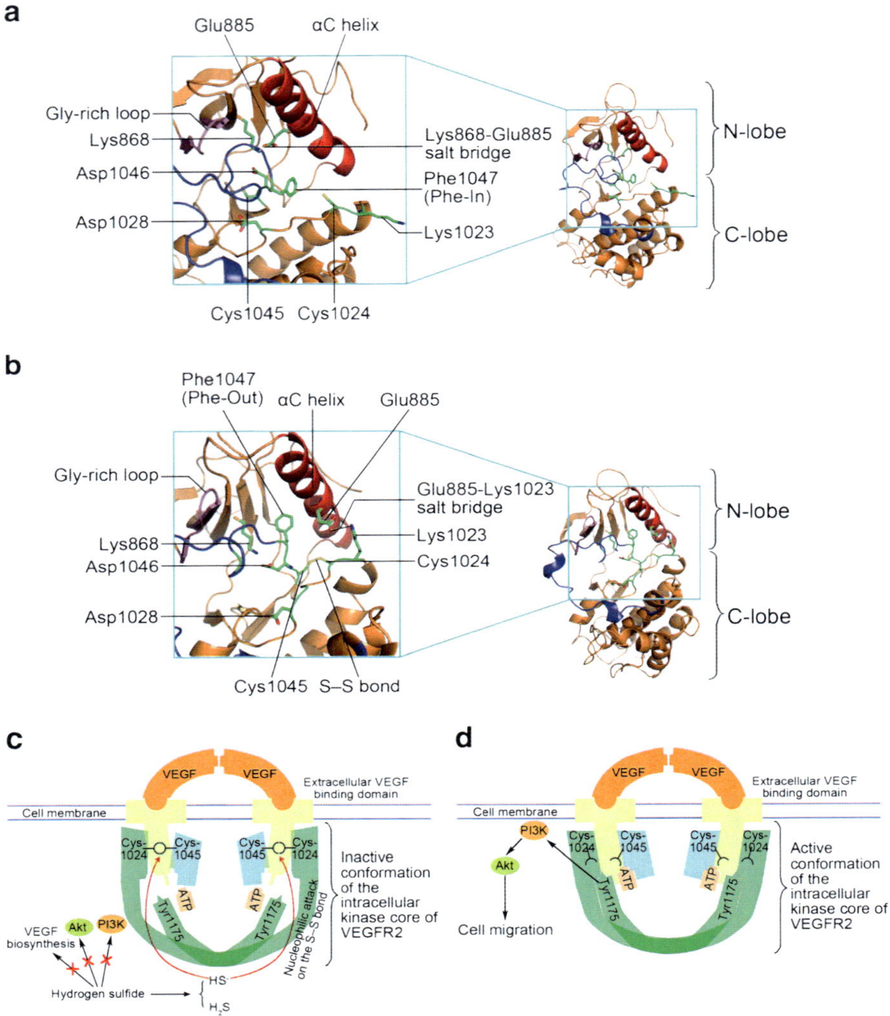

Fig. 4.7 Molecular dynamics simulation showing conformation of the kinase core of VEGFR2. (**a**) There is a conserved salt bridge between Lys868 and Glu885 in VEGFR2 without the Cys1045-Cys1024 S–S bond. The DFG motif is in the Phe-In conformation. (**b**) The Lys868-Glu885 salt bridge is disrupted and the DFG motif rotates anticlockwise from the Phe-In conformation in VEGFR2 with the Cys1045-Cys1024 S–S bond. (**c**, **d**) Schematic illustration of the main finds of the present study. Hydrogen sulfide directly acts on VEGFR2, but not on PI3K or on Akt in promoting vascular endothelial cell migration. There is a disulfide bond between Cys1045 and Cys1024 serving as an intrinsic inhibitory motif that disrupts the integrity of the VEGFR2 kinase core, which is essential for precise coordination of ATP for the phosphotransfer reaction. (**c**) Hydrogen sulfide yields HS^-, which breaks the Cys1045-Cys1024 disulfide bond and thus recovers the integrity of the VEGFR2 kinase core resulting in an increase in VEGFR2 kinase activity (**d**). Consequently, VEGFR2 (at the Tyr1175 site), PI3K and Akt are phosphorylated and cell migration is promoted. Although the migration-promoting effects of hydrogen sulfide were not mediated by an increase in VEGF biosynthesis, extracellular VEGF binding (to dimerize VEGFR2) is a premise for hydrogen sulfide-induced VEGFR2 activation (Reproduced from Tao et al. 2013)

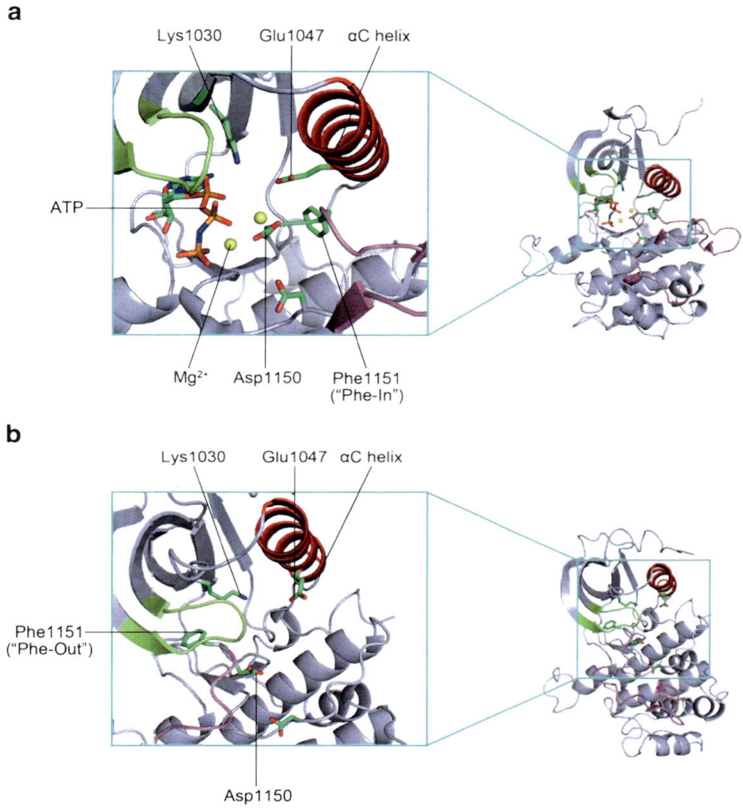

Fig. 4.8 Conformation of the kinase core of the insulin receptor. (**a**) The active conformation showing making of the Lys1030-Glu1047 salt bridge resulting proper positioning and stabilization of Lys1030 where the α and β phosphates of ATP are coordinated precisely for the phosphotransfer reaction. The DFG motif is also in its active form, which is characterized with the Phe-In conformation of Phe1151. (**b**) The inactive conformation showing a flip of Glu1047 disrupting proper positioning of Lys1030, which is essential for ATP coordination. The DFG motif is also in its inactive conformation, which is characterized with the Phe-Out conformation of Phe1151 where ATP binding is hindered (Reproduced from Tao et al. 2013)

Though the members of the receptor tyrosine kinase family have different extracellular ligand domains which bind with their ligands such as VEGF or insulin, the structure of their intracellular kinase core are similar. For example, the insulin receptor has an intracellular kinase core which also switches between its active "Phe-in" and inactive "Phe-out" conformation (Fig. 4.8) being very similar to that of VEGFR2. This gives rise to an idea that some members of the receptor tyrosine kinase family may have a common mechanism for the regulation of their similar intracellular kinase core. Therefore, H_2S directly targets at the intrinsic inhibitory disulfide bond molecular switches contained in the intracellular kinase core of the receptor tyrosine kinases and activates them despite the fact that the extracellular domains of these tyrosine kinases are distinguished from each other. The receptor tyrosine kinases with a similar intracellular kinase core are potential "receptors" for H_2S and the intrinsic inhibitory disulfide bond serves as molecular switch for H_2S regulation. To date, only one such a molecular switch, the Cys1045-Cys1024 disulfide bond, has been successfully indentified (Tao et al. 2013). Future works are required to identify potential disulfide bond molecular switches in other receptor tyrosine kinase family members such as the insulin receptor. Interestingly, we have recently shown that H_2S directly activates the insulin receptor and subsequently increase glucose uptake in insulin targeting cells such as myotubes and adipocytes (Xue et al. 2013). This new evidence supports the idea that additional members of the receptor tyrosine kinase family may also serve as "receptor" for H_2S.

In H_2S-induced activation of VEGFR2 and subsequent increase in migration of vascular endothelial cells, basal VEGF is required for H_2S to activate VEGFR2. H_2S relieves the monomers of VEGFR2 from its intrinsic inhibition by breaking the Cys1045-Cys1024 disulfide bond to allow ATP binding in each monomer. However, full activation of VEGFR2 requires two monomers to be dimerized upon extracellular VEGF binding which could not be induced by H_2S. On the other hand, endogenous H_2S is required for VEGF to activate VEGFR2 and promote migration of vascular endothelial cells (Tao et al. 2013). These data uncover a new mechanism underlying activation of the receptor tyrosine kinase. Without H_2S-induced turn on of its molecular switch, the endogenous ligand, VEGF, is not able to activate its receptor (Tao et al. 2013). Therefore, H_2S-induced activation of its "receptor" is carried out with a mechanism beyond the typical ligand-receptor binding mechanism which is based on conformational matching between the ligand and the receptor. H_2S is a molecule too small to have a conformation essential for ligand-receptor docking. The specificity of the interaction between H_2S and its "receptor" is based on the geometry and energy of the electron orbitals of the sulfur atom of H_2S and the sulfur atom of the disulfide bond which is under attack. In particular, the highest occupied molecular orbital (HOMO) of the HS^- anion specifically reacts with the lowest unoccupied molecular orbital (LUMO) of the disulfide bond under attack (Fig. 4.9). This is a nucleophilic attack where two HS^- anions (yielded from H_2S in aquatic solution) are required to break one disulfide bond. Quantum chemical calculations reveal a two-step reaction (Fig. 4.10):

Reaction 1: $HS^- + C_3H_6NO_2–S–S–C_3H_6NO_2 \rightarrow C_3H_6NO_2–S^- + HS–S–C_3H_6NO_2$

Reaction 2: $HS^- + HS–S–C_3H_6NO_2 \rightarrow C_3H_6NO_2–S^- + HS–SH$

These theoretical calculations are supported by our electrospray ionization mass spectrometry (ESI-MS) experiments where Cys–S–SH (a product of S-sulfhydration of cysteine) is identified transiently before its subsequent disappearance nearing completion of the reactions. These data support the idea that Cys–S–SH is a product of reaction 1 and serves as a reactant for reaction 2.

The data indicate that the "receptor" of H_2S is not a typical concept of receptor, but a specific motif (a disulfide bond) within a protein molecule. Any protein molecules containing a functional disulfide bond may serve as a "receptor" of H_2S and the disulfide bond serves as a molecular switch for H_2S.

One question arises that do other endogenous biological thiols such as glutathione (GSH), cysteine (Cys), cysteinylglycine (Cys-Gly) and homocysteine may also

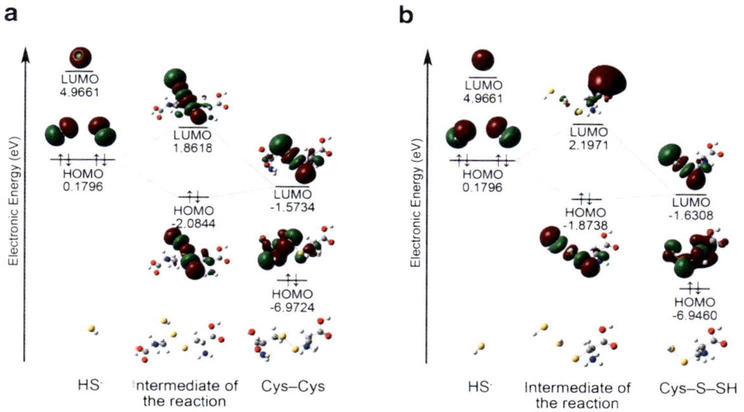

Fig. 4.9 Quantum chemical mechanisms underlying hydrogen sulfide-induced breaking of the S–S bond with the model chemical Cys–Cys and comparison of the S–S bond cleaving effect of H_2S with a group of common biological thiols. The specificity of the interaction between HS^- and the S–S bond is based on the energy and geometry of the frontier molecular orbitals involved in reaction 1 (**a**) and reaction 2 (**b**) which are required to break an S–S bond. The HOMO of HS^- reacts with the LUMO of Cys–Cys resulting in the intermediate of reaction 1 (**a**). In the consequent reaction 2, the HOMO of HS^- reacts with the LUMO of Cys–S–SH (**b**) (Reproduced from Tao et al. 2013)

break the disulfide bonds which are labile to H_2S-induced reduction? In other words, is H_2S a unique signaling molecule being distinguished from other biological thiols? Using mass spectrometry, we find that H_2S is most potent in breaking the disulfide bond in comparison with these biological thiols including GSH, Cys, Cys-Gly and homocysteine (Tao et al. 2013). The data indicate that H_2S breaks the disulfide bonds with some mechanisms beyond its reducing property. In addition, H_2S has a smallest molecular size among all known biological thiols. This gives rise to a hypothesis that H_2S is most probably to penetrate into the protein molecules to attack a disulfide bond which is deeply embedded in a protein molecule and the thiols with a larger molecular size may be not able to reach such a disulfide bond. On the other hand, the cellular distribution of different thiols including H_2S may be different from each other. In this context, each thiol may exert their biological roles in the cells/tissues or even in certain intracellular compartments where some specific thiol is distributed. Therefore, H_2S is a unique gas signaling molecule which has specific untypical "receptors" and specific molecular switch.

4.6 Two Current Theories Which Interpret the Interaction Between H_2S and Its Receptor: The Disulfide Bond Breaking Theory Versus the Cysteine Sulfhydration Theory

As mentioned above, we have recently identified VEGFR2 as a "receptor" for H_2S to stimulate angiogenesis and have found a disulfide bond serving as a molecular switch for H_2S regulation. Two H_2S molecules are required to break one disulfide

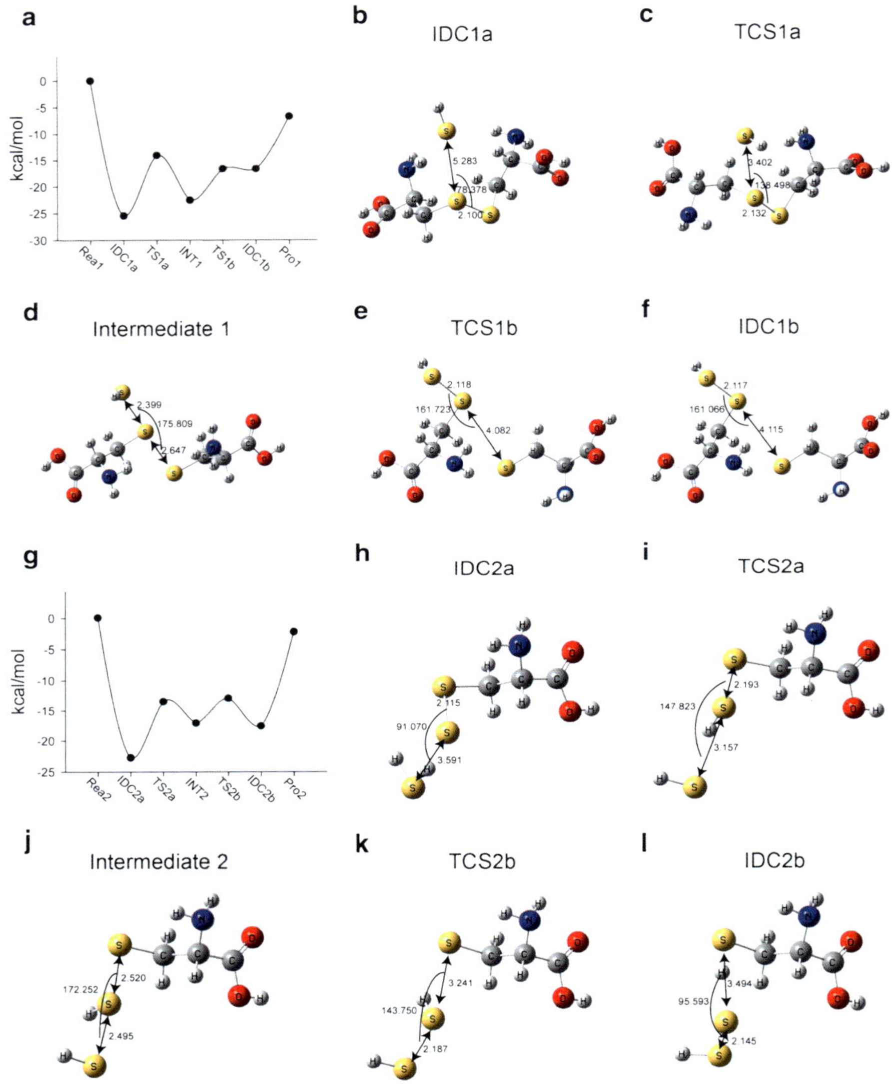

Fig. 4.10 The energy and geometries of the two-step nucleophilic attack of HS^- to break an S–S bond. Potential-energy surface (**a**) and optimized key geometries (**b–f**) for the nucleophilic attack of HS^- on Cys–S–S–Cys (reaction 1). (**b**) The ion-dipole complex 1a (*IDC1a*); (**c**) transition state 1a (*TS1a*); (**d**) intermediate 1; (**e**) transition state 1b (*TS1b*); F, ion-dipole complexe 1b (*IDC1b*). Potential-energy surface (**g**) and optimized key geometries (**h–l**) for the nucleophilic attack of HS^- on Cys–S–SH (reaction 2). (**h**) the ion-dipole complex 2a (IDC2a); (**i**) transition state 2a (*TS2a*); (**j**) intermediate 2; (**k**) transition state 2b (*TS2b*); (**l**) ion-dipole complexe 2b (*IDC2b*) (Reproduced from Tao et al. 2013)

bond that results in a shift in the conformation and function of the receptor tyrosine kinase (Tao et al. 2013). In our experimental settings, H_2S treatment does not cause any chemical modification in any of the 20 free amino acides including cysteine, nor in model peptides containing the cysteine residue using mass spectrometry analysis. The only chemical modification induced by H_2S is breaking of the disulfide bond contained in synthesized hexapeptide and the broken disulfide bond is reformed when H_2S is washed out. It is interesting that we do identify S-sulfhydration of cysteine as described by Mustafa et al. (Mustafa et al. 2009), however, this is an intermediate which appears transiently during the two-step reaction of H_2S-induced breaking of the disulfide bonds. The S-sulfhydrated intermediate is soon attacked by a second HS^- and is further reduced to cysteine (Fig. 4.10) (Tao et al. 2013).

All there mass spectrometry data demonstrate that H_2S does not react with the free –SH groups of the unbond cysteine residues. H_2S only targets at pair of cysteine residues which bond each other with a disulfide bond. H_2S breaks this disulfide molecular switch to change the conformation and function of its "receptor" (Tao et al. 2013).

One the other hand, according to the cysteine S-sulfhydration theory, H_2S targets at the free –SH groups of the unbond cysteine residues. In case of VEGFR2, there are 10 cysteines residues (Cys862, Cys905, Cys919, Cys1007, Cys1024, Cys1045, Cys1116, Cys1142, Cys1201 and Cys1208) in its intracellular kinase domain (Mctigue et al. 1999). If the cysteine S-sulfhydration theory applies to H_2S-induced activation of VEGFR2, a question is which of the 10 cysteine residues of the intracellular kinase domain of VEGFR2 would be S-sulfhydrated by H_2S? To date, it remains unknown about the quantum chemical mechanisms underlying the possible S-sulfhydration of the free –SH group of the cysteine residues. In the first report of the S-sulfhydration theory, Mustafa et al. show that the free –SH group of the cysteine residue of glyceraldehyde-3-phosphate dehydrogenase (GAPDH) is S-sulfhydrated by H_2S and this chemical modification caused an increase in the activity of GAPDH (Mustafa et al. 2009).

If H_2S-induced S-sulfhydration is the mechanism underlying the interaction between H_2S and its "receptors", all protein molecules would function as potential "receptors" for H_2S since nearly every protein in mammalian cells contain at least one cysteine residues. In this context, the S-sulfhydration theory may not be able to explain the specific biological effects which have been reported by numerous independent groups (Li et al. 2011; Kimura et al. 2012). Though H_2S has been shown to exert various biological effects in many cell types such as vascular endothelial cells (Cai et al. 2007), vascular smooth muscle cells (Cindrova-Davies et al 2013; Coletta et al. 2012), cardiomyocytes (Yao et al. 2010; Hu et al. 2011; Pan et al. 2011; Sun et al. 2012), neurons (Shibuya et al. 2013; Xie et al. 2013) and keratinocytes (Mirandola et al. 2011), these H_2S effects do not seem to be non-specific. For example, in vascular endothelial cells where H_2S activates the VEGFR2/PI3K/Akt signaling pathway, VEGFR2 is directly activated by H_2S while neither PI3K nor Akt directly reacts with H_2S (Tao et al. 2013). H_2S specifically reacts with only one of these three kinases though all of these three protein kinases contain some unbond cysteine residues with free –SH group which is supposed to be S-sulfhydrated by H_2S according to the cysteine S-sulfhydration theory. However, H_2S did not directly react with PI3K nor Akt (Tao et al. 2013). From chemical point of view, cysteine S-sulfhydration results from H_2S-induced oxidation of the –SH groups of unbond cysteine residues, however, breaking of the disulfide bond is obviously a reduction reaction. The cysteine S-sulfhydration theory and the disulfide bond breaking theory seem to be opposite. Is it possible that H_2S can exert two sides of chemical property, i.e. being either oxidant or antioxidant? Indeed, H_2S exerts its action by mechanisms beyond its antioxidant property. The microenvironment around each unbond cysteine residue and a disulfide bond may be different. The chemical nature of the amino acid residues neighboring the cysteine residues may affect the microenvironment of these bound or unbound cysteine residues. In big protein molecules which serve as potential receptors for H_2S, protein conformation may also have some impact on the microenvironment.

Much future works are required to testify these two theories. Moreover, if there is some additional unknown mechanisms underlying the interaction between H_2S and its "receptors" also remains to be investigated. The major challenge for "the cysteine S-sulfhydration theory" is to clarify if H_2S has some specificity for certain cysteine residues since most protein molecules have multiple free cysteine sites. If H_2S can really cause cysteine S-sulfhydration, how such a chemical modification can take place? Does it happen at all cysteine sites or at some of cysteine sites? And why such a chemical modification can change the conformation and function of a certain protein molecule? On the other hand, the major challenge for the disulfide bond breaking theory is to identify disulfide bonds that may serve as potential targeting motif for H_2S. It may be too early to conclude which of the two theories is true. Some additional unknown mechanisms also remain to be discovered. Identification of H_2S "receptors" as well as mechanisms underlying "receptor" activation and inactivation would be a landmark to move forward the field of H_2S biology and also help scientists to understand how a small gas molecule can regulate the structure and function of a big protein molecule.

References

Abe K, Kimura H (1996) The possible role of hydrogen sulfide as an endogenous neuromodulator. J Neurosci 16(3):1066–1071

Bian JS, Yong QC, Pan TT et al (2006) Role of hydrogen sulfide in the cardioprotection caused by ischemic preconditioning in the rat heart and cardiac myocytes. J Pharmacol Exp Ther 316(2): 670–678

Bir SC, Kolluru GK, McCarthy P et al (2012) Hydrogen sulfide stimulates ischemic vascular remodeling through nitric oxide synthase and nitrite reduction activity regulating hypoxia-inducible factor-1α and vascular endothelial growth factor-dependent angiogenesis. J Am Heart Assoc 1(5):e004093

Buckler KJ (2012) Effects of exogenous hydrogen sulphide on calcium signalling, background (TASK) K channel activity and mitochondrial function in chemoreceptor cells. Pflugers Arch 463(5):743–754

Burke MA, Mutharasan RK, Ardehali H (2008) The sulfonylurea receptor, an atypical ATP-binding cassette protein, and its regulation of the K_{ATP} channel. Circ Res 102(2):164–176

Cai WJ, Wang MJ, Moore PK et al (2007) The novel proangiogenic effect of hydrogen sulfide is dependent on Akt phosphorylation. Cardiovasc Res 76(1):29–40

Cindrova-Davies T, Herrera EA, Niu Y et al (2013) Reduced cystathionine γ-lyase and increased miR-21 expression are associated with increased vascular resistance in growth-restricted pregnancies: hydrogen sulfide as a placental vasodilator. Am J Pathol 182(4):1448–1458

Coletta C, Papapetropoulos A, Erdelyi K et al (2012) Hydrogen sulfide and nitric oxide are mutually dependent in the regulation of angiogenesis and endothelium-dependent vasorelaxation. Proc Natl Acad Sci USA 109(23):9161–9166

Hosoki R, Matsuki N, Kimura H (1997) The possible role of hydrogen sulfide as an endogenous smooth muscle relaxant in synergy with nitric oxide. Biochem Biophys Res Commun 237(3): 527–531

Hu LF, Li Y, Neo KL et al (2011) Hydrogen sulfide regulates Na+/H + exchanger activity via stimulation of phosphoinositide 3-kinase/Akt and protein kinase G pathways. J Pharmacol Exp Ther 339(2):726–735

Jackson-Weaver O, Paredes DA, Gonzalez Bosc LV et al (2011) Intermittent hypoxia in rats increases myogenic tone through loss of hydrogen sulfide activation of large-conductance Ca(2+)-activated potassium channels. Circ Res 108(12):1439–1447

Jain SK, Kahlon G, Morehead L et al (2012) The effect of sleep apnea and insomnia on blood levels of leptin, insulin resistance, IP-10, and hydrogen sulfide in type 2 diabetic patients. Metab Syndr Relat Disord 10(5):331–336

Kimura H, Shibuya N, Kimura Y (2012) Hydrogen sulfide is a signaling molecule and a cytoprotectant. Antioxid Redox Signal 17(1):45–57

Kombian SB, Reiffenstein RJ, Colmers WF (1993) The actions of hydrogen sulfide on dorsal raphe serotonergic neurons in vitro. J Neurophysiol 70(1):81–96

Lee HJ, Mariappan MM, Feliers D et al (2012) Hydrogen sulfide inhibits high glucose-induced matrix protein synthesis by activating AMP-activated protein kinase in renal epithelial cells. J Biol Chem 287(7):4451–4461

Li L, Bhatia M, Zhu YZ et al (2005) Hydrogen sulfide is a novel mediator of lipopolysaccharide-induced inflammation in the mouse. FASEB J 19(9):1196–1198

Li L, Rose P, Moore PK (2011) Hydrogen sulfide and cell signaling. Annu Rev Pharmacol Toxicol 51:169–187

Liang GH, Xi Q, Leffler CW et al (2012) Hydrogen sulfide activates Ca^{2+} sparks to induce cerebral arteriole dilatation. J Physiol 590(Pt 11):2709–2720

Matsunami M, Kirishi S, Okui T et al (2012) Hydrogen sulfide-induced colonic mucosal cytoprotection involves T-type calcium channel-dependent neuronal excitation in rats. J Physiol Pharmacol 63(1):61–68

McTigue MA, Wickersham JA, Pinko C et al (1999) Crystal structure of the kinase domain of human vascular endothelial growth factor receptor 2: a key enzyme in angiogenesis. Structure 7(3):319–330

Mirandola P, Gobbi G, Micheloni C et al (2011) Hydrogen sulfide inhibits IL-8 expression in human keratinocytes via MAP kinase signaling. Lab Invest 91(8):1188–1194

Mustafa AK, Gadalla MM, Sen N et al (2009) H_2S signals through protein S-sulfhydration. Sci Signal 2(96):ra72. doi:10.1126/scisignal.2000464

Mustafa AK, Sikka G, Gazi SK et al (2011) Hydrogen sulfide as endothelium-derived hyperpolarizing factor sulfhydrates potassium channels. Circ Res 109(11):1259–1268

Noma A (1983) ATP-regulated K^+ channels in cardiac muscle. Nature 305(5930):147–148

Okubo K, Takahashi T, Sekiguchi F et al (2011) Inhibition of T-type calcium channels and hydrogen sulfide-forming enzyme reverses paclitaxel-evoked neuropathic hyperalgesia in rats. Neuroscience 188:148–156

Okubo K, Matsumura M, Kawaishi Y et al (2012) Hydrogen sulfide-induced mechanical hyperalgesia and allodynia require activation of both Cav3.2 and TRPA1 channels in mice. Br J Pharmacol 166(5):1738–1743

Padiya R, Khatua TN, Bagul PK et al (2011) Garlic improves insulin sensitivity and associated metabolic syndromes in fructose fed rats. Nutr Metab (Lond) 8:53. doi:10.1186/1743-7075-8-53

Pan TT, Neo KL, Hu LF et al (2008) H_2S preconditioning-induced PKC activation regulates intracellular calcium handling in rat cardiomyocytes. Am J Physiol Cell Physiol 294(1): C169–C177

Pan LL, Liu XH, Gong QH et al (2011) S-Propargyl-cysteine (SPRC) attenuated lipopolysaccharide-induced inflammatory response in H9c2 cells involved in a hydrogen sulfide-dependent mechanism. Amino Acids 41(1):205–215

Papapetropoulos A, Pyriochou A, Altaany Z et al (2009) Hydrogen sulfide is an endogenous stimulator of angiogenesis. Proc Natl Acad Sci USA 106(51):21972–21977

Patel M, Shah G (2010) Possible role of hydrogen sulfide in insulin secretion and in development of insulin resistance. J Young Pharm 2(2):148–151

Qipshidze N, Metreveli N, Mishra PK et al (2012) Hydrogen sulfide mitigates cardiac remodeling during myocardial infarction via improvement of angiogenesis. Int J Biol Sci 8(4):430–441

Shi YX, Chen Y, Zhu YZ et al (2007) Chronic sodium hydrosulfide treatment decreases medial thickening of intramyocardial coronary arterioles, interstitial fibrosis, and ROS production in spontaneously hypertensive rats. Am J Physiol Heart Circ Physiol 293(4):H2093–H2100

Shibuya N, Mikami Y, Kimura Y et al (2009) Vascular endothelium expresses 3-mercaptopyruvate sulfurtransferase and produces hydrogen sulfide. J Biochem 146(5):623–626

Shibuya N, Koike S, Tanaka M et al (2013) A novel pathway for the production of hydrogen sulfide from D-cysteine in mammalian cells. Nat Commun 4:1366. doi:10.1038/ncomms2371

Si YF, Wang J, Guan J et al (2013) Treatment with hydrogen sulfide alleviates streptozotocin-induced diabetic retinopathy in rats. Br J Pharmacol. doi:10.1111/bph.12163

Streeter E, Hart J, Badoer E (2012) An investigation of the mechanisms of hydrogen sulfide-induced vasorelaxation in rat middle cerebral arteries. Naunyn Schmiedebergs Arch Pharmacol 385(10):991–1002

Sun YG, Cao YX, Wang WW et al (2008) Hydrogen sulphide is an inhibitor of L-type calcium channels and mechanical contraction in rat cardiomyocytes. Cardiovasc Res 79(4):632–641

Sun WH, Liu F, Chen Y et al (2012) Hydrogen sulfide decreases the levels of ROS by inhibiting mitochondrial complex IV and increasing SOD activities in cardiomyocytes under ischemia/reperfusion. Biochem Biophys Res Commun 421(2):164–169

Suzuki K, Olah G, Modis K et al (2011) Hydrogen sulfide replacement therapy protects the vascular endothelium in hyperglycemia by preserving mitochondrial function. Proc Natl Acad Sci USA 108(33):13829–13834

Tao BB, Liu SY, Zhang CC et al (2013) VEGFR2 functions as an H_2S-targeting receptor protein kinase with its novel Cys1045-Cys1024 disulfide bond serving as a specific molecular switch for hydrogen sulfide actions in vascular endothelial cells. Antioxid Redox Signal. doi:10.1089/ars.2012.4565

Tian XY, Wong WT, Sayed N et al (2012) NaHS relaxes rat cerebral artery in vitro via inhibition of l-type voltage-sensitive Ca^{2+} channel. Pharmacol Res 65(2):239–246

Wang MJ, Cai WJ, Li N et al (2010) The hydrogen sulfide donor NaHS promotes angiogenesis in a rat model of hind limb ischemia. Antioxid Redox Signal 12(9):1065–1077

Whiteman M, Gooding KM, Whatmore JL et al (2010) Adiposity is a major determinant of plasma levels of the novel vasodilator hydrogen sulphide. Diabetologia 53(8):1722–1726

Wu L, Yang W, Jia X et al (2009) Pancreatic islet overproduction of H_2S and suppressed insulin release in Zucker diabetic rats. Lab Invest 89(1):59–67

Xie L, Hu LF, Teo XQ et al (2013) Therapeutic effect of hydrogen sulfide-releasing L-dopa derivative ACS84 on 6-OHDA-induced Parkinson's disease rat model. PLoS One 8(4):e60200. doi:10.1371/journal.pone.0060200

Xue R, Hao DD, Sun JP et al (2013) Hydrogen sulfide treatment promotes glucose uptake by increasing insulin receptor sensitivity and ameliorates kidney lesions in type 2 diabetes. Antioxid Redox Signal. doi:10.1089/ars.2012.5024

Yang G, Tang G, Zhang L et al (2011) The pathogenic role of cystathionine γ-lyase/hydrogen sulfide in streptozotocin-induced diabetes in mice. Am J Pathol 179(2):869–879

Yao LL, Huang XW, Wang YG et al (2010) Hydrogen sulfide protects cardiomyocytes from hypoxia/reoxygenation-induced apoptosis by preventing GSK-3beta-dependent opening of mPTP. Am J Physiol Heart Circ Physiol 298(5):H1310–H1319

Yuan P, Xue H, Zhou L et al (2011) Rescue of mesangial cells from high glucose-induced over-proliferation and extracellular matrix secretion by hydrogen sulfide. Nephrol Dial Transplant 26(7):2119–2126

Zhang L, Yang G, Tang G et al (2011) Rat pancreatic level of cystathionine γ-lyase is regulated by glucose level via specificity protein 1 (SP1) phosphorylation. Diabetologia 54(10):2615–2625

Zhao W, Zhang J, Lu Y et al (2001) The vasorelaxant effect of H_2S as a novel endogenous gaseous K(ATP) channel opener. EMBO J 20(21):6008–6016

Hydrogen Sulfide: Its Production, Release and Functions

5

Therapeutic Applications of Hydrogen Sulfide

Kyle L. Flannigan and John L. Wallace

Abstract

The first descriptions of physiological effects of hydrogen sulfide (H_2S) were published in the mid-1990s, but the therapeutic benefits of this gaseous mediator have been appreciated for centuries, through bathing in sulfide-containing hot springs and through the consumption of foods that can release H_2S, such as garlic. In the past decade, with the marked increase in our knowledge about the contribution of H_2S to many physiological and pathophysiological processes, efforts have been made to exploit this endogenous substance in drug design. In this chapter we review some of the key roles played by H_2S in disease processes, strategies that have been employed to design novel H_2S-releasing drugs, and the evidence that these drugs exert beneficial effects.

Keywords

Inflammation • Pain • Ulceration • Oxidative stress • Chemoprevention • Neuromodulation • Drug development

Abbreviations

AMP	Adenosine monophosphate
ATP	Adenosine triphosphate
COX	Cyclooxygenase
GI	Gastrointestinal
H_2S	Hydrogen sulfide

K.L. Flannigan • J.L. Wallace (✉)
Farncombe Institute for Digestive Health Research, McMaster University, Hamilton, ON, Canada
e-mail: flannik@mcmaster.ca; altapharm@hotmail.com

H. Kimura (ed.), *Hydrogen Sulfide and its Therapeutic Applications*,
DOI 10.1007/978-3-7091-1550-3_5,

IL	Interleukin
iNOS	Inducible nitric oxide synthase
Na_2S	Sodium sulfide
NaHS	Sodium hydrosulfide
NSAID	Nonsteroidal anti-inflammatory drug
TNF	Tumor necrosis factor

5.1 H_2S Actions and Mechanisms

Many actions have been described for H_2S and are reviewed in more detail in other chapters in this volume. With respect to the therapeutic applications of H_2S, many attempted thus far involve exploitation of the ability of this gaseous mediator to lessen, prevent or promote the repair of injury (e.g., ischemia/reperfusion-induced injury), while many others focus on enhancement of anti-inflammatory activity. Some of the key mechanisms through which H_2S modulates inflammation are depicted in Fig. 5.1. One of the first-described and possibly most significant anti-inflammatory effects of H_2S is its ability to inhibit leukocyte adherence to the vascular endothelium (Zanardo et al. 2006). This appears to be a tonic, physiological role of H_2S, since suppression of endogenous H_2S synthesis leads rapidly to the adherence of leukocytes (mainly neutrophils) to the vascular endothelium. Administration of H_2S donors was shown to suppress leukocyte adherence stimulated by a formylated peptide as well as to reduce leukocyte accumulation and edema formation in various inflammatory models (Zanardo et al. 2006; Andruski et al. 2008; Li et al. 2009; Benetti et al. 2013).

In the gastrointestinal (GI) tract, which is said to be in a constant state of low-grade inflammation because of the multitude of microbes within the lumen, suppression of H_2S synthesis leads to mucosal inflammation and loss of tissue integrity (Wallace et al. 2009). Oral or intracolonic administration of iodoacetamide, which binds to L-cysteine and thereby makes it unavailable for conversion to H_2S (Wallace 2010), leads to inflammation in the stomach and colon, respectively (Barnett et al. 2000; Szabo et al. 1997). On the other hand, administration of H_2S donors can accelerate the resolution of GI inflammation, increase the resistance of the mucosa to damage, and enhance the healing of pre-existing ulcers (Wallace et al. 2007b, 2009). Some of these beneficial effects of H_2S may be due to a promotion of a pro-resolution phenotype of macrophages (Dufton et al. 2012), induction of neutrophil apoptosis (Mariggio et al. 1998), and through stimulation of angiogenesis, a crucial element of wound healing (Papapetropoulos et al. 2009). Moreover, H_2S can be utilized by mitochondria to generate ATP, particularly in settings of hypoxia (Goubern et al. 2007; Kimura et al. 2010). In these circumstances, H_2S may act to reduce tissue injury and enhance repair processes (Kimura et al. 2005, 2010; Elrod et al. 2007). H_2S can also trigger the Nrf2 stress response pathway, up-regulating a battery of detoxifying proteins and anti-oxidant enzymes (Yang et al. 2013). This effect may be achieved via H_2S sulfhydration of a protein that tonically suppresses Nrf2 activity (Yang et al. 2013).

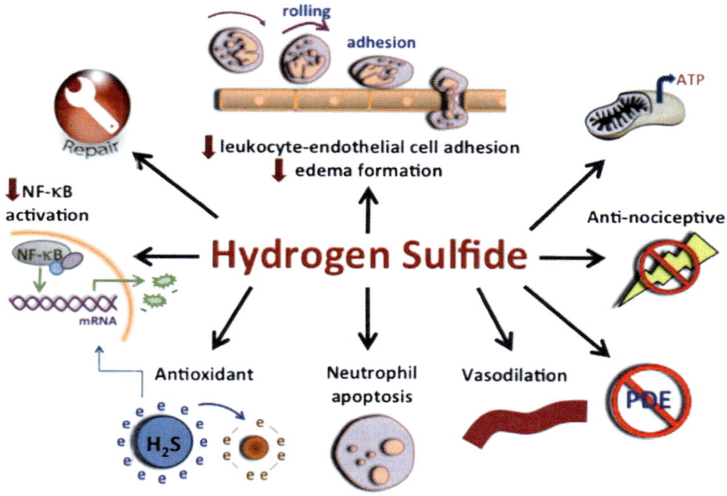

Fig. 5.1 Anti-inflammatory actions of hydrogen sulfide. H_2S can act on multiple targets to reduce inflammation and promote repair. H_2S inhibits leukocyte adherence to the vascular endothelium by down-regulating leukocyte and endothelial adhesion molecule expression (Zanardo et al. 2006). H_2S donors can reduce edema formation (Zanardo et al. 2006). H_2S can act as a 'metabolic fuel', substituting for oxygen in mitochondrial respiration (Goubern et al. 2007; Kimura et al. 2010). This may be particularly important in situations of low oxygen (e.g., ischemia). H_2S reduces visceral pain (Distrutti et al. 2006a), can inhibit phosphodiesterase activity (Bucci et al. 2010) and can relax vascular smooth muscle tone (Wang 2009). Promotion of resolution of inflammation by H_2S is achieved in part via induction of apoptosis in neutrophils (Mariggio et al. 1998) and causing a shift in macrophage phenotype to 'anti-inflammatory' (not shown) (Dufton et al. 2012). H_2S can scavenge free radicals (Whiteman et al. 2005). It can also inhibit activation of the transcription factor, NF-kB (Li et al. 2009), and in doing so H_2S reduces production of several pro-inflammatory cytokines. H_2S has been shown to promote healing of ulcers in the gastrointestinal tract (Wallace et al. 2007), which may be related to its ability to stimulate angiogenesis (Papapetropoulos et al. 2009)

The mechanisms of action of H_2S are reviewed in detail elsewhere in this volume. Much of the drug design around H_2S has not been 'mechanism-based'; rather, it has been based on observed effects of inhibitors of H_2S synthesis or of H_2S donors (generally simple ones such as NaHS and Na_2S). Where data are available relevant to therapeutic applications of H_2S, we have referred below to the studies that suggest a specific mechanism of action.

5.2 Examples of H_2S-Based Therapeutics in Development

Over the past two decades, a number of H_2S-based experimental drugs have been developed and tested in various animal models. Some of these compounds are already entering clinical trials.

5.2.1 Arthritis

Nonsteroidal anti-inflammatory drugs (NSAIDs) are the most commonly used drugs for treatment of the symptoms of osteoarthritis. The major limitation to their use is their untoward effects on the gastrointestinal tract, most notably the induction of bleeding ulcers (Wallace 2008). Although most attention has been focused on ulcers in the stomach and proximal duodenum, most NSAID-induced injury and bleeding occurs in the more distal small intestine (Wallace 2013). Suppression of gastric acid secretion with proton pump inhibitors or histamine H_2 receptor antagonists can significantly reduce the incidence of upper GI ulceration induced by NSAIDs, but these drugs do not protect the lower intestine. Indeed, there is recent evidence that suppression of gastric acid secretion leads to more severe NSAID-induced enteropathy (Wallace et al. 2011; Satoh et al. 2012; Sakai et al. 2012; Watanabe et al. 2013). The discovery that almost all NSAIDs, most notably the cyclooxygenase(COX)-2-selective NSAIDs (e.g., celecoxib, rofecoxib, etoricoxib), increased the risk of serious cardiovascular events has led to co-prescription of low-dose aspirin to chronic NSAID users, to achieve cardioprotection. However, this further exacerbates NSAID-induced ulceration and bleeding throughout the GI tract (Wallace 2013). Thus, there is a need for NSAIDs that can provide symptomatic relief to patients with chronic diseases such as osteoarthritis, but without the detrimental effects on the GI tract.

H_2S was shown to be effective in reducing the severity of acute NSAID injury to the gastric mucosa of rats (Fiorucci et al. 2005), so two groups of investigators independently undertook to determine if H_2S-releasing derivatives of an NSAID would be more "GI safe" than the NSAID itself. Li et al. (2007) and Wallace et al. (2007a) demonstrated that an H_2S-releasing derivative of diclofenac produced less acute gastric damage than diclofenac itself. In one study, significantly less intestinal damage was observed after repeated administration of the H_2S-releasing diclofenac than with the parent drug. Moreover, there was no decrease in hematocrit (indicative of bleeding) with the H_2S-releasing diclofenac group, in contrast to a significant decrease in rats treated with diclofenac (Wallace et al. 2007a). Interestingly, the H_2S-releasing diclofenac also exhibited enhanced anti-inflammatory activity in a carrageenan-induced paw edema model in rats (Wallace et al. 2007a), consistent with a previous demonstration that H_2S donors were anti-inflammatory in the model (Zanardo et al. 2006). Other NSAIDs, including naproxen and indomethacin, have similarly been covalently linked to an H_2S-releasing moiety and demonstrated markedly reduced gastrointestinal toxicity (Wallace 2007; Wallace et al. 2010). The gastrointestinal-sparing properties of H_2S-NSAIDs are not mediated via nitric oxide synthesis, ATP-sensitive potassium channels or sensory afferent nerves (Wallace et al. 2010).

NSAID-induced small intestinal damage and bleeding are major clinical concerns, in part because the damage often goes undetected (there is a poor correlation between tissue damage and symptoms). Also, there are no effective treatments of NSAID-enteropathy. One of the most important features of the H_2S-releasing NSAIDs is that they produce little, if any, small intestinal damage, despite exhibiting comparable inhibitory effects on COX-1 and COX-2 as the parent drugs (Wallace et al. 2010; Blackler et al. 2012). The H_2S-releasing NSAIDs were found to be completely GI safe even when given to animals with compromised mucosal defence, or animals with co-morbidities that increase susceptibility to ulceration (e.g., old age, obesity, arthritis, combined use with low-dose aspirin and proton pump inhibitors) (Wallace et al. 2010; Blackler et al. 2012). Moreover, when administered to mice with pre-existing gastric ulcers, the H_2S-releasing derivative of naproxen (ATB-346) accelerated healing of the ulcers. This was in contrast to significant inhibition of ulcer healing with naproxen or celecoxib. (Wallace et al. 2010). H_2S donors had previously been shown to accelerate ulcer healing, while inhibitors of H_2S synthesis delayed ulcer healing (Wallace et al. 2007b).

H_2S donors may have therapeutic benefit in arthritis without being coupled to an NSAID. Fox et al. (2012) demonstrated that H_2S synthesis was induced in chondrocytes and mesenchymal progenitor cells in vitro when exposed to several pro-inflammatory cytokines, and suggested that it acted as a protective mediator in this setting. Elevated levels of H_2S in plasma and synovial fluid were found to be elevated in rheumatoid arthritis and osteoarthritis patients as compared to healthy controls, and the H_2S levels correlated with clinical indices of disease activity (Whiteman et al. 2010). However, it was not clear from this clinical study if H_2S was acting as a pro-inflammatory or anti-inflammatory mediator. In an in vivo study of mice with adjuvant-induced arthritis, the H_2S donor GYY4137 was able to produce beneficial effects when administered after the adjuvant, but not given as a pre-treatment (Li et al. 2013).

5.2.2 Inflammatory Bowel Disease

Mesalamine (5-amino salicylic acid) is the first-line therapy for treatment of ulcerative colitis and Crohn's disease. It is a relatively safe drug, but it also lacks potency, so patients may take as much as 6 g a day divided over 3–4 doses. Mesalamine appears to act topically in the GI tract. There are various formulations of mesalamine designed to prevent the drug being absorbed in the upper GI tract, so that it will reach the sites of inflammation that are most commonly in the terminal ileum and colon. The mechanism of action of mesalamine is not known, but it is a potent anti-oxidant.

The development of a H_2S-releasing derivative of mesalamine was based on the notion that the anti-inflammatory and ulcer-healing of effects of this gaseous mediator would be beneficial in treating colitis. Indeed, H_2S has been shown to contribute significantly to the resolution of colitis in animal models. Suppression of endogenous H_2S synthesis results in exacerbation of colitis, while local administration of H_2S donors can accelerate healing (Wallace et al. 2009). As shown in Fig. 5.2, an H_2S-releasing derivative of mesalamine (ATB-429) was significantly more effective than mesalamine in reducing disease activity in mice with hapten-induced colitis (Fiorucci et al. 2007). ATB-429 reduced tissue numbers of granulocytes to normal levels, even when given at a dose that contained only half

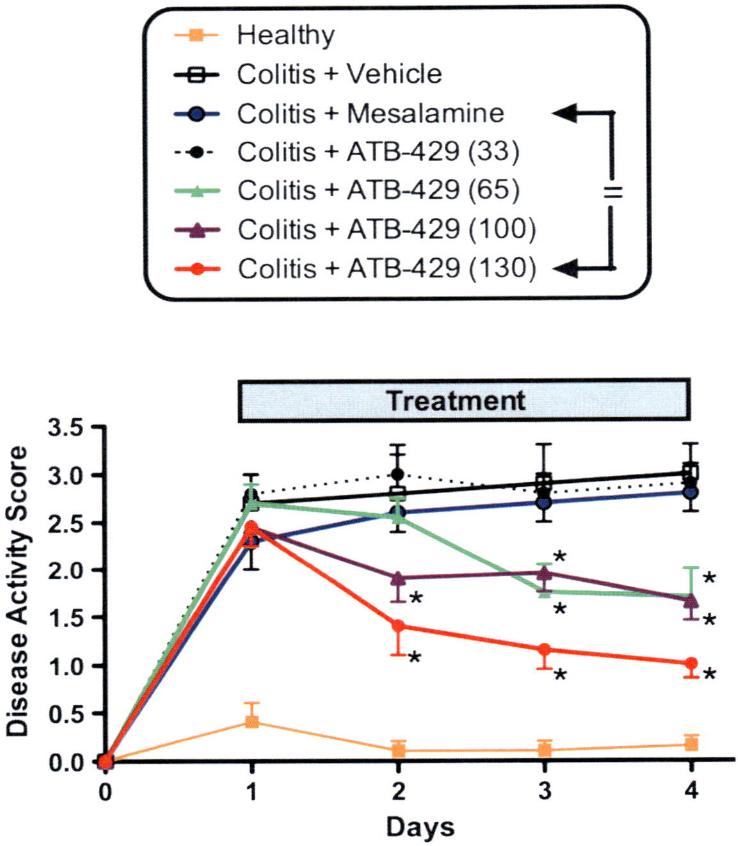

Fig. 5.2 Improved therapeutic effect of an H_2S-releasing derivative of mesalamine. Mice with colitis induced by a hapten were treated twice-daily with mesalamine or ATB-429 (H_2S-releasing derivative). Mesalamine (50 mg/kg) was ineffective in this model, but ATB-429 produced a dose-dependent improvement of disease activity. ATB-429 at 130 mg/kg is the molar equivalent of mesalamine at 50 mg/kg. The Disease Activity Score is a blindly evaluated composite score of incidence and severity of rectal bleeding and diarrhea, plus weight loss. $*p < 0.05$ compared to the corresponding vehicle-treated group

as much mesalamine as that in the mesalamine-treated group. Interestingly, administration of mesalamine together with the H_2S-releasing moiety of ATB-429 had no benefit over administration of mesalamine alone. Thus, the improved activity of ATB-429 was dependent upon the linkage of mesalamine to the H_2S-releasing moiety. This may be related to the rate of release of H_2S from the moiety alone being less than that when it is conjoined to mesalamine (Wallace et al. 2008).

There are likely multiple mechanisms of action of ATB-429 in reducing the severity of colitis. The marked reduction of myeloperoxidase activity in the colonic tissue is consistent with the well-characterized ability of H_2S to reduce leukocyte extravasation (Zanardo et al. 2006). As mentioned above, H_2S has also been shown enhance resolution of inflammation through numerous mechanisms, and to reduce oxidative stress (Whiteman et al. 2005; Elrod et al. 2007; Kimura et al. 2010). The ability of H_2S to accelerate ulcer healing, possibly by stimulating angiogenesis, likely also contributes to the enhanced activity of ATB-429 as compared to mesalamine. When given orally, ATB-429 was shown to significantly accelerate the healing of gastric ulcers in rats, while mesalamine itself had no effect (Wallace 2010).

In addition to accelerating the repair of ulcerated and inflamed tissue, ATB-429 has been shown to significantly reduce visceral pain. Treatment with this compound markedly reduces colonic distention-induced pain in rats, including rats with colitis (Distrutti et al. 2006b). These effects appeared to be mediated via ATP-sensitive potassium channels, since they were blocked by glibenclamide. Abdominal pain is one of the most common symptoms of inflammatory bowel disease, with limited options for treatment. Opioids are effective in reducing visceral pain, but their inhibitory effects on intestinal motility limits their use in patients with IBD. As described in more detail below, an H_2S-releasing salt of trimebutine is in clinical trials for its use as a colonic analgesic, to facilitate colonoscopy (Cukier-Meisner 2013).

Unpublished studies performed in healthy dogs have shown that ATB-429 is not absorbed to any significant extent after oral administration (plasma mesalamine levels are very low). If this is also the case in a setting of colitis, it has some important implications. First, it means that ATB-429 could be used to treat inflammation throughout the GI tract without need of the formulations that are typically required to prevent mesalamine absorption in the upper GI tract. This is important because Crohn's disease can occur anywhere in the GI tract, but current mesalamine preparations deliver the drug only to the lower part of the GI tract. Second, ATB-429 will not produce significant mesalamine-related systemic adverse effects.

5.2.3 Oxidative Stress-Induced Injury

Perhaps the most investigated type of injury with respect to the potential therapeutic value of H_2S-releasing drugs is that characterized by ischemia-reperfusion and oxidative stress. For example, Elrod et al. (2007) demonstrated that rats subjected to cardiac ischemia-reperfusion developed significant infarcts and loss of left ventricular function. However, these detrimental changes could be significantly attenuated when H_2S was delivered during the reperfusion of the cardiac tissue. In addition to reduced inflammation, there was significant preservation of cardiac mitochondrial function. Cardiac tissue injury and dysfunction could also be attenuated through cardiac-specific up-regulation of endogenous H_2S synthesis. Sivarajah et al. (2009) also demonstrated beneficial effects of H_2S in experimental myocardial infarction, with marked reductions in inflammation of the tissue. Kondo et al. (2013) reported that some of the beneficial effects of H_2S in experimental heart failure are attributable to up-regulation of endothelial nitric oxide synthase. Peake et al. (2013) demonstrated beneficial effects of Na_2S against ischemia-reperfusion-induced damage to the heart in a diabetic mouse model. They further demonstrated that the effect was mediated via activation of Erk-dependent Nrf2 signaling. Diabetes was found not to alter the ability of H_2S to stimulate the nuclear localization of Nrf2, but it was associated with an impairment of certain aspects of Nrf2 signaling.

These results suggest there could be significant therapeutic value of H_2S donors in terms of limiting cardiac tissue damage in acute myocardial infarction and heart failure. There are also a number of studies suggesting that H_2S-releasing drugs could exert significant therapeutic benefits in oxidative stress-induced injury in other organs, most notably in the central nervous system (Kimura and Kimura 2004; Kimura et al. 2006), the eye (Mikami and Kimura 2012) the kidney (Liu et al. 2011; Simon et al. 2011) and the lung (Benetti et al. 2013). In many cases, these effects of H_2S are likely mediated via its ability to preserve mitochondrial respiratory function, thereby reducing oxidative stress (Goubern et al. 2007; Kimura et al. 2010)

5.2.4 Cancer Chemoprevention

Several studies have highlighted the potential utility of H_2S-releasing drugs as chemopreventative agents, possibly attributable to a combination of induction of apoptosis and cell cycle arrest in cancer cells (Lee et al. 2011). GYY4137, a slow-releasing H_2S donor, was found to concentration-dependently kill seven different human cancer cell lines, but not to kill normal human lung fibroblasts (Lee et al. 2011). A structural analogue of GYY4137 (ZYJ1122) lacking the ability to release H_2S was inactive. These studies were extended by the same investigators to an in vivo model, where GYY4137 significantly reduced tumour growth.

NSAIDs have been shown to reduce the growth of several types of tumours (Vendramini-Costa and Carvalho 2012), but their ability to trigger GI bleeding significantly limits their use for this purpose. Since addition of an H_2S-releasing moiety to NSAIDs greatly reduces the GI toxicity, H_2S-releasing NSAIDs have been explored for their use as chemopreventative agents. The potential chemopreventative effects of the Antibe Therapeutics' naproxen derivative, ATB-346, were assessed in a mouse model of colon cancer (Elsheikh and Wallace 2012). Azoxymethane was administered to the mice weekly for 4 weeks. During the first 2 weeks, ATB-346, naproxen or vehicle were administered once daily. When examined 5 weeks after the first azoxymethane treatment, extensive aberrant crypt foci were evident in the colon (these are a precancerous lesion). Both naproxen and ATB-346 dose-dependently reduced the number of aberrant crypt foci, but ATB-346 was significantly more effective, particularly at the lower doses. In contrast, administration of equimolar doses of the H_2S-releasing moiety alone did not significantly affect the number of aberrant crypt foci (Elsheikh and Wallace 2012).

Chattopadhyay et al. (2012) also studied the anti-cancer effects of H_2S-releasing NSAIDs, though with a different H_2S-releasing moiety to that in the study of Elsheikh and Wallace (2012). They examined the effects of four different

H_2S-releasing NSAIDs on the growth properties of 11 different human cancer cell lines (from six different types of tissues). Human colon, breast, pancreatic, prostate, lung, and leukemia cancer cell lines were exposed to H_2S-releasing derivatives of aspirin, sulindac, ibuprofen, naproxen, and to the corresponding NSAIDs themselves. The H_2S-releasing derivatives of each of the NSAIDs inhibited the growth of all cancer cell lines studied to a greater extent than the respective parent NSAID. Thus, the effects of the H_2S-releasing NSAIDs appeared to be a general property, rather than compound-specific, and they were also not tissue type-specific. Increases in potencies over those of the parent NSAID ranged between 28- and >3,000-fold, with the aspirin derivative consistently being the most potent. The H_2S-releasing NSAIDs induced apoptosis, inhibited cell proliferation, and caused G_0/G_1 cell cycle block. Importantly, the administration of aspirin and the H_2S-releasing moiety as separate entities was found to be no more effective in reducing cell growth than aspirin alone. Thus, as was the case for the GI safety of H_2S-releasing NSAIDs (Wallace et al. 2007a, 2010), and the enhanced anti-inflammatory activity of H_2S-releasing mesalamine (Fiorucci et al. 2005), the activity of a conjoined H_2S-releasing drug is substantially greater than when the two elements are administered simultaneously, but as separate entities.

Moody et al. (2010) also described a marked improvement of the chemopreventative effects of H_2S-releasing derivatives of NSAIDs and of valproic acid as compared to the parent drugs. In contrast to the above-mentioned studies, the H_2S-releasing moieties alone (dithiolethiones) also exhibited significant beneficial effects in models of lung and breast cancer (Moody et al. 2010; Switzer et al. 2012).

5.2.5 Erectile Dysfunction

Early studies exploring the possibility that H_2S could be important in erectile dysfunction yielded disappointing results, with an H_2S-releasing derivative of sildenafil not demonstrating superior relaxant effects on the corpus cavernosum as compared to sildenafil itself (Shukla et al. 2009). However, a series of elegant studies by Cirino and colleagues have convincingly demonstrated an important role of H_2S in erectile function. They demonstrated that H_2S is an endogenous inhibitor of phosphodiesterase activity, and it contributed to physiological relaxation of the corpus cavernosum (d'Emmanuele di Villa Bianca et al. 2009; Bucci et al. 2010). They also demonstrated the importance of cGMP-dependent protein kinase activity in the actions of H_2S (Bucci et al. 2012). Consistent with a role of H_2S in erectile function, Srilatha et al. (2012) have demonstrated that an insufficiency of H_2S is a predictor of sexual dysfunction in aging rats.

Table 5.1 Hydrogen sulfide-releasing drugs in development

Company	Drug name	Description	Primary target indication
Antibe Therapeutics	ATB-346	Naproxen derivative	Osteoarthritis
	ATB-344	Indomethacin derivative	Gout
Antibe Holdings	ATB-429	Mesalamine derivative	Inflammatory Bowel Disease
CTG Pharma	ACS14[a]	Aspirin derivative	Anti-thrombotic
GIcare Pharma	GIC-1001	Trimebutine salt	Colonic analgesia
Ikaria	IK-1001[b]	Sodium sulfide	Myocardial Infarction
National University of Singapore	GYY4137	Slow-releasing H_2S donor	Unknown
Sulfagenix	SG-1002	Precursor to natural substance	Heart failure

[a]CTG Pharma have published data on a number of drugs, but it is not clear if any of the drugs have progressed beyond academic characterization. The company's website has been deactivated
[b]Ikaria appears to have ceased the development of IK-1001, as two clinical trials of this compound were halted (http://clinicaltrials.gov/ct2/show/nct01007461 and http://clinicaltrials.gov/ct2/show/nct00858936)

5.3 Development Programs for H_2S-Based Drugs

As outlined in Table 5.1, several companies and universities have, or have had, H_2S-based compounds in development for a range of disorders.

5.3.1 Antibe Holdings Ltd.

Antibe Holdings is a privately held company based in Calgary, Canada. This company is developing ATB-429, an H_2S-derivative of mesalamine, for treatment of inflammatory bowel disease. It is also developing gastrointestinal-sparing NSAIDs for veterinary indications through a subsidiary (Antibe Animal Health). Antibe Holdings has out-licensed compounds to GIcare Pharma and to Antibe Therapeutics Inc. (see below).

5.3.2 Antibe Therapeutics Inc.

Antibe Therapeutics is a public company (Toronto Venture Exchange) based in Toronto, Canada (www.antibethera.com). Antibe's lead drug is ATB-346, an H_2S-releasing derivative of naproxen, for treatment of osteoarthritis. This drug has been extensively evaluated in models of arthritis and gastrointestinal ulceration. ATB-346 exhibits comparable anti-inflammatory and analgesic activity to naproxen, but does not produce gastrointestinal damage. Extensive studies have been performed in animals with compromised mucosal defence and in animal models mimicking the co-morbidities commonly seen in osteoarthritis patients

(Wallace et al. 2010; Blackler et al. 2012). There is also evidence that, unlike naproxen, ATB-346 does not elevate blood pressure in hypertensive rats (Wallace et al. 2010). Antibe Therapeutics is also developing a GI-safe derivative of indomethacin for treatment of gout, and it has intellectual property protection of H_2S-releasing statins.

5.3.3 CTG Pharma

Based in Milan, Italy, CTG Pharma has developed a number of H_2S-releasing derivatives of other drugs (e.g., NSAIDs, sildenafil, L-DOPA), and there have been numerous publications describing the effects of these drugs (Lee et al. 2011; Muzaffar et al. 2008; Li et al. 2007; Pircher et al. 2012). For example, ACS15, an H_2S-releasing derivative of diclofenac, exhibited comparable anti-inflammatory effects to diclofenac, but with greatly reduced gastric damage (Li et al. 2007). ACS14 is an H_2S-releasing derivative of aspirin that is more effective than aspirin at blocking thrombus formation (Pircher et al. 2012). This is due to its ability to impair activation of the fibrinogen receptor via a cyclic AMP-dependent mechanism. The website for this company has not been active since 2011, so it is unclear if the company is still active.

5.3.4 GIcare Pharma

GIcare Pharma is a gastrointestinal-focused, privately held company based in Montreal, Canada (www.gicarepharma.com). They have acquired patents covering a group of gastrointestinal compounds from Antibe Holdings Ltd. The patents cover non-centrally acting opioid agonists formulated as salts. They are developing a non-sedating colonic analgesic compound (GIC-1001) to provide more convenient and less expensive alternatives for pain management during colonoscopy. GIC-1001 is a H_2S-releasing derivative of trimebutine, a weak peripheral kappa and mu opioid receptor agonist. Sedation during endoscopy has become routine practice, but it can be expensive and inconvenient. Patients treated with GIC-1001 may recover faster than sedated patients, enabling them to consult with the physician and drive themselves home after a colonoscopy, as well as return to work sooner.

Unpublished data from a mouse study demonstrated that GIC-1001 significantly decreased nociceptive responses to colorectal distention and was more effective than the parent opioid (Cukier-Meisner 2013). The company has also completed a Phase I trial in healthy volunteers that showed no significant adverse events.

5.3.5 Ikaria Therapeutics LLC

Ikaria Therapeutics is a private company based in Clinton, NJ, USA (www.ikaria.com). They are focused on critical care applications, mainly with delivery of nitric oxide administered via inhalation. Ikaria was developing sodium sulfide (IK-1001) for reduction of myocardial injury and other hypoxic/ischemic conditions (Szabo 2007). However, two clinical trials were halted in 2011 (http://clinicaltrials.gov/ct2/show/nct01007461 and http://clinicaltrials.gov/ct2/show/nct00858936) and there does not appear to be any further activity with respect to development of H_2S-based therapies by this company.

5.3.6 National University of Singapore

GYY4137 is a slow-releasing H_2S donor developed at and patented by the National University of Singapore (www.nus.edu.sg). It was first described in 2008 (Li et al. 2008). As well as releasing H_2S more slowly, it appears to have a more prolonged action than conventional H_2S donors. It is not clear if this compound is in development, or is just a prototype of an H_2S-based therapeutic. GYY4137 has been shown to induce ATP-sensitive K+ channel-dependent vasodilation in rats and thereby to exert anti-hypertensive actions (Li et al. 2008). Consistent with the actions of other H_2S donors (Fiorucci et al. 2007; Gao et al. 2012; Sen et al. 2012), GYY4137 has been shown to reduce production or expression of a number of pro-inflammatory cytokines and mediators (e.g., TNF-α, IL-1β, IL-6, iNOS), as well as inhibiting activation of NF-κB (Li et al. 2009). In an endotoxic shock model, GYY4137 also significantly reduced neutrophil infiltration into the lung and several systemic markers of inflammation (Li et al. 2009). GYY4137 exerted anti-inflammatory effects in a mouse adjuvant arthritis model, but this was dependent upon the time of administration (Li et al. 2013). Only when administered after the adjuvant were the beneficial effects observed: reduction of synovial fluid granulocyte numbers (myeloperoxidase activity) and N-acetyl-β-D-glucosaminidase activity, as well as decreased concentrations of TNF-α, IL-1β, IL-6 and IL-8. GYY4137 was also found to consistently reduce pro-inflammatory mediator generation by human joint cells in vitro (Fox et al. 2012).

5.3.7 Sova Pharmaceuticals Inc.

Sova Pharmaceuticals is a privately held company based in La Jolla, CA, USA (www.sovapharm.com). They are developing inhibitors of cystathionine-γ-lyase "and other enzymes", based on the premise that over-production of H_2S contributes to the pathogenesis of several diseases. They are particularly focused on neuropathic and neurodegenerative diseases, as well as diseases characterized by inflammation and inflammatory pain. No specific compounds are identified on the company's website, and no research publications could be found that were affiliated with this company or with the principals of the company listed on their website.

5.3.8 Sulfagenix

Sulfagenix is a privately held company located in Cleveland, USA (www.sulfagenixinc.com). Its lead product, SG1002, is being developed to treat problems associated with congestive heart failure, based largely on the work of Lefer, Elrod, Calvert and colleagues that demonstrated beneficial effects of H_2S in cardiac ischemia-reperfusion (Elrod et al. 2007; Kondo et al. 2013; Peake et al. 2013). On the company's website, SG1002 is described as "a precursor to a natural-occurring molecule for which deficits have been shown to exist in a number of serious diseases", including cardiovascular disease, type II diabetes, cancer, hypertension, Alzheimer's Disease, asthma, & wound healing. While there are no publications specifically on SG1002, the Sulfagenix website states that the compound has been demonstrated to produce a number of effects in clinically relevant cardiovascular disease models: decrease infarct size, improve cardiac function, increase angiogenesis, down-regulate oxidative stress and decrease inflammation. It also states that SG1002 has been evaluated in a double-blind human trial for male infertility, with significant efficacy and safety. Moreover, patient studies in Mexico have demonstrated safety and provided encouraging results in several disease conditions including cancer, hypercholesterolemia, and Down's syndrome.

5.3.9 Sulfidris S.R.L

Sulfidris S.R.L. is a spin-off drug discovery company from the University of Milan (www.sulfidris.com) that is focused on development of H_2S-releasing drugs for the treatment of cancer and inflammation (Moody et al. 2010; Switzer et al. 2012). Their patents cover novel thiosulfonate derivatives for preventing, treating and/or reducing inflammation-associated diseases in the cardiovascular, connective tissue, pulmonary, gastrointestinal, respiratory, urogenital, nervous or cutaneous systems. They have also developed derivatives of the NSAID pro-drug, sulindac, for treatment/prevention of cancer. The founders of Sulfidris S.R.L. are the same as those for CTG Pharma. As in the case for the latter, the website for Sulfidris is no longer active.

5.3.10 University of Exeter

The University of Exeter, UK, have developed and patented several H_2S releasing compounds, including AP39 and AP123 (www.exeter.ac.uk). These compounds consist of a mitochondrion-targeting group linked to a moiety capable of releasing hydrogen sulfide. The compounds show a range of activities, including lowering of blood pressure, reduction of oxidative stress and various anti-inflammatory activities.

5.4 Future Perspectives

The use of H_2S as the basis for novel drugs was initially met with skepticism, because of the well-known toxicity of this substance in industrial settings. However, the realization that H_2S is produced throughout the body, and performs a wide range of physiological functions, led to a realization that like other gaseous mediators (nitric oxide, carbon monoxide), there was considerable potential for H_2S-based therapies. There is now overwhelming evidence for roles of H_2S in a number of diseases, and considerable data from experimental models to support the hypothesis that drug design focused on H_2S is rational. There is still interest in suppression of H_2S synthesis in some circumstances, but most of the drug development around this molecule is focused on donors that will release appropriate amounts of H_2S at the appropriate target for a desirable period of time. There are significant challenges in achieving these objectives, but increasingly it is becoming clear that H_2S-releasing drugs are promising. As these novel compounds enter the clinical trial phase of development, as is occurring with several at present, we will gain a better understanding of true potential of the use of H_2S as a therapeutic modality.

References

Andruski B, McCafferty DM, Ignacy T, Millen B, McDougall JJ (2008) Leukocyte trafficking and pain behavioural responses to a hydrogen sulfide donor in acute monoarthritis. Am J Physiol Regul Integr Comp Physiol 295:R814–R820

Barnett K, Bell CJ, McKnight W, Dicay M, Sharkey KA, Wallace JL (2000) Role of cyclooxygenase-2 in modulating gastric acid secretion in the normal and inflamed rat stomach. Am J Physiol 279:G1292–G1297

Benetti LR, Campos D, Gurgueira SA, Vercesi AE, Guedes CE, Santos KL, Wallace JL, Teixeira SA, Florenzano J, Costa SK, Muscará MN, Ferreira HH (2013) Hydrogen sulfide inhibits oxidative stress in lungs from allergic mice in vivo. Eur J Pharmacol 698:463–469

Blackler R, Syer S, Bolla M, Ongini E, Wallace JL (2012) Gastrointestinal-sparing effects of novel NSAIDs in rats with compromised mucosal defence. PLoS One 7:e35196

Bucci M, Papapetropoulos A, Vellecco V, Zhou Z, Pyriochou A, Roussos C, Roviezzo F, Brancaleone V, Cirino G (2010) Hydrogen sulfide is an endogenous inhibitor of phosphodiesterase activity. Arterioscler Thromb Vasc Biol 30:1998–2004

Bucci M, Papapetropoulos A, Vellecco V, Zhou Z, Zaid A, Giannogonas P, Cantalupo A, Dhayade S, Karalis KP, Wang R, Feil R, Cirino G (2012) cGMP-dependent protein kinase contributes to hydrogen sulfide-stimulated vasorelaxation. PLoS One 7:e53319

Chattopadhyay M, Kodela R, Nath N, Dastagirzada YM, Velasquez-Martinez CA, Boring D, Kashfi K (2012) Hydrogen sulfide-releasing NSAIDs inhibit the growth of human cancer cells: a general property and evidence of a tissue type-independent effect. Biochem Pharmacol 83:715–722

Cukier-Meisner E (2013) GIcare: colonoscopy comfort. BioCentury A13, March 18

d'Emmanuele di Villa Bianca R, Sorrentino R, Maffia P, Mirone V, Imbimbo C, Fusco F, De Palma R, Ignarro LJ, Cirino G (2009) Hydrogen sulfide as a mediator of human corpus cavernosum smooth-muscle relaxation. Proc Nat Acad Sci U S A 106:4513–4518

Distrutti E, Sediari L, Mencarelli A, Renga B, Orlandi S, Antonelli E, Roviezzo F, Morelli A, Cirino G, Wallace JL, Fiorucci S (2006a) Evidence that hydrogen sulphide exerts antinociceptive effects in the gastrointestinal tract by activating KATP channels. J Pharmacol Exp Ther 316:325–335

Distrutti E, Sediari L, Mencarelli A, Renga B, Orlandi S, Russo G, Caliendo G, Santagada V, Cirino G, Wallace JL, Fiorucci S (2006b) 5-Amino-2-hydroxybenzoic acid 4-(5-thioxo-5H-[1,2]dithiol-3yl)-phenyl ester (ATB-429), a hydrogen sulfide releasing derivative of mesalamine, exerts antinociceptive effects in a model of postinflammatory hypersensitivity. J Pharmacol Exp Ther 319:447–458

Dufton N, Natividad J, Verdu EF, Wallace JL (2012) Hydrogen sulfide and resolution of acute inflammation: a comparative study utilizing a novel fluorescent probe. Sci Rep 2:499

Elrod JW, Calvert JW, Morrison J, Doeller JE, Kraus DW, Tao L, Jiao X, Scalia R, Kiss L, Szabo C, Kimura H, Chow CW, Lefer DJ (2007) Hydrogen sulfide attenuates myocardial ischemia-reperfusion injury by preservation of mitochondrial function. Proc Natl Acad Sci U S A 104:15560–15565

Elsheikh W, Wallace JL (2012) Chemo-preventative effects of a hydrogen sulfide-releasing naproxen derivative (ATB-346) in murine colorectal cancer. Nitric Oxide 27(Suppl 2):S14

Fiorucci S, Antonelli E, Distrutti E, Rizzo G, Mencarelli A, Orlandi S, Antonelli E, Roviezzo F, Morelli A, Cirino G, Wallace JL, Fiorucci S (2005) Inhibition of hydrogen sulfide generation contributes to gastric injury caused by anti-inflammatory nonsteroidal drugs. Gastroenterology 129:1210–1224

Fiorucci S, Orlandi S, Mencarelli A, Caliendo G, Santagada V, Distrutti E, Santucci L, Cirino G, Wallace JL (2007) Enhanced activity of a hydrogen sulphide-releasing derivative of mesalamine (ATB-429) in a mouse mode of colitis. Br J Pharmacol 150:996–1002

Fox B, Schantz JT, Haigh R, Wood ME, Moore PK, Viner N, Spencer JP, Winyard PG, Whiteman M (2012) Inducible hydrogen sulfide synthesis in chondrocytes and mesenchymal progenitor cells: is H_2S a novel cytoprotective mediator in the inflamed joint? J Cell Mol Med 16:896–910

Gao C, Xu DQ, Gao CJ, Ding Q, Yao LN, Li ZC, Chai W (2012) An exogenous hydrogen sulphide donor, NaHS, inhibits the nuclear factor κB inhibitor kinase/nuclear factor κB inhibitor/nuclear factor-κB signaling pathway and exerts cardioprotective effects in a rat hemorrhagic shock model. Biol Pharm Bull 35:1029–1034

Goubern M, Andriamihaja M, Nubel T, Blachier F, Bouillaud F (2007) Sulfide, the first inorganic substrate for human cells. FASEB J 21:1699–1706

Kimura Y, Kimura H (2004) Hydrogen sulfide protects neurons from oxidative stress. FASEB J 18:1165–1167

Kimura H, Nagai Y, Umemura K, Kimura Y (2005) Physiological roles of hydrogen sulfide: synaptic modulation, neuroprotection, and smooth muscle relaxation. Antioxid Redox Signal 7:795–803

Kimura Y, Dargusch R, Schubert D, Kimura H (2006) Hydrogen sulfide protects HT22 neuronal cells from oxidative stress. Antioxid Redox Signal 8:661–670

Kimura Y, Goto Y, Kimura H (2010) Hydrogen sulfide increases glutathione production and suppresses oxidative stress in mitochondria. Antioxid Redox Signal 12:1–13

Kondo K, Bhushan S, King AL, Prabhu SD, Hamid T, Koenig S, Murohara T, Predmore BL, Gojon G Sr, Gojon G Jr, Wang R, Karusula N, Nicholson CK, Calvert JW, Lefer DJ (2013) H_2S protects against pressure overload-induced heart failure via upregulation of endothelial nitric oxide synthase. Circulation 127:1116–1127

Lee ZW, Zhou J, Chen CS, Zhao Y, Tan CH, Li L, Moore PK, Deng LW (2011) The slow-releasing hydrogen sulfide donor, GYY4137, exhibits novel anti-cancer effects in vitro and in vivo. PLoS One 6:e21077

Li L, Rossoni G, Sparatore A, Lee LC, Del Soldato P, Moore PK (2007) Anti-inflammatory and gastrointestinal effects of a novel diclofenac derivative. Free Radic Biol Med 42:706–719

Li L, Whiteman M, Guan YY, Neo KL, Cheng Y, Lee SW, Zhao Y, Baskar R, Tan CH, Moore PK (2008) Characterization of a novel, water-soluble hydrogen sulfide-releasing molecule (GYY4137): new insights into the biology of hydrogen sulfide. Circulation 117:2351–2360

Li L, Salto-Tellez M, Tan CH, Whiteman M, Moore PK (2009) GYY4137, a novel hydrogen sulfide-releasing molecule, protects against endotoxic shock in the rat. Free Radic Biol Med 47:103–113

Li L, Fox B, Keeble J, Salto-Tellez M, Winyard PG, Wood ME, Moore PK, Whiteman M (2013) The complex effects of the slow-releasing hydrogen sulfide donor GYY4137 in a model of acute joint inflammation and in human cartilage cells. J Cell Mol Med 17:365–376

Liu YH, Lu M, Bian JS (2011) Hydrogen sulfide and renal ischemia. Expert Rev Clin Pharmacol 4:49–61

Mariggio MA, Minunno V, Riccardi S, Santacroce R, De Rinaldis P, Fumarulo R (1998) Sulfide enhancement of PMN apoptosis. Immunopharmacol Immunotoxicol 20:399–408

Mikami Y, Kimura H (2012) A mechanism of retinal protection from light-induced degeneration by hydrogen sulfide. Commun Integr Biol 5:169–171

Moody TW, Switzer C, Santana-Flores W, Ridnour LA, Berna M, Thill M, Jensen RT, Sparatore A, Del Soldato P, Yeh GC, Robers DD, Giacone G, Wink DA (2010) Dithiolethione modified valproate and diclofenac increase E-cadherin expression and decrease proliferation of non-small cell lung cancer cells. Lung Cancer 68:154–160

Muzaffar S, Jeremy JY, Sparatore A, Del Soldato P, Angelini GD, Shukla N (2008) H_2S-donating sildenafil (ACS6) inhibits superoxide formation and gp91phox expression in arterial endothelial cells: role of protein kinases A and G. Br J Pharmacol 155:984–994

Papapetropoulos A, Pyriochou A, Altaany Z, Yang G, Marazioti A, Zhou Z, Jeschke MG, Branski LK, Herndon DN, Wang R, Szabo C (2009) Hydrogen sulfide is an endogenous stimulator of angiogenesis. Proc Natl Acad Sci U S A 106:21972–21977

Peake BF, Nicholson CK, Lambert JP, Hood RL, Amin H, Amin S, Calvert JW (2013) Hydrogen sulfide preconditions the db/db diabetic mouse heart against ischemia- reperfusion injury by activating Nrf2 signaling in an Erk-dependent manner. Am J Physiol Heart Circ Physiol 304(9):H1215–H1224

Pircher J, Fochler F, Czermak T, Mannell H, Kraemer BF, Wörnle M, Sparatore A, Del Soldato P, Pohl U, Krötz F (2012) Hydrogen sulfide-releasing aspirin derivative ACS14 exerts strong antithrombotic effects in vitro and in vivo. Arterioscler Thromb Vasc Biol 32:2884–2891

Sakai E, Endo H, Taniguchi L, Hata Y, Ezuka A, Nagase H, Yamada E, Ohkubo H, Higurashi T, Sekino Y, Koide T, Iida H, Hosono K, Nonaka T, Takahashi H, Inamori M, Maeda S, Nakajima A (2012) Factors predicting the presence of small bowel lesions in patients with obscure gastrointestinal bleeding. Dig Endosc. doi:10.1111/den.12002, Dec 17, [Epub ahead of print]

Satoh H, Amagase K, Takeuchi K (2012) Exacerbation of NSAID-induced small intestinal lesions by anti-secretory drugs in rats: the role of intestinal motility. J Pharmacol Exp Therap 343:270–277

Sen N, Paul BD, Gadalla MM, Mustafa AK, Sen T, Xu R, Kim S, Snyder SH (2012) Hydrogen sulfide-linked sulfhydration of NF-κB mediates its antiapoptotic actions. Mol Cell 45:13–24

Shukla N, Rossoni G, Hotston M, Sparatore A, Del Soldato P, Tazzari V, Persad R, Angelini GD, Jeremy JY (2009) Effect of hydrogen sulphide-donating sildenafil (ACS6) on erectile function and oxidative stress in rabbit isolated corpus cavernosum and in hypertensive rats. BJU Int 103:1522–1529

Simon F, Scheuerle A, Gröger M, Stahl B, Wachter U, Vogt J, Speit G, Hauser B, Möller P, Calzia E, Szabó C, Schelzig H, Georgieff M, Radermacher P, Wagner F (2011) Effects of intravenous sulfide during porcine aortic occlusion-induced kidney ischemia/reperfusion injury. Shock 35:156–163

Sivarajah A, Collino M, Yasin M, Benetti E, Gallicchio M, Mazzon E, Cuzzocrea S, Fantozzi R, Thiemermann C (2009) Anti-apoptotic and anti-inflammatory effects of hydrogen sulfide in a rat model of regional myocardial I/R. Shock 31:267–274

Srilatha B, Muthulakshmi P, Adaikan PG, Moore PK (2012) Endogenous hydrogen sulfide insufficiency as a predictor of sexual dysfunction in aging rats. Aging Male 15:153–158

Switzer CH, Cheng RY, Ridnour LA, Murray MC, Tazzari V, Sparatore A, Del Soldato P, Hines HB, Glynn SA, Ambs S, Wink DA (2012) Dithiolethiones inhibit NF-κB activity via covalent modification in human estrogen receptor-negative breast cancer. Cancer Res 72:2394–2404

Szabo C (2007) Hydrogen sulphide and its therapeutic potential. Nat Rev Drug Discov 6:917–935

Szabo H, Sato F, Takami K, Szabo S (1997) New ulcerative colitis model induced by sulfhydryl blockers in rats and the effects of antiinflammatory drugs on the colitis. Jpn J Pharmacol 73:299–309

Vendramini-Costa DB, Carvalho JE (2012) Molecular link mechanisms between inflammation and cancer. Curr Pharm Des 18:3831–3852

Wallace JL (2007) Hydrogen sulfide-releasing anti-inflammatory drugs. Trends Pharmacol Sci 28:501–505

Wallace JL (2008) Prostaglandins, NSAIDs, and gastric mucosal protection: why doesn't the stomach digest itself? Physiol Rev 88:1547–1565

Wallace JL (2010) Physiological and pathophysiological roles of hydrogen sulfide in the gastrointestinal tract. Antioxid Redox Signal 12:1125–1133

Wallace JL (2013) Mechanisms, prevention and clinical implications of nonsteroidal anti-inflammatory drug-enteropathy. World J Gastroenterol 19:1861–1876

Wallace JL, Caliendo G, Santagada V, Cirino G, Fiorucci S (2007a) Gastrointestinal safety and anti-inflammatory effects of a hydrogen sulfide-releasing diclofenac derivative in the rat. Gastroenterology 132:261–271

Wallace JL, Dicay M, McKnight W, Martin GR (2007b) Hydrogen sulfide enhances ulcer healing in rats. FASEB J 21:4070–4076

Wallace JL, Cirino G, Santagada V, Caliendo G (2008) Hydrogen sulfide derivatives of nonsteroidal anti-inflammatory drugs. US Patent Application No. WO/2008/009127

Wallace JL, Vong L, McKnight W, Dicay M, Martin GR (2009) Endogenous and exogenous hydrogen sulfide promotes resolution of colitis in rats. Gastroenterology 137:569–578

Wallace JL, Caliendo G, Santagada V, Cirino G (2010) Markedly reduced toxicity of a hydrogen sulfide-releasing derivative of naproxen (ATB-346). Br J Pharmacol 159:1236–1246

Wallace JL, Syer S, Denou E, de Palma G, Vong L, McKnight W, Jury J, Bolla M, Bercik P, Collins SM, Verdu E, Ongini E (2011) Proton pump inhibitors exacerbate NSAID-induced small intestinal injury by inducing dysbiosis. Gastroenterology 141:1314–1322

Wang R (2009) Hydrogen sulfide: a new EDRF. Kidney Int 76:700–704

Watanabe T, Tanigawa T, Nadatani Y, Nagami Y, Sugimori S, Okazaki H, Yamagami H, Watanabe K, Tominaga K, Fujiwara Y, Koike T, Arakawa T (2013) Risk factors for severe nonsteroidal anti-inflammatory drug-induced small intestinal damage. Dig Liver Dis 45:390–395

Whiteman M, Cheung NS, Zhu YZ, Chu SH, Siau JL, Wong BS, Armstrong JS, Moore PK (2005) Hydrogen sulphide: a novel inhibitor of hypochlorous acid-mediated oxidative damage in the brain? Biochem Biophys Res Commun 326:794–798

Whiteman M, Haigh R, Tarr JM, Gooding KM, Shore AC, Winyard PG (2010) Detection of hydrogen sulfide in plasma and knee-joint synovial fluid from rheumatoid arthritis patients: relation to clinical and laboratory measures of inflammation. Ann N Y Acad Sci 1203:146–150

Yang G, Zhao K, Ju Y, Mani S, Cao Q, Puukila S, Khaper N, Wu L, Wang R (2013) Hydrogen sulfide protects against cellular senescence via S-sulfhydration of Keap1 and activation of Nrf2. Antioxid Redox Signal 18:1906–1919

Zanardo RC, Brancaleone V, Distrutti E, Fiorucci S, Cirino G, Wallace JL (2006) Hydrogen sulfide is an endogenous modulator of leukocyte-mediated inflammation. FASEB J 20:2118–2120

Hydrogen Sulfide: Physiological and Pathophysiological Functions

6

Yi Tong Liu and Jin-Song Bian

Abstract

Hydrogen sulfide (H_2S) has been recognized as an endogenous gaseous mediator. The past decade has seen an exponential growth of scientific interest in the physiological and pathological significance of H_2S especially with respect to its roles in the central nervous and the cardiovascular systems. In cardiovascular system, H_2S regulates heart contractile function and may serve as a cardioprotectant for treating ischemic heart diseases and heart failure. Alterations of the endogenous H_2S level have been found in animal models with various pathological conditions such as myocardial ischemia, spontaneous hypertension, and hypoxic pulmonary hypertension. In the central nervous system, H_2S facilitates long-term potentiation and regulates intracellular calcium concentration in brain cells. Intriguingly, H_2S produces antioxidant, anti-inflammatory, and anti-apoptotic effects that may be of relevance to neurodegenerative disorders. Abnormal generation and metabolism of H_2S have been reported in the pathogenesis of ischemic stroke, Alzheimer's disease, Parkinson's disease, and recurrent febrile seizure. Exogenously applied H_2S is demonstrated to be valuable in the treatment against febrile seizure and Parkinson's disease. In addition, H_2S also regulates the physiological and pathological functions of kidney, pancreas and bone. Exogenously applied H_2S may protect against ischemic kidney injuries and osteoporosis. This article surveys the growing recognition of H_2S as an endogenous signaling molecule in mammals and its functions in different biological systems. We will emphasize on its physiological and pathological functions in the cardiovascular, central nervous and renal systems.

Keywords

Gasotransmitter • Hydrogen sulphide • Pharmacology • Physiology • Pathology

Y.T. Liu • J.-S. Bian (✉)
Department of Pharmacology, Yong Loo Lin School of Medicine, National University of Singapore, 18 Medical Drive, Singapore 117597, Singapore
e-mail: phcbjs@nus.edu.sg

H. Kimura (ed.), *Hydrogen Sulfide and its Therapeutic Applications*,
DOI 10.1007/978-3-7091-1550-3_6, © Springer-Verlag Wien 2013

Abbreviations

$[Ca^{2+}]_i$	Intracellular Ca^{2+}
$\pm LVdp/dt_{max}$	Maximal/minimal left ventricular pressure development
1-K	Uninephrectomy
2K1C	2-Kidneys-1-clip
3-MST	Mercaptopyruvate sulfurtransferase
6-OHDA	6-Hydroxydopamine
AC	Adenylyl cyclase
ACE	Angiotensin-converting enzyme
AD	Alzheimer's disease
Ang II	Angiotensin II
AOAA	Aminooxyacetic acid
APD	Action potential duration
ApoE	Apolipoprotein E
ATN	Acute tubular necrosis
ATP	Adenosine triphosphate
AVF	Arteriovenous fistula
BACE-1	Beta-site amyloid precursor protein cleaving enzyme 1
BK_{Ca}	Large Conductance Ca^{2+}–activated potassium channels
BP	Blood pressure
cAMP	Cyclic adenosine monophosphate
CAT	Cysteine aminotransferase
CBS	Cystathionine β-synthase
cGMP	Cyclic guanosine monophosphate
CKD	Chronic kidney disease
CNS	Central nervous system
CO	Carbon monoxide
COX-2	Cyclooxygenase 2
CSE	Cystathionine γ-lyase
DA	Dopamine/dopaminergic
DEANO	Diethylamine nitric oxide
ECs	Endothelial cells
EDHF	Endothelium derived hyperpolarizing factor
ER	Endoplasmic reticulum
ERK (MAPK)	Extracellular signal-regulated kinase
FE_{Na}	Fractional excretion of Na^+
FE_K	Fractional excretion of K^+
FF	Filtration rate
FS	Febrile seizures
GABA	Gamma-aminobutyric acid
GFR	Glomerular filtration rate
GLT1	Glial glutamate transporter 1

GSH	Glutathione
GSK-3β	Glycogen synthase kinase-3
GSSG	Oxidized glutationine
H_2S	Hydrogen sulfide
HF	Heart failure
HIF	Hypoxia-inducible factors
HMC1.1	Human mast cell line 1.1
HNO	Nitroxyl anion
Hsp	Heat shock protein
HUVECs	Human umbilical vein endothelial cells
I/R	Ischemia/reperfusion
ICAM-1	Intercellular adhesion molecule 1
IK_{Ca}	Intermediate conductance Ca^{2+}-activated potassium channels
IL	Interleukin
iNOS	Inducible NO synthase
IPreC	Ischemic preconditioning
IRR	Intrarenal resistance
ISO	Isoproterenol
JNK	c-Jun N-terminal kinases
K_{ATP}	ATP-sensitive potassium channel
LCA	Left coronary artery
LDL	Low-density lipoprotein
LPO	Lipid hydroperoxidation
LPS	Lippolysaccharide
LTCC $I_{Ca, L}$	L-type Ca^{2+} channels
MAP	Mean arteriole pressure
MCAO	Middle cerebral artery occlusion
MDA	Malondialdehyde
MEK	ERK Kinase
MI	Myocardial infarction
MMP	Matrix metalloproteinase
MPP^+	1-Methyl-4-phenylpyridine
mPTP	Mitochondrial permeability transition pore
MR	Mental retardation
Na_2S	Sodium sulfide
NADPH	Nicotinamide adenine dinucleotide phosphate
NaHS	Sodium hydrosulfide
NF-E2	Nuclear factor-erythroid-derived 2
NF-κB	Nuclear factor kappa-light-chain-enhancer of activated B cells
NHE	Na^+/H^+ exchanger
NO	Nitric oxide
NOS	Nitric oxide synthase
NRF-1	Nuclear respiratory factor 1
Nrf2	NF-E2 related factor 2
NSAIDs	Non-steroidal anti-inflammatory drugs

$ONOO^-$	Peroxynitrite
PAG	DL-Propargylglycine
PARP	Poly (ADP-ribose) polymerase
p-CREB	Phosphorlyated cAMP response element-binding protein
PD	Parkinson's disease
pH_i	Intracellular pH
PI3K	Phosphoinositide 3-kinase
PKA	Protein kinase A
PKC	Protein kinase C
PGE	Prostaglandin E2
PKG	Protein kinase G
p-NR1	Phosphorylated N-methyl-D-aspartate receptor 1 subunit
p-NR2A	Phosphorylated N-methyl-D-aspartate receptor 2A subunit
p-NR2B	Phosphorylated N-methyl-D-aspartate receptor 2B subunit
RAS	Renin-angiotensin system
RBF	Renal blood flow
ROS	Reactive oxygen species
SBP	Systolic blood pressure
SHR	Spontaneously hypertensive rats
SIN-1	3-Morpholinosydnonimine
SK_{Ca}	Small conductance Ca^{2+}-activated potassium channels
SMCs	Smooth muscle cells
SNAP	S-Nitroso-N-acetylpenicillamine
SNP	Sodium nitroprusside
SOD	Superoxide dismutase
SPreC	H_2S preconditioning
STAT	Signal transducer and activator of transcription
TH	Tyrosine hydroxylase
TNF	Tumor necrosis factor
TUNEL	Terminal deoxynucleotidyl transferase dUTP Nick end labeling
U.V	Urine flow rate
UCP2	Uncoupling protein 2
$U_K.V$	Urinary K^+ excretion
$U_{Na}.V$	Urinary Na^+ excretion
VCAM-1	Vascular cell adhesion molecule-1
VEGF	Vascular endothelial growth factor
WT	Wild type

6.1 Introduction

The physiologic importance of H_2S was first reported by Abe and Kimura in 1996, when H_2S was found to act as a novel neuromodulator (Abe and Kimura 1996). H_2S is now commonly regarded as the third 'gasotransmitter' subsequent to nitric oxide (NO) and carbon monoxide (CO) (Wang 2002). Similar to NO and CO, H_2S can be endogenously synthesized by several enzymes. It has also been well demonstrated to influence a wide range of physiological and pathological processes. In the heart, H_2S has been recognized to induce protective effects (Johansen et al. 2006; Elrod et al. 2007). In vascular tissues, H_2S induces both blood vessel relaxation (Hosoki et al. 1997; Zhao et al. 2001; Zhao and Wang 2002; Cheng et al. 2004; Ali et al. 2006; Kiss et al. 2008; Webb et al. 2008; Yang et al. 2008) as well as constriction (Ali et al. 2006; Kiss et al. 2008; Lim et al. 2008; Webb et al. 2008), depending on the concentration of H_2S administered and the type of vessels involved. In the nervous system, H_2S has been found to mediate neurotransmission (Abe and Kimura 1996) and induces both neuroprotection and neurotoxicity (Hu et al. 2010; Kida et al. 2010). H_2S has also been reported to regulate inflammation (Li et al. 2006a; Hu et al. 2007) and insulin release. In this chapter, we present current knowledge of H_2S to facilitate better understanding of its biological functions in both health and disease, with a special emphasis on its protective effects in cardiovascular, central nervous and renal systems.

Under physiological conditions, H_2S is present in plasma and organ systems as ~14 % H_2S, 86 % HS^- and a trace of S^{2-} (Giggenbach 1971; Hvitved-Jacobsen 2002; Dombkowski et al. 2004). Since these species coexist in aqueous solution together, it is difficult to identify the biologically active species that underlie the effects observed. Hence, the terminology "H_2S" refers to the sum of H_2S, HS^- and S^{2-}in the context of this chapter unless otherwise specified. Till date, most researchers utilize NaHS or Na_2S (or their hydrous forms) as exogenous sources of H_2S. In aqueous solution, both release rapid bolus of H_2S which triggers downstream mechanisms. More recently, a handful of slow-releasing H_2S compounds have been developed (Li et al. 2007, 2008; Sidhapuriwala et al. 2007; Lee et al. 2010b; Xie et al. 2013). The effects of these exogenous H_2S donors in different systems are discussed in this chapter.

6.2 Physiological and Pathological Functions of H_2S in the Cardiovascular System

6.2.1 H_2S Biosynthesis in the Cardiovascular System

The most important mammalian enzymes that are responsible for the synthesis of H_2S are cystathionine β-synthase (CBS, EC 4.2.1.22), cystathionine γ-lyase (cystathionase, CSE, EC 4.4.1.1), mercaptopyruvate sulfurtransferase (3-MST, EC 2.8.1.2) and cysteine aminotransferase (CAT, EC 2.6.1.3). Recently, a novel H_2S

biosynthetic pathway from D-cysteine involving 3-MST and D-amino acid oxidase has been unveiled (Shibuya et al. 2013).

Among these enzymes, CSE is the main H_2S-generating enzyme that is expressed in the cardiovascular system (Zhao et al. 2001; Bian et al. 2006) and various vascular tissues (Chen et al. 1999; Zhao et al. 2001). CSE mRNA expression has been detected in the myocardium (Geng et al. 2004b), endothelial cells (ECs) (Yang et al. 2008) and smooth muscle cells (SMCs) (Zhao et al. 2001), and the intensity rank of CSE mRNA expression in various vascular tissues is as follows:

Pulmonary artery > aorta > tail artery > mesenteric artery (Zhao et al. 2001).

6.2.2 Physiological Functions of H_2S in the Cardiovascular System

6.2.2.1 Physiological Functions of H_2S in the Heart

H_2S may markedly reduce action potential duration (APD) and decelerate sinus rhythm, while having no significant effect on the amplitude of action potential and resting potential (Sun et al. 2008). HERG/I_{kr} and KvLQT1/I_{ks} are two important potassium channels that control APD. Till date, H_2S has not been reported to affect the function of these channels in the heart. Therefore, the effect of H_2S on APD is probably attributed to the opening of K_{ATP} channels (Abramochkin et al. 2009). H_2S is capable of opening K_{ATP} channels directly, first demonstrated by Wang and coworkers (Tang et al. 2005; Jiang et al. 2010). Furthermore, H_2S may also activate K_{ATP} channels indirectly by inducing intracellular acidosis (Cuevas et al. 1991; Koyano et al. 1993; Bethell et al. 1998; Lee et al. 2007) and other potassium channels (Martelli et al. 2013). However, the involvement of these channel activations towards shortening of APD is yet to be clearly understood and warrants further research.

H_2S produces negative inotropic effect in rat hearts. In isolated rat ventricular myocytes, H_2S decreased the amplitudes of myocyte twitch and electrically-induced calcium transients upon stimulation of β_1-adrenergic receptors with isoproterenol (ISO) (Yong et al. 2008b). Using isolated heart, perfusion with H_2S inhibited maximal/minimal left ventricular pressure development ($\pm$LVdp/dt_{max}) (Geng et al. 2004b). H_2S perfusion in vivo via femoral vein produced a similar effect on the cardiodynamics of anesthetized rats (Geng et al. 2004b). However, H_2S at concentration up to 100 μM NaHS had no significant effect on heart rate in isolated rat hearts (Zhong et al. 2003; Minamishima et al. 2009).

Different mechanisms have been implicated in the inhibitory effect of H_2S on heart contractility. Firstly, H_2S opens K_{ATP} channels. Secondly, H_2S may inhibit adenylyl cyclase (AC)/ cyclic adenosine monophosphate (cAMP) pathway to suppress β-adrenoceptor system, thereby producing negative inotropic effects (Yong et al. 2008b). Thirdly, H_2S reduced peak current of L-type Ca^{2+} channels (LTCC; $I_{Ca,\ L}$) which is important in controlling heart contractility and cardiac rhythm (Sun et al. 2008). Interestingly, in various brain cell types, H_2S (100–300 μM) has been reported to increase $[Ca^{2+}]_i$ via opening (instead of closing)

LTCC (Nagai et al. 2004; Garcia-Bereguiain et al. 2008; Yong et al. 2010a). It is intriguing that H_2S directly blocks LTCC in cardiomyocytes, but opens the same channels in neurons. One possibility accounting for such phenomenon is that the effect of H_2S on LTCC may be secondary to other signaling pathways. For instance, reduction of Ca^{2+} current through LTCC could have resulted from hyperpolarization caused by opening of K_{ATP} channels (Tang et al. 2005; Jiang et al. 2010) or the suppression of cAMP/PKA pathway (Yong et al. 2008b). More evidence, including single channel recording, is needed to conclude whether H_2S is a direct LTCC blocker.

It should be noted, however, that the significance of findings mentioned above may require further valiadation since H_2S concentrations administered in those experiments are much higher than physiological, which is now generally regarded to be less than 0.1–1 μM (Whitfield et al. 2008; Levitt et al. 2011). Nonetheless, sulfides may bind to proteins in plasma and tissues, and is released in response to stimuli (Liu et al. 2012b). For instance the concentration of acid-labile sulfur in the heart was reported to be about 300 μM (Levitt et al. 2011). Free and bound sulfide originates from the action of enzymes that synthesize H_2S. It is therefore unclear if H_2S administered at concentrations between 100 and 500 μM plays physiological roles in heart functions.

6.2.2.2 Physiological Functions of H_2S in the Vascular System

Regulation of Vascular Tone

The biological function of H_2S on vascular tissue is biphasic. H_2S induces vasorelaxation at a higher concentration range (NaHS 100–1,600 μM), but causes concentration-dependent constriction at lower concentrations (NaHS 10–100 μM) (Ali et al. 2006; Kubo et al. 2007; Lim et al. 2008). H_2S induced vasodilation has been reported in thoracic aorta, mesenteric arteries, pulmonary artery, tail artery and other types of vascular tissues (Hosoki et al. 1997; Zhao et al. 2001). H_2S-induced vasorelaxation is mainly underlied by opening of K_{ATP} channels (Zhao et al. 2001; Cheng et al. 2004; Kubo et al. 2007) and partially mediated by endothelium-dependent mechanism(s) (Zhao et al. 2001). Other signaling mechanisms involved includes intracellular acidosis (Lee et al. 2007), depletion of intracellular ATP levels (Szabo 2007; Kiss et al. 2008; Webb et al. 2008) and elevations in cyclic guanosine monophosphate (cGMP)/protein kinase G (PKG) (Bucci et al. 2010). More recent studies refer H_2S as an endothelium derived hyperpolarizing factor (EDHF) (Mustafa et al. 2011). This is supported by findings that intermediate and small conductance Ca^{2+}-activated potassium channels (IK_{Ca}/SK_{Ca}) channels underlie H_2S effect, and IK_{Ca}, but not K_{ATP} and large conductance Ca^{2+}-activated potassium channels (BK_{Ca}) channels, mediate H_2S-induced hyperpolarization in cultured human aortic ECs (Mustafa et al. 2011). Taken together, these studies are suggestive that H_2S plays important roles in mediating vascular responses of small and intermediate resistance vessels.

H_2S-induced vasoconstrictive effects are also mediated by multiple mechanisms. It has been found that H_2S may reduce NO synthesis in endothelium (Kubo et al. 2007), or interact with NO to form a nitrosothiol compound, which itself has no effect on vascular activity (Ali et al. 2006). However, H_2S-induced vasoconstriction is not completely abolished in the presence of NO synthase (NOS) inhibitor or removal of endothelium, suggesting that other NO-independent mechanisms might be implicated. One possibility is the downregulation of cAMP level in VSMCs (Lim et al. 2008), which then upregulates the activation of myosin light chain kinase to induce vasoconstriction.

Angiogenesis

Current evidence suggests that H_2S promotes angiogenesis and cell growth. H_2S enhances cell migration, growth and proliferation in endothelial cells (Cai et al. 2007; Papapetropoulos et al. 2009). Under hypoxic conditions, H_2S-induced angiogenesis is probably hypoxia-inducible factors (HIF)-1α/ vascular endothelial growth factor (VEGF)-dependent (Liu et al. 2010).

H_2S also promotes vascular network formation under pathological situations. A hindlimb ischemic model was established in rats that were subjected to unilateral femoral artery ligation. NaHS at 50 μmol/kg/day, but not (200 μmol/kg/day), promoted collateral vessel growth in ischemic hindlimbs, along with increased regional blood flow and increased capillary density (Wang et al. 2010). This implies that H_2S may promote vascular network formation in vivo at near physiological concentrations. The signaling mechanisms for the angiogenic effect of H_2S involve activation of Akt (Cai et al. 2007), extracellular signal-regulated (ERK)-kinase (MEK) (Papapetropoulos et al. 2009) and heat shock protein (Hsp)-27 (Papapetropoulos et al. 2009).

6.2.2.3 Interaction Among Gasotransmitters in the Cardiovascular System

Under physiological conditions, gaseous mediators (i.e. H_2S, NO and CO) might be present at the same time, and accumulating evidence now suggests that the interaction among gaseous mediators may influence or alter overall biological effects, in contrast to their individual effects (Kashiba et al. 2002; Fukuto and Collins 2007; Li et al. 2009; Olson and Donald 2009; Kajimura et al. 2010). Interaction between H_2S and NO may also regulate heart function. Yong et al. first reported that a mixture of NO donor and H_2S (100 μM) produces positive isotropic effect in the heart whereas H_2S and NO alone produces opposite effect. The effect of interaction could be abolished by thiols, suggesting that a new molecule that is thiol sensitive could have been formed. Nitroxyl (HNO) was proposed to be the product (Yong et al. 2010b) due to the strong reducing capability of H_2S (Warenycia et al. 1989b; Wang 2002; Szabo 2007) and the structural and pharmacological similarities with HNO (Yong et al. 2010b). The formation of HNO as an end-product of H_2S and NO donor (sodium nitroprusside; SNP) interaction was further supported by Filipovic et al. under physiological cellular conditions and in isolated mouse heart (Filipovic et al. 2013). Filipovic et al. proposed that the interaction is independent of NO released from SNP, but rather a direct effect between H_2S and SNP. This is in contrast with Yong et al.'s observations in which various types of NO donors such as L-arginine (NOS substrate) or diethylamine NO (DEANO) were also used and similar effect to that of SNP was found (Yong et al. 2010b, 2011). Nevertheless, the formation of HNO as a result of H_2S and NO or SNP interaction warrants further in depth studies to be fully resolved.

In the vascular system, interaction between NO and H_2S is controversial. Hosoki et al. first reported that NO and H_2S act synergistically in vasorelaxations (Hosoki et al. 1997). On the contrary, later studies reported that H_2S pretreatment inhibited SNP-induced vasorelaxations (Zhao and Wang 2002). Ali et al. showed that mixing NO donors (SNP, SIN-1 or SNAP) with NaHS (100 μM) reduced the extent of vasorelaxation compared to the relaxation with NO donors alone, further indicating inactivation of NO by H_2S (Ali et al. 2006). The authors ascribed these observations to formation of a nitrosothiol compound (Ali et al. 2006), which is still unidentified till date. It is highly likely that this new compound is HNO, as mentioned above, instead of a nitrosothiol (Yong et al. 2010b, 2011; Filipovic et al. 2013).

6.2.3 Pathological Functions of H_2S in the Cardiovascular System

6.2.3.1 Acute Ischemic Heart Diseases

Endogenous H_2S Level Under Ischemic Conditions

Accumulating evidence now suggests that under ischemic conditions, endogenous H_2S production in the heart is reduced. In ventricular myocytes, for example, treatment with ischemic solution reduced endogenous H_2S level (Bian et al. 2006). Under ischemic conditions, both in vivo and in vitro studies showed that CSE activity (Yong et al. 2008a) and mRNA gene expression (Zhu et al. 2007) were downregulated.

In an in vivo animal study, rats that were injected with ISO to produce "infarct-like" myocardial necrosis were found to have reduced H_2S levels in myocardium (Rona et al. 1959). Geng et al. further confirmed that plasma H_2S level dropped by 66 % (from 60 to 20 μM) in an ISO-induced myocardial ischemic rat model (Geng et al. 2004a). Consistent with this, a clinical observational study showed that plasma H_2S concentration in patients with coronary diseases is significantly lower compared with control subjects (26 μM vs. 52 μM), suggesting that the decreased plasma H_2S levels may correlate with severity of coronary diseases (Jiang et al. 2005). These observations suggest that plasma H_2S level has the potential to be used as a biomarker for ischemic heart diseases.

Ischemia/reperfusion (I/R)-induced arrhythmias may develop as a result of free radical species (ROS) production and accumulation in the myocardium during reperfusion. Since H_2S production is markedly decreased during ischemia (Geng et al. 2004a; Jiang et al. 2005; Bian et al. 2006; Yong et al. 2008a, b), ROS may therefore be increased. Excessive free radicals may react with proteins, lipids and nucleic acids, thereby disrupting myocardium structure and functions.

Therapeutic Effects of H_2S Against Ischemic Heart Diseases

Exogenously applied H_2S may reduce myocardial infarction (MI) size in rats (Johansen et al. 2006; Zhu et al. 2007; Pan et al. 2009), mice (Elrod et al. 2007) and pigs (Sodha et al. 2008; Osipov et al. 2009; Sodha et al. 2009). Treatment with H_2S also significantly protected heart against I/R-induced arrhythmias (Bian et al. 2006; Zhang et al. 2007) and improved myocardial contractile function in ISO-induced ischemic rat heart (Geng et al. 2004a) and I/R-induced ischemic porcine heart (Sodha et al. 2008). Endogenous H_2S level is vital to protect heart against ischemic injuries. Inhibition of endogenous H_2S production significantly increased infarct size (Sivarajah et al. 2006; Bliksoen et al. 2008), whereas stimulation of endogenously produced H_2S by overexpression of CSE reduced infarct size (Elrod et al. 2007).

H_2S inhibits the progression of apoptosis subsequent to I/R injury. H_2S treatment suppressed the activation of caspase-3, poly (ADP-ribose) polymerase (PARP) and/or terminal deoxynucleotidyl transferase dUTP nick end labeling (TUNEL)-positive nuclei in mice (Elrod et al. 2007) and swine (Sodha et al. 2008). It also suppressed the expression of pro-apoptotic proteins via caspase-independent cell death through phosphorylation of glycogen synthase kinase-3 (GSK-3β) (Osipov et al. 2009). Similarly, Yao et al. also demonstrated that H_2S increased phosphorylation of GSK-3β (Ser9) and thus inhibited the opening of mitochondrial permeability transition pore (mPTP) (Yao et al. 2010). H_2S also improved cardiac ATP pools (Szabo et al. 2011) and reduced mitochondrial oxygen consumption (Elrod et al. 2007). It preserves mitochondrial function by increasing complex I and II efficiency (Alves et al. 2011) and inhibiting respiration and thus limiting the generation of ROS (Elrod et al. 2007). Therefore, the cardioprotective effects of H_2S also involve its anti-oxidative function (Sojitra et al. 2011; Szabo et al. 2011).

Anti-inflammatory effect of H_2S also contributes to its cardioprotection. H_2S decreased the number of leukocytes within the ischemic zone by inhibition of leukocyte-EC interactions (Elrod et al. 2007). It also decreased myocardial interleukin (IL)-1β (Elrod et al. 2007), TNF-α, IL-6 and IL-8 levels (Sodha et al. 2009). Therefore, inhibition of leukocyte transmigration and inhibition of cytokine release are possible mechanisms for the anti-inflammatory and cardioprotective effects of H_2S. Other cardioprotective mechanisms of H_2S may include suppression of β-adrenergic function (Yong et al. 2008b), inhibition of Na^+/H^+ exchanger (NHE) activity (Hu et al. 2011a), opening of K_{ATP} channels (Johansen et al. 2006), blockade of LTCC (Sun et al. 2008), attenuation of endoplasmic reticulum (ER) stress (Wei et al. 2010) and preservation of endothelial function (Szabo et al. 2011) etc.

Ischemic preconditioning (IPreC) is a powerful natural cardioprotective mechanism. H_2S preconditioning (SPreC) produces cardioprotective effects (Bian et al. 2006; Pan et al. 2006, 2008, 2009; Hu et al. 2008a, b; Calvert et al. 2009). Interestingly, SPreC produces stronger effect than post-ischemic H_2S treatment (Pan et al. 2009). The protective effects of direct H_2S treatment may rely mainly on the ability of sulfide to reduce inflammatory responses (Zanardo et al. 2006) and to neutralize cytotoxic ROS such as peroxynitrite ($ONOO^-$) (Whiteman et al. 2004), which may relieve oxidative stress partly, but not enough to salvage infarcted myocardium. SPreC is more likely to protect the heart by switching it to a defensive mode against ischemic insult.

SPreC may trigger a series of signaling proteins including opening K_{ATP} channels (Pan et al. 2006) activation of Protein Kinase C (PKC, especially ε-isoform) (Pan et al. 2008), ERK1/2-MAPK (Hu et al. 2008b) and PI3K/Akt pathways (Hu et al. 2008b). By activating pro-survival pathways, SPreC may stimulate cells to counteract stressful conditions. These pathways result in the production of various molecules (e.g. HSPs, GSH, and bilirubin) endowed with antioxidant and antiapoptotic activities (Calvert et al. 2009). SPreC also activates signal transducer and activator of transcription (STAT)-3, which prevents cleavage of caspase-3, inhibits translocation of cytochrome C and reduces the number of TUNEL-positive nuclei (Calvert et al. 2009). The anti-apoptotic actions are found to be, at least partially, mediated by inhibition of pro-apoptotic factor Bad, upregulation of pro-survival factors Bcl-2 and Bcl-xL, and an upregulation of Hsps.

In addition, COX-2/PGE2 pathway (Hu et al. 2008a; Sojitra et al. 2011), prevention of intracellular calcium overload and hypercontracture (Pan et al. 2008), NO (Pan et al. 2006) and nuclear factor-erythroid-derived 2 (NF-E2) related factor 2 (Nrf2)/anti-oxidative stress (Calvert et al. 2009) have all been implicated in SPreC-induced cardioprotection (Liu et al. 2012b). These results suggest that H_2S therapy may enhance endogenous antioxidant defense of myocytes and create an environmental resistance to the oxidative stress associated with myocardial I/R injury, as evidenced by the preservation of redox state and a reduction in lipid peroxidation.

6.2.3.2 Hypertrophy, Cardiomyopathy and Heart Failure (HF)

Myocardial infarction (MI) is the leading cause of HF. Cardiac hypertrophy as a result of sustained overload can lead to progression of HF. Plasma H_2S level was found to be decreased in MI-induced (Wang et al. 2011), and arteriovenous fistula (AVF)-induced CHF model (Sen et al. 2008; Mishra et al. 2010). In addition, endogenous H_2S synthesis in the heart was also found to be lowered in adriamycin -induced cardiomyopathy model (Su et al. 2009). This was further supported by transgenic mice overexpressing CSE as excessive H_2S production protected against CHF injuries in both permanent LCA ligation model as well as LCA I/R model (Calvert et al. 2010).

H_2S pretreatment prevented cardiomyocyte hypertrophy by lowering intracellular ROS, upregulating microRNA-133a and suppressing microRNA-21 in rat primary cultures (Liu et al. 2011a). Overexpression of CSE reduces left ventricle dilation and cardiac hypertrophy (Calvert et al. 2010). Exogenous application of H_2S attenuated the development of hypertrophy in spontaneously hypertensive rats (SHR) (Shi et al. 2007). H_2S also attenuated development of adriamycin-induced cardiomyopathy (Su et al. 2009).

Anti-oxidative effect of H_2S is the main mechanism for its therapeutic effect on CHF. Application with H_2S inhibited lipid hydroperoxidation (LPO) and increased superoxide dismutase (SOD) and GSH peroxidase activities. Therefore, treatment with H_2S stimulated the activity of anti-oxidant enzymes (Zhang et al. 2011b). H_2S also reduced LPO and protected heart against HF injury via stimulation of Akt and nuclear localization of nuclear respiratory factor 1 (NRF-1) and NF-E2 related factor 2 (Nrf2) (Calvert et al. 2010). H_2S also decreased the number of apoptotic cells through promoting the expression of anti-apoptotic factor Bcl-2 while suppressing expressions of pro-apoptotic factors Bax and caspase-3. The release of cytochrome c from mitochondria was reduced. These anti-apoptotic effects therefore mediated the cardioprotective effects of H_2S (Wang et al. 2011). In addition, H_2S may also protect against HF via promoting angiogenesis (Mishra et al. 2010; Givvimani et al. 2011).

H_2S was also found recently to prevent HF progression via attenuating mast cell accumulation and degranulation in response to toxic cardiomyopathy (Liu et al. 2013). The inhibition of mast cell number increments is probably due to downregulation of leukotriene A4 hydrolase protein expression and leukotriene B4 level, which acts as chemoattractant in the recruitment of mast cell uptake into tissue. In addition, H_2S treatment inhibited forskolin-induced renin degranulation mast cell line (HMC1.1) via lowering of intracellular cAMP level (Liu et al. 2013).

6.2.3.3 Atherosclerosis

H_2S level was found to be significantly reduced in either vascular beds or plasma during the development of atherosclerosis. This is probably due to the inhibition of CSE expression and activity (Wu et al. 2006; Meng et al. 2007). In apolipoprotein E knockout ($apoE^{-/-}$) mice, plasma H_2S and aortic H_2S synthesis were decreased. However, CSE mRNA in aorta was found to be elevated, probably due to the existence of a positive compensatory feedback mechanism (Wang et al. 2009).

Exogenously administered H_2S suppressed the development of neointima hyperplasia (Meng et al. 2007), decreased vascular calcium content, calcium overload and alkaline phosphatase activity in calcified vessels (Wu et al. 2006) and reduced atherosclerotic plaque size and improved aortic ultrastructure (Wang et al. 2009). The anti-atherosclerotic effects involve anti-inflammatory (Wang et al. 2009) and anti-apoptotic (Yang et al. 2006) effects on SMCs, cytoprotective effects in ECs (Jeney et al. 2009) and inhibition of LDL modifications and oxidation (Laggner et al. 2007b; Jeney et al. 2009).

6.2.3.4 Hypertension

The role of endogenous H_2S in blood pressure (BP) regulation is still controversial. Pharmacological blockade of endogenous H_2S production with hydroxylamine hydrochloride, a non-specific inhibitor of both CSE and CBS, for 4 weeks failed to influence systolic BP in rats (Lu et al. 2010b). In contrast, Yan et al. found that administration of PAG, an inhibitor of CSE, to rats for 5 weeks significantly elevated blood pressure (Yan et al. 2004). The discrepancy was also observed in CSE-knockout mice. Yang et al. reported that CSE knockouts exhibited pronounced hypertension (Yang et al. 2008), whereas Ishii et al. did not find hypertension in these mice (Ishii et al. 2010).

Plasma level of H_2S and the expression of CSE mRNA was significantly lowered in spontaneously hypertensive rats (SHR) (Yan et al. 2004) and hypoxic pulmonary hypertensive rats (Zhang et al. 2003). These findings suggest that the hypertension in SHR involves a reduction in the production and function of H_2S (Yan et al. 2004).

Treatment with H_2S can significantly lower BP in different hypertensive animal models, such as SHR (Yan et al. 2004), renovascular hypertension (Lu et al. 2010b) and pulmonary hypertension (Zhang et al. 2003). The mechanisms for its anti-hypertensive effects probably implicate the inhibition of renin-angiotensin system (RAS) (Lu et al. 2011), attenuation of vascular remodeling (Zhao et al. 2008) and activation of K_{ATP} channels (Li et al. 2008).

6.3 Physiological and Pathological Function of H_2S in the Central Nervous System

Accumulating evidence establishes that H_2S is a neuromodulator in CNS and regulation of H_2S synthetic system may be a promising therapeutic approach for CNS diseases. Therefore H_2S confers pathophysiological regulatory functions in brain, instead of being a 'mere' environmental toxin.

6.3.1 H_2S Biosynthesis in Brain

Earlier reports from various groups showed high concentrations of H_2S in brain (ranging 50–160 μM) in a variety of mammalian species including rat, bovine, mouse and human (Goodwin et al. 1989; Warenycia et al. 1989a; Savage and Gould 1990). Recent works suggest that the concentration of H_2S in brain may be in the nanomolar range (Furne et al. 2008; Ishigami et al. 2009). Determination of H_2S in biological samples is often influenced by a number of factors such as its instability, high volatility, great susceptibility to oxidation, and release of sulfide out of the commonly used reagent dithiothreitol. Therefore, without a reliable and well-validated method with high sensitivity at the nanomolar range, it is difficult to determine the actual value of H_2S level in the brain.

All the three H_2S biosynthesis enzymes, namely, CBS, CSE and 3-MST, are expressed in the brain. CBS is the primary physiologic source of H_2S in the CNS (Abe and Kimura 1996). CBS protein is predominantly localized in most areas of the brain, especially in hippocampus and cerebellum (Robert et al. 2003). It was found to be preferentially expressed in astrocytes rather than in neurons (Enokido et al. 2005; Lee et al. 2009). CSE is expressed in brain and was found to be predominantly present in neurons. It is critical for maintaining GSH homeostasis in brain (Diwakar and Ravindranath 2007). Furthermore, an intact transsulfuration pathway in the brain mediated by both CBS and CSE links to GSH homeostasis, which greatly contributes to the redox-buffering capacity in brain (Vitvitsky et al. 2006). 3-MST in combination with CAT produces H_2S from L-cysteine (Shibuya et al. 2009). 3-MST is localized to mitochondria and nerve endings. However, the contributions of CBS and 3-MST with respect to H_2S generation under different physiological and pathological conditions are still not clearly understood. Detailed biosynthesis and metabolism of H_2S in CNS was described in a previous publication (Hu et al. 2011b). A novel pathway for the production of H_2S from D-cysteine was recently reported in mammalian cells (Shibuya et al. 2013). Unlike the L-cysteine pathway, this D-cysteine-dependent pathway operates predominantly in the cerebellum. This study presents a novel pathway of H_2S production and provides a new therapeutic approach to deliver H_2S.

6.3.2 Physiological Function of H_2S in Brain

H_2S may serve as a neuromodulator based on the following evidence. H_2S modulates LTP in active synapses. It facilitates the induction of LTP in the presence of a weak tetanic stimulation (Abe and Kimura 1996) and reversibly inhibits both fast and slow synaptic responses in dorsal raphe serotonergic neurons (Kombian et al. 1993). As H_2S upregulates the expression of γ-aminobutyric acid (GABA) B receptor (Han et al. 2005a), it is therefore critical in maintaining the excitatory/inhibitory balance. H_2S also induces astrocytic glutamate uptake (Lu et al. 2008), which removes excessive glutamate from synaptic clefts and maintains normal neurotransmission between neurons. These observations indicate that H_2S plays an important modulatory role in CNS.

Intracellular calcium ($[Ca^{2+}]_i$) is vital in regulating various brain functions. H_2S increases $[Ca^{2+}]_i$ in neurons, astrocytes and microglia (Nagai et al. 2004; Lee et al. 2006; Yong et al. 2010a), therefore plays important regulatory roles in synaptic activity and plasticity, as well as signal transmission between neuron and glial cells. H_2S also regulates intracellular pH (pH_i) in microglia and astrocytes (Lu et al. 2010a). Taken together, these findings suggest that H_2S modulates cell function via changes in ion channel conductance, synaptic transmission as well as gap junctions.

6.3.3 Pathological Functions of H_2S in the Central Nervous System

6.3.3.1 Neuroprotective Effects of H_2S

At micromolar range, H_2S may produce neuroprotective effects via its anti-inflammatory, anti-apoptotic and anti-oxidative actions.

Microglia cells are the resident macrophages of the brain, and thus act as the first and main form of active immune defense in CNS. H_2S inhibits production and release of NO and TNF-α in microglia and astrocytes when these cells are treated with lipopolysaccharide (LPS) (Hu et al. 2007). This is further confirmed by different groups of scientists with different H_2S-releasing compounds (Lee et al. 2010a; Yin et al. 2013). H_2S may exert anti-neuroinflammatory actions via inhibiting the production of pro-inflammatory factors and enhancing the production of anti-inflammatory cytokines. Inhibition of p38/JNK MAPK and NF-κB signalling pathways are recognized as possible mechanisms by which H_2S restrains the extent of neuroinflammation and thereby limits the extent of neuronal injury.

H_2S, by itself, may act as a poor reductant (Kabil and Banerjee 2010). Physiological relevance of the antioxidant properties of H_2S probably rely more on other mechanisms. H_2S stimulates glutamate uptake in astrocytes by enhancing the trafficking of glial glutamate transporter GLT-1 (Lu et al. 2008). The enhanced glutamate uptake lowers extracellular glutamate and relieves the inhibition by glutamate on cystine transportation. This produces the driving force for cystine/glutamate antiporter Xc^- which transports cystine into cells, thereby an increase in intracellular L-cysteine followed by an increase in intracellular GSH. Moreover, H_2S may also increase GSH levels both directly or indirectly (Kimura and Kimura 2004; Whiteman et al. 2005; Umemura and Kimura 2007). These findings are strongly suggestive of the powerful anti-oxidative actions of H_2S in CNS.

H_2S has anti-apoptotic property in neuronal cells. H_2S protects hippocampal neurons against vascular dementia-induced cell apoptosis (Zhang et al. 2009), and inhibits apoptosis of neuronal cells induced by various toxins that are commonly used in establishing in vivo and in vitro models for PD and AD. These toxins include 1-methyl-4-phenylpyridine (MPP^+), 6-hydroxydopamine (6-OHDA), rotenone and β amyloid (Tang et al. 2008; Yin et al. 2009; Tiong et al. 2010). Preservation of mitochondrial integrity is the main mechanism for the anti-apoptotic effects of H_2S (Hu et al. 2009; Yin et al. 2009). H_2S prevents formation and opening of mitochondrial permeability transition pore, the subsequent release of cytochrome c from mitochondria to cytosol and the activation of caspase cascades. H_2S exerts these effects via opening of mitochondrial K_{ATP} channels and suppression of p38-MAPK (Hu et al. 2009).

6.3.3.2 H_2S in CNS Diseases

H_2S at normal level is important in brain physiology. Abnormal H_2S biosynthesis may contribute towards the progression of CNS diseases. Deficiency of CBS in humans, for example, results in higher plasma levels of homocysteine and methionine along with decreased level of L-cysteine. Patients with Alzheimer's Disease (AD) or Parkinson's Disease (PD) commonly show significantly increased homocysteine level in their cerebrospinal fluid (Isobe et al. 2005). This indicates that alterations of H_2S level in brain may contributes to pathophysiology of CNS diseases.

Alzheimer's Disease (AD)

The role of H_2S in AD development is incompletely understood. The level of S-adenosylmethionine, a CBS activator, is largely reduced in the brain of AD patients (Morrison et al. 1996). Furthermore, the serum level of homocysteine, a precursor of L-cysteine, is elevated in AD patients (Clarke et al. 1998). One possible explanation is that the transsulfuration pathway linking homocysteine and GSH metabolism, mediated by CBS and CSE, is disrupted.

There are various pieces of evidence that suggest H_2S treatment is capable of eliciting neuroprotective effects against pathological progression of AD. For example,

H_2S may decrease **β**-site amyloid precursor protein cleaving enzyme 1 (BACE-1) mRNA and protein expression and Aβ1-42 release in PC-12 neuronal cells (Zhang et al. 2011a). In addition, H_2S ameliorates β amyloid-induced damage in microglial (Liu and Bian 2010) and neuronal cells (Tang et al. 2008). Furthermore, H_2S attenuated LPS-induced cognitive impairment through reducing the overproduction of pro-inflammatory mediators via inhibition of NF-κB pathways in rats (Gong et al. 2010). These data imply that H_2S would be beneficial for AD treatment. However, more direct evidence for the potential benefits of H_2S or its donors in AD animal models is lacking at present.

Parkinson's Disease (PD)

The therapeutic effect of H_2S on PD has been well studied by several groups. Endogenous H_2S levels in substantia nigra and striatum were found to be reduced in PD animal models created by 6-OHDA or rotenone (Hu et al. 2010). This suggests that endogenous H_2S is likely to play a role in the development of PD. H_2S treatment was found to inhibit microglial activation in the substantia nigra and inflammation in the striatum. Since neuroinflammation is considered to be a critical factor in the pathogenesis of PD, these findings may suggest a therapeutic effect of H_2S. In separate animal models, H_2S treatment has been shown to inhibit loss of tyrosine hydroxylase positive (TH^+)-neurons in substantia nigra, and progression of movement dysfunction in these PD models was attenuated (Hu et al. 2010; Kida et al. 2010; Lu et al. 2012).

The mechanisms underlying therapeutic effects of H_2S on PD include anti-oxidative stress (Hu et al. 2010; Kida et al. 2010), anti-inflammation (Hu et al. 2010), anti-apoptosis (Hu et al. 2009) and anti-ER stress (Xie et al. 2012). Interestingly, Lu et al. reported that H_2S induced protection in dopaminergic (DA) neurons against neurodegeneration is independent of K_{ATP} activation (Lu et al. 2012), but mediated through a uncoupling protein 2 (UCP2) dependent mechanism. A very recent study demonstrated that H_2S may also induce S-sulfhydration of neuroprotective ubiquitin E3 ligase, parkin, to enhance its catalytic activity. Moreover, Parkin sulfhydration is markedly depleted in the brains of PD patients (Vandiver et al. 2013).

Taken together, these data further confirm that H_2S donors may be of high therapeutic value in the treatment of PD. ACS84 is a hydrogen sulfide-releasing-L-Dopa derivative compound. ACS84 has been found to prevent neurodegeneration via an anti-oxidative mechanism, and shown to have potential therapeutic values against PD (Xie et al. 2013).

Ischemic Stroke

High plasma level of L-cysteine correlates with poor clinical outcome 3 months post stroke in acute stroke patients (Wong et al. 2006). L-cysteine loading increases infarct volume in rats after middle cerebral artery occlusion (MCAO), and this effect can be reversed by inhibition of H_2S synthesis (Wong et al. 2006) and mimicked by exogenous application of H_2S (Qu et al. 2006).

However, under in vitro conditions, H_2S protects neurons against hypoxic injury (Tay et al. 2010; L. et al. 2011; Yin et al. 2013). The protective effects were mediated by anti-inflammatory, anti-oxidative and anti-apoptotic properties of H_2S. The discrepancy in observations could have resulted from differing concentrations of H_2S used in these studies. It is highly likely that physiological level of H_2S exerts a protective effect on cells against insults, such as hypoxia. During stroke, however, over-production of H_2S may facilitate cell death through enhancing excitotoxicity induced by excessive accumulation of extracellular glutamate.

Other CNS Diseases

Down syndrome is the most common chromosomal abnormality in humans. It is typically associated with a delay in cognitive ability (mental retardation, or MR) and physical growth, and a particular set of facial characteristics. High level of thiosulfate – a catabolite of H_2S, was found in the urine of Down syndrome patients (Belardinelli et al. 2001). Overproduction of endogenous H_2S was also found in Down syndrome patients and thus established a correlation between Down syndrome and chronic H_2S poisoning. Excessive H_2S may account for many clinical features of Down syndrome such as MR (Kamoun 2001).

There is an interaction between CBS and Huntington disease. Deficiency of CBS causes homocystinuria, as homocysteine is a substrate of CBS. The plasma homocysteine levels are also reported to be higher in patients with Huntington disease (Boutell et al. 1998; Andrich et al. 2004). Homocysteine is metabolized to homocysteate and homocysteine sulphinate, both are known to be powerful excitotoxic amino acids. It has been suggested that Huntington disease involves the action of excitotoxic amino acids and this interaction with CBS may suggest a mechanism for H_2S in this disorder.

Recurrent febrile seizures (FS) is the most common seizure type in children, often causing hippocampal damage. H_2S treatment may alleviate hippocampal damage induced by recurrent FS whereas inhibition of H_2S synthesis aggravates this damage (Han et al. 2005b). However, in a rat models of recurrent FS, the plasma level of H_2S and expressions of CBS in hippocampus were dramatically increased (Han et al. 2005b, 2006). As a result, the elevated H_2S concentration and CBS expression during recurrent FS may be a compensatory response to suppress neuronal hyperexcitability and thus alleviate neuronal damage in hippocampus.

Repeated exposure to opioids leads to development of addiction dependence, which can be assessed by observing emergence of withdrawal syndromes subsequent to discontinuation of chronic opioid administration or the administration of a competitive opioid antagonist such as naloxone (Maldonado and Koob 1993). Withdrawal-induced symptoms are the main cause to keep drug-dependent individuals craving continued opioids. It was found that exogenous administration of H_2S alleviates morphine and heroin withdrawal symptoms. This was mediated by suppression of supersensitivity of AC/cAMP/p-CREB pathway and modification of the levels of p-NR1, p-NR2A and p-NR2B levels (Jiang et al. 2012; Yang et al. 2013).

6.4 Physiological and Pathological Function of H_2S in the Kidneys

6.4.1 H_2S Synthesis in Kidneys

Endogenous H_2S plays an important role in mediating both glomerular and tubular functions of the kidneys. H_2S synthesizing enzymes are highly expressed in renal tissues, especially in proximal tubules (House et al. 2003; Ishii et al. 2004; Li et al. 2006b). This results in high amount of H_2S production when renal tissues are incubated with L-cysteine, the H_2S synthesizing enzyme substrate. Moreover, blockade of endogenous H_2S production with PAG (a CSE inhibitor) and aminooxyacetic acid (AOAA, a CBS inhibitor) reduced H_2S synthesis completely (Xia et al. 2009), suggesting that both enzymes contribute towards H_2S production in kidneys (Stipanuk and Beck 1982; House et al. 2003). The involvement of 3-MST and CAT in H_2S synthesis has yet to be characterized in renal tissues, hence warrants future research to understand the full picture of H_2S generation in kidneys.

6.4.2 Physiology Function of H_2S in Kidney

When H_2S is exogenously infused into renal artery, vascular activity of kidney such as renal blood flow (RBF), glomerular filtration rate (GFR) and filtration rate (FF) are significantly increased. However, there is no change in mean arteriole blood pressure (MAP), suggesting that H_2S may produce greater vasodilation in preglomerular arterioles than in postglomerular arterioles (Xia et al. 2009). H_2S also increased urine flow rate (U.V), urinary Na^+ and K^+ excretion (U_{Na}.V; U_K.V), fractional excretion of Na^+ and K^+ (FE_{Na}, FE_K) (Xia et al. 2009), suggesting that H_2S infusion altered renal tubular function (Xia et al. 2009).

Consistent effects on renal hemodynamics and excretory functions were observed when L-cysteine was infused into renal artery. These effects could be abolished by a combination of PAG and AOAA, but not by either of these alone. These observations bespeak physiological importance of endogenous H_2S, produced by CBS and CSE, in the basal regulation of renal filtration and tubular functions (Xia et al. 2009).

6.4.3 H_2S and Renal Ischemic Injury

6.4.3.1 Endogenous H_2S Production in Renal I/R

The effect of I/R on the level of endogenous H_2S is unclear due to controversies in research findings. Xu et al. reported that renal and plasma H_2S level in rats subjected to unilateral renal occlusion were significantly decreased (Xu et al. 2009), an effect due to reduced CBS activity in the ischemic kidney (Prathapasinghe et al. 2008;

Xu et al. 2009). Wu et al. also noticed a significant decrease in CBS enzyme activity during renal I/R, and they postulated that this effect was underlied by a decrease in Sp1 transcriptional activity (Wu et al. 2010).

CSE activity in the kidneys, however, was reported to be unaffected by Xu et al. (2009). This is in direct contrast to Tripatara et al.'s findings which suggest that renal H_2S production rate and plasma H_2S concentration were markedly elevated in mice subjected to bilateral renal occlusion due to an upregulation of CSE expression (Tripatara et al. 2009).

Nevertheless, both groups believed that endogenous H_2S protects against I/R injury. Xu et al. postulated that H_2S production by CBS is compromised in the kidneys during renal I/R, and the resultant reduction in H_2S leads to renal injuries (Xu et al. 2009). Tripatara et al., on the other hand, proposed that the elevated CSE activity and endogenous H_2S level act as a defensive mechanism against I/R induced injuries (Tripatara et al. 2008; Liu et al. 2011b).

6.4.3.2 Protective Effects of H_2S Against Renal Ischemia Injury

H_2S protects against I/R-induced renal injury, reperfusion injury, glomerular dysfunction and tubular dysfunction (Tripatara et al. 2009; Xu et al. 2009). H_2S administration also decreased elevated FE_{Na} during I/R, but had no significant effect on urine flow (Tripatara et al. 2009). In a large animal model of non-heart beating donor kidneys, H_2S protected the kidneys against I/R injuries, probably through improvements in RBF and decrease in intrarenal resistance (IRR) of kidneys (Hosgood and Nicholson 2010).

Renal I/R leads to both necrotic and apoptotic forms of cell death (Prathapasinghe et al. 2007). Rats that were subjected to I/R displayed severe acute tubular damage. Treatment with H_2S markedly reduced these histological signs and histological score for acute tubular necrosis (ATN), indicating that H_2S protects against I/R induced structural injuries (Tripatara et al. 2008).

The protective effects of H_2S involve its anti-oxidant effects. H_2S significantly reduce urinary 8-isoprostane, indicative of reduced extent of lipid peroxidation (Hosgood and Nicholson 2010). H_2S also reduces MDA level in the kidney as compared to I/R model rats (Xu et al. 2009), indicating that H_2S provides protective effect against IR-induced lipid peroxidation.

H_2S also significantly reduced the number of TUNEL-positive cells in renal tissues subjected to I/R (Xu et al. 2009). Furthermore, H_2S injection reduced the number of propidium iodide-positive cells, an index for necrotic cells, in the kidney tissues of rats subjected to I/R (Xu et al. 2009). Administration of H_2S into kidneys subjected to renal I/R also prevented caspase-3 activation (Bos et al. 2009). Tripatara et al. showed that NaHS administration attenuated I/R-induced Bid translocation and activation, which prevented I/R induced decrease in Bcl-2 protein levels (Tripatara et al. 2008). On the contrary, Bos et al. failed to observe change in Bcl-2 mRNA expression among treatment groups. Instead, they showed that H_2S

pretreatment decreased IR-induced elevation of Bax (Bos et al. 2009). These data suggest that H_2S produces anti-apoptotic effects to against ischemic injury.

In addition, H_2S also produced anti-inflammatory effect. H_2S lowered total NO level, a marker for tubular cell inflammation, in urine of pigs subjected to I/R (Hosgood and Nicholson 2010). Triparata et al. reported that H_2S attenuates NF-κB activation and expression of its dependent proteins, iNOS, COX-2 and ICAM-1 in the kidneys. These results clearly demonstrated the anti-inflammatory effects of H_2S in renal I/R model (Tripatara et al. 2008). Immunohistochemical staining of inflammatory components were assessed by Bos et al. H_2S pretreatment, but not post-treatment, significantly reduced the influx of Mac-1 (present on macrophages, monocytes, granulocytes and natural killer cells) and Ly-6G-positive cells (expressed on mature granulocytes) (Bos et al. 2009).

6.4.4 Role of H_2S in Other Renal Diseases

6.4.4.1 Renovascular Hypertension

In a renovascular hypertensive model established by 2-kidneys-1-clip (2K1C), Lu et al. found that H_2S exerted antihypertensive effects via inhibition of plasma renin activity and Ang II production in plasma (Lu et al. 2010b). This effect was underlied by downregulation of elevated cAMP by H_2S in kidney tissue (Lu et al. 2010b). In fact, using primary cultured renin-rich kidney cells, H_2S has also been proven to inhibit renin release by decreasing intracellular cAMP levels. Interestingly, H_2S was reported to inhibit angiotensin-converting enzyme (ACE) activity in human umbilical vein endothelial cells (HUVECs) (Laggner et al. 2007a). Lu and colleagues, however, observed no such inhibitory effect of H_2S on ACE activity of rat aortic endothelial cells (Lu et al. 2010b).

6.4.4.2 Chronic Kidney Disease (CKD)

In an experiment conducted on humans, Perna et al. found that endogenous production of H_2S was lowered in uraemic patients due to downregulation of CSE (Perna et al. 2009). Interestingly, 3-MST was found to be upregulated despite an overall decrease in plasma H_2S concentration, suggesting a predominant role of CSE in producing H_2S.

CBS heterozygous (CBS +/−) mice and/or uninephrectomy (1-K) were used as models of HHcy-associated end stage renal failure (Sen et al. 2009). H_2S supplementation prevents apoptosis of glomerular cells, macrophage infiltration, excessive superoxide production and decrease in glutathione (GSH) -to-oxidized glutationine (GSSG) ratio of CBS (+/−) 2-K, 1-K mice and WT 1-K mice. H_2S treatment also rectifies the expressions of desmin, nephrin, pro- and active forms of matrix metalloproteinase (MMP) −2 and −9, collagen IV, NAD(P)H oxidase p47 ^{phox} subunit, inflammatory molecules ICAM-1 and VCAM-1 (Sen et al. 2009, 2010).

6.5 Physiological and Pathological Functions of H_2S in Other Systems

6.5.1 Gastrointestinal System

In an in vivo model of hepatic I/R injury, exogenously applied H_2S significantly reduced elevations in serum alanine aminotransferase (Jha et al. 2008), suggesting that H_2S protects against liver injury.

H_2S also protects gastric mucosa against injury caused by NSAIDs or ischemic injury. This effect was mediated by suppression of leukocyte adherence (Fiorucci et al. 2005) and stimulation of production of antioxidant enzymes like SOD-1 and GSH (Liu et al. 2012a; Cui et al. 2013). In addition, H_2S may also improve blood flow to the injured gastric mucosal (Fiorucci et al. 2005; Henderson et al. 2010; Liu et al. 2012a).

6.5.2 Lungs

In a pulmonary I/R model, pretreatment of isolated rat lung with H_2S attenuated I/R injury, indicated by improvements in lung histological change, perfusion flow rate, ratio of lung wet weight to dry weight and lung compliance (Fu et al. 2008). In separate experiments, perfusion of lungs with PAG, a CSE inhibitor, showed aggravated lung I/R injury (Fu et al. 2008). Therefore, it is likely that endogenous H_2S is involved in the pathogenesis of lung I/R injury, whereas exogenous H_2S may be of clinical benefit to lung I/R injury.

6.5.3 Bones

Using an in vitro osteoblastic cell system, Xu et al. demonstrated that H_2S may be of potential therapeutic value for treatment against osteoporosis (Xu et al. 2011). In osteoblastic cells treated with H_2O_2, H_2S treatment stimulated osteoblast proliferation by enhancing both transcription and activity of alkaline phosphatase, and stimulated the transcriptional level of osteocalcin, the main bone matrix protein, and protein expression of collagen, a major constituent of bone tissue. These effects were mediated by antioxidant and anti-inflammatory effects of H_2S via a MAPK (p38 and ERK1/2)-dependent mechanism (Xu et al. 2011).

6.5.4 Diabetes

CSE/H_2S system plays an important role in regulating β-cell functions (Yang et al. 2011). H_2S can promote glucose uptake and produce insulin-sensitizing effects in type 2 diabetes (Xue et al. 2013). In vitro experiments demonstrated that H_2S enhances glucose uptake in both myotubes and adipocytes. These effects are mediated by upregulated phosphorylation of insulin receptors, PI3K and Akt (Xue et al. 2013). In Goto-Kakizaki diabetic rats, chronic H_2S treatment decreased fasting blood glucose, increased insulin sensitivity, and increased glucose tolerance with increased phosphorylation of PI3K and Akt in muscles (Xue et al. 2013). Henceforth, H_2S holds promise as a new therapeutic drug against insulin resistance. The role of H_2S in the pathogenesis of diabetes mellitus has been extensively discussed in a recent review (Szabo 2012).

6.6 Concluding Remarks

In summary, H_2S acts as a gaseous modulator in many mammalian tissues and is likely to be involved in the pathogenesis of many diseases. Knowledge of H_2S biology in mammalian systems raises the possibility of manipulating H_2S system for therapeutic benefits to patients suffering from various diseases. The narrow therapeutic window of H_2S, however, continues to pose as a major challenge for its widespread utilization as a therapeutic drug despite its potent and beneficial effects seen in many systems and diseased conditions. Slow-releasing H_2S donors that mimic endogenous H_2S synthesis release are therefore urgently sought. Research efforts in recent years has focused on the development of a variety of such novel donors, including GYY4137, S-diclofenac and S-dopa (Li et al. 2007, 2008; Xie et al. 2013). Moreover, currently available H_2S biosynthesis inhibitors, such as AOAA and hydroxylamine, are nonspecific for their actions. More potent and specific inhibitors are deemed to be developed for us to have a better understanding of endogenous H_2S functions.

At present, most of these slow-releasing H_2S donors and H_2S biosynthesis inhibitors have poor solubility in physiological medium, and the adverse side effects of most novel compounds yet to be fully explored. There is still a long way to go before H_2S-releasing compound can be exploited for clinical usage; the biology of H_2S still beholds lots of mysteries and excitements waiting to be unveiled.

Acknowledgements This work was supported by research grants from Singapore National Medical Research Council (NMRC 1183 ⁄ 2008, 1219/2009) and Singapore Kidney foundation (NKFRC/2009/01/10).

References

Abe K, Kimura H (1996) The possible role of hydrogen sulfide as an endogenous neuromodulator. J Neurosci 16(3):1066–1071

Abramochkin DV, Moiseenko LS et al (2009) The effect of hydrogen sulfide on electrical activity of rat atrial myocardium. Bull Exp Biol Med 147(6):683–686

Ali MY, Ping CY et al (2006) Regulation of vascular nitric oxide in vitro and in vivo; a new role for endogenous hydrogen sulphide? Br J Pharmacol 149(6):625–634

Alves MG, Soares AF et al (2011) Sodium hydrosulfide improves the protective potential of the cardioplegic histidine buffer solution. Eur J Pharmacol 654(1):60–67

Andrich J, Saft C et al (2004) Hyperhomocysteinaemia in treated patients with Huntington's disease homocysteine in HD. Mov Disord 19(2):226–228

Belardinelli MC, Chabli A et al (2001) Urinary sulfur compounds in Down syndrome. Clin Chem 47(8):1500–1501

Bethell HW, Vandenberg JI et al (1998) Changes in ventricular repolarization during acidosis and low-flow ischemia. Am J Physiol 275(2 Pt 2):H551–H561

Bian JS, Yong QC et al (2006) Role of hydrogen sulfide in the cardioprotection caused by ischemic preconditioning in the rat heart and cardiac myocytes. J Pharmacol Exp Ther 316(2):670–678

Bliksoen M, Kaljusto ML et al (2008) Effects of hydrogen sulphide on ischaemia-reperfusion injury and ischaemic preconditioning in the isolated, perfused rat heart. Eur J Cardiothorac Surg 34(2):344–349

Bos EM, Leuvenink HG et al (2009) Hydrogen sulfide-induced hypometabolism prevents renal ischemia/reperfusion injury. J Am Soc Nephrol 20(9):1901–1905

Boutell JM, Wood JD et al (1998) Huntingtin interacts with cystathionine beta-synthase. Hum Mol Genet 7(3):371–378

Bucci M, Papapetropoulos A et al (2010) Hydrogen sulfide is an endogenous inhibitor of phosphodiesterase activity. Arterioscler Thromb Vasc Biol 30(10):1998–2004

Cai WJ, Wang MJ et al (2007) The novel proangiogenic effect of hydrogen sulfide is dependent on Akt phosphorylation. Cardiovasc Res 76(1):29–40

Calvert JW, Jha S et al (2009) Hydrogen sulfide mediates cardioprotection through Nrf2 signaling. Circ Res 105(4):365–374

Calvert JW, Elston M et al (2010) Genetic and pharmacologic hydrogen sulfide therapy attenuates ischemia-induced heart failure in mice. Circulation 122(1):11–19

Chen P, Poddar R et al (1999) Homocysteine metabolism in cardiovascular cells and tissues: implications for hyperhomocysteinemia and cardiovascular disease. Adv Enzyme Regul 39:93–109

Cheng Y, Ndisang JF et al (2004) Hydrogen sulfide-induced relaxation of resistance mesenteric artery beds of rats. Am J Physiol Heart Circ Physiol 287(5):H2316–H2323

Clarke R, Smith AD et al (1998) Folate, vitamin B12, and serum total homocysteine levels in confirmed Alzheimer disease. Arch Neurol 55(11):1449–1455

Cuevas J, Bassett AL et al (1991) Effect of H+ on ATP-regulated K+ channels in feline ventricular myocytes. Am J Physiol 261(3 Pt 2):H755–H761

Cui J, Liu L et al (2013) Protective effect of endogenous hydrogen sulfide against oxidative stress in gastric ischemia-reperfusion injury. Exp Ther Med 5(3):689–694

Diwakar L, Ravindranath V (2007) Inhibition of cystathionine-gamma-lyase leads to loss of glutathione and aggravation of mitochondrial dysfunction mediated by excitatory amino acid in the CNS. Neurochem Int 50(2):418–426

Dombkowski RA, Russell MJ et al (2004) Hydrogen sulfide as an endogenous regulator of vascular smooth muscle tone in trout. Am J Physiol Regul Integr Comp Physiol 286(4): R678–R685

Elrod JW, Calvert JW et al (2007) Hydrogen sulfide attenuates myocardial ischemia-reperfusion injury by preservation of mitochondrial function. Proc Natl Acad Sci U S A 104(39):15560–15565

Enokido Y, Suzuki E et al (2005) Cystathionine beta-synthase, a key enzyme for homocysteine metabolism, is preferentially expressed in the radial glia/astrocyte lineage of developing mouse CNS. FASEB J 19(13):1854–1856

Filipovic MR, Eberhardt M et al (2013) Beyond H(2)S and NO Interplay: hydrogen sulfide and nitroprusside react directly to give nitroxyl (HNO). A new pharmacological source of HNO. J Med Chem 56(4):1499–1508

Fiorucci S, Antonelli E et al (2005) Inhibition of hydrogen sulfide generation contributes to gastric injury caused by anti-inflammatory nonsteroidal drugs. Gastroenterology 129(4):1210–1224

Fu Z, Liu X et al (2008) Hydrogen sulfide protects rat lung from ischemia-reperfusion injury. Life Sci 82(23–24):1196–1202

Fukuto JM, Collins MD (2007) Interactive endogenous small molecule (gaseous) signaling: implications for teratogenesis. Curr Pharm Des 13(29):2952–2978

Furne J, Saeed A et al (2008) Whole tissue hydrogen sulfide concentrations are orders of magnitude lower than presently accepted values. Am J Physiol Regul Integr Comp Physiol 295(5):R1479–R1485

Garcia-Bereguiain MA, Samhan-Arias AK et al (2008) Hydrogen sulfide raises cytosolic calcium in neurons through activation of L-type Ca^{2+} channels. Antioxid Redox Signal 10(1):31–42

Geng B, Chang L et al (2004a) Endogenous hydrogen sulfide regulation of myocardial injury induced by isoproterenol. Biochem Biophys Res Commun 318(3):756–763

Geng B, Yang J et al (2004b) H_2S generated by heart in rat and its effects on cardiac function. Biochem Biophys Res Commun 313(2):362–368

Giggenbach W (1971) Optical spectra of highly alkaline sulfide solutions and the second dissociation constant of hydrogen sulfide. Inorg Chem 10:1333–1338

Givvimani S, Munjal C et al (2011) Hydrogen sulfide mitigates transition from compensatory hypertrophy to heart failure. J Appl Physiol 110(4):1093–1100

Gong QH, Wang Q et al (2010) Hydrogen sulfide attenuates lipopolysaccharide-induced cognitive impairment: a pro-inflammatory pathway in rats. Pharmacol Biochem Behav 96(1):52–58

Goodwin LR, Francom D et al (1989) Determination of sulfide in brain tissue by gas dialysis/ion chromatography: postmortem studies and two case reports. J Anal Toxicol 13(2):105–109

Han Y, Qin J et al (2005a) Modulating effect of hydrogen sulfide on gamma-aminobutyric acid B receptor in recurrent febrile seizures in rats. Neurosci Res 53(2):216–219

Han Y, Qin J et al (2005b) Hydrogen sulfide may improve the hippocampal damage induced by recurrent febrile seizures in rats. Biochem Biophys Res Commun 327(2):431–436

Han Y, Qin J et al (2006) Hydrogen sulfide and carbon monoxide are in synergy with each other in the pathogenesis of recurrent febrile seizures. Cell Mol Neurobiol 26(1):101–107

Henderson PW, Weinstein AL et al (2010) Hydrogen sulfide attenuates ischemia-reperfusion injury in in vitro and in vivo models of intestine free tissue transfer. Plast Reconstr Surg 125(6):1670–1678

Hosgood SA, Nicholson ML (2010) Hydrogen sulphide ameliorates ischaemia-reperfusion injury in an experimental model of non-heart-beating donor kidney transplantation. Br J Surg 97(2):202–209

Hosoki R, Matsuki N et al (1997) The possible role of hydrogen sulfide as an endogenous smooth muscle relaxant in synergy with nitric oxide. Biochem Biophys Res Commun 237(3):527–531

House JD, O'Connor CP et al (2003) Plasma homocysteine and glycine are sensitive indices of folate status in a rodent model of folate depletion and repletion. J Agric Food Chem 51(15):4461–4467

Hu LF, Wong PT et al (2007) Hydrogen sulfide attenuates lipopolysaccharide-induced inflammation by inhibition of p38 mitogen-activated protein kinase in microglia. J Neurochem 100(4):1121–1128

Hu LF, Pan TT et al (2008a) Cyclooxygenase-2 mediates the delayed cardioprotection induced by hydrogen sulfide preconditioning in isolated rat cardiomyocytes. Pflugers Arch 455(6):971–978

Hu Y, Chen X et al (2008b) Cardioprotection induced by hydrogen sulfide preconditioning involves activation of ERK and PI3K/Akt pathways. Pflugers Arch 455(4):607–616

Hu LF, Lu M et al (2009) Hydrogen sulfide inhibits rotenone-induced apoptosis via preservation of mitochondrial function. Mol Pharmacol 75(1):27–34

Hu LF, Lu M et al (2010) Neuroprotective effects of hydrogen sulfide on Parkinson's disease rat models. Aging Cell 9(2):135–146

Hu LF, Li Y et al (2011a) Hydrogen sulfide regulates Na+/H+ exchanger activity via stimulation of Phosphoinositide 3-kinase/Akt and protein kinase G pathways. J Pharmacol Exp Ther 339(2): 726–735

Hu LF, Lu M et al (2011b) Hydrogen sulfide: neurophysiology and neuropathology. Antioxid Redox Signal 15(2):405–419

Hvitved-Jacobsen T (2002) Sewer processes: microbial and chemical process engineering of sewer networks, vol 70. CRC Press, Boca Raton

Ishigami M, Hiraki K et al (2009) A source of hydrogen sulfide and a mechanism of its release in the brain. Antioxid Redox Signal 11(2):205–214

Ishii I, Akahoshi N et al (2004) Murine cystathionine gamma-lyase: complete cDNA and genomic sequences, promoter activity, tissue distribution and developmental expression. Biochem J 381 (Pt 1):113–123

Ishii I, Akahoshi N et al (2010) Cystathionine gamma-Lyase-deficient mice require dietary cysteine to protect against acute lethal myopathy and oxidative injury. J Biol Chem 285(34):26358–26368

Isobe C, Murata T et al (2005) Increase of total homocysteine concentration in cerebrospinal fluid in patients with Alzheimer's disease and Parkinson's disease. Life Sci 77(15):1836–1843

Jeney V, Komodi E et al (2009) Supression of hemin-mediated oxidation of low-density lipoprotein and subsequent endothelial reactions by hydrogen sulfide (H(2)S). Free Radic Biol Med 46(5):616–623

Jha S, Calvert JW et al (2008) Hydrogen sulfide attenuates hepatic ischemia-reperfusion injury: role of antioxidant and antiapoptotic signaling. Am J Physiol Heart Circ Physiol 295(2): H801–H806

Jiang HL, Wu HC et al (2005) Changes of the new gaseous transmitter H_2S in patients with coronary heart disease. Di Yi Jun Yi Da Xue Xue Bao 25(8):951–954

Jiang B, Tang G et al (2010) Molecular mechanism for H(2)S-induced activation of K(ATP) channels. Antioxid Redox Signal 12(10):1167–1178

Jiang LH, Wang J et al (2012) Exogenous sodium hydrosulfide can attenuate naloxone-precipitated withdrawal syndromes and affect cAMP signaling pathway in heroin-dependent rat's nucleus accumbens. Eur Rev Med Pharmacol Sci 16(14):1974–1982

Johansen D, Ytrehus K et al (2006) Exogenous hydrogen sulfide (H_2S) protects against regional myocardial ischemia-reperfusion injury–Evidence for a role of K ATP channels. Basic Res Cardiol 101(1):53–60

Kabil O, Banerjee R (2010) The redox biochemistry of hydrogen sulfide. J Biol Chem 285(29): 21903–21907

Kajimura M, Fukuda R et al (2010) Interactions of multiple gas-transducing systems: hallmarks and uncertainties of CO, NO, and H_2S gas biology. Antioxid Redox Signal 13(2):157–192

Kamoun P (2001) Mental retardation in Down syndrome: a hydrogen sulfide hypothesis. Med Hypotheses 57(3):389–392

Kashiba M, Kajimura M et al (2002) From O_2 to H_2S: a landscape view of gas biology. Keio J Med 51(1):1–10

Kida K, Yamada M et al (2010) Inhaled hydrogen sulfide prevents neurodegeneration and movement disorder in a mouse model of Parkinson's disease. Antioxid Redox Signal 15:343–352

Kimura Y, Kimura H (2004) Hydrogen sulfide protects neurons from oxidative stress. FASEB J 18(10):1165–1167

Kiss L, Deitch EA et al (2008) Hydrogen sulfide decreases adenosine triphosphate levels in aortic rings and leads to vasorelaxation via metabolic inhibition. Life Sci 83(17–18):589–594

Kombian SB, Reiffenstein RJ et al (1993) The actions of hydrogen sulfide on dorsal raphe serotonergic neurons in vitro. J Neurophysiol 70(1):81–96

Koyano T, Kakei M et al (1993) ATP-regulated K+ channels are modulated by intracellular H+ in guinea-pig ventricular cells. J Physiol 463:747–766

Kubo S, Doe I et al (2007) Direct inhibition of endothelial nitric oxide synthase by hydrogen sulfide: contribution to dual modulation of vascular tension. Toxicology 232(1–2):138–146

Laggner H, Hermann M et al (2007a) The novel gaseous vasorelaxant hydrogen sulfide inhibits angiotensin-converting enzyme activity of endothelial cells. J Hypertens 25(10):2100–2104

Laggner H, Muellner MK et al (2007b) Hydrogen sulphide: a novel physiological inhibitor of LDL atherogenic modification by HOCl. Free Radic Res 41(7):741–747

Lee SW, Hu YS et al (2006) Hydrogen sulphide regulates calcium homeostasis in microglial cells. Glia 54(2):116–124

Lee SW, Cheng Y et al (2007) Hydrogen sulphide regulates intracellular pH in vascular smooth muscle cells. Biochem Biophys Res Commun 358(4):1142–1147

Lee M, Schwab C et al (2009) Astrocytes produce the antiinflammatory and neuroprotective agent hydrogen sulfide. Neurobiol Aging 30(10):1523–1534

Lee M, Sparatore A et al (2010a) Hydrogen sulfide-releasing NSAIDs attenuate neuroinflammation induced by microglial and astrocytic activation. Glia 58(1):103–113

Lee M, Tazzari V et al (2010b) Effects of hydrogen sulfide-releasing L-DOPA derivatives on glial activation: potential for treating Parkinson disease. J Biol Chem 285(23):17318–17328

Levitt MD, Abdel-Rehim MS et al (2011) Free and acid-labile hydrogen sulfide concentrations in mouse tissues: anomalously high free hydrogen sulfide in aortic tissue. Antioxid Redox Signal 15(2):373–378

Li L, Bhatia M et al (2006a) Hydrogen sulphide–a novel mediator of inflammation? Curr Opin Pharmacol 6(2):125–129

Li N, Chen L et al (2006b) Hyperhomocysteinemia associated with decreased renal transsulfuration activity in Dahl S rats. Hypertension 47(6):1094–1100

Li L, Rossoni G et al (2007) Anti-inflammatory and gastrointestinal effects of a novel diclofenac derivative. Free Radic Biol Med 42(5):706–719

Li L, Whiteman M et al (2008) Characterization of a novel, water-soluble hydrogen sulfide-releasing molecule (GYY4137): new insights into the biology of hydrogen sulfide. Circulation 117(18):2351–2360

Li L, Hsu A et al (2009) Actions and interactions of nitric oxide, carbon monoxide and hydrogen sulphide in the cardiovascular system and in inflammation–a tale of three gases! Pharmacol Ther 123(3):386–400

Li Z, Wang Y et al (2011) Protective effects of exogenous hydrogen sulfide on neurons of hippocampus in a rat model of brain ischemia. Neurochem Res 36(10):1840–1849

Lim JJ, Liu YH et al (2008) Vasoconstrictive effect of hydrogen sulfide involves downregulation of cAMP in vascular smooth muscle cells. Am J Physiol Cell Physiol 295(5):C1261–C1270

Liu YY, Bian JS (2010) Hydrogen sulfide protects amyloid-beta induced cell toxicity in microglia. J Alzheimers Dis 22(4):1189–1200

Liu X, Pan L et al (2010) Hypoxia-inducible factor-1alpha is involved in the pro-angiogenic effect of hydrogen sulfide under hypoxic stress. Biol Pharm Bull 33(9):1550–1554

Liu J, Hao DD et al (2011a) Hydrogen sulphide inhibits cardiomyocyte hypertrophy by up-regulating miR-133a. Biochem Biophys Res Commun 413(2):342–347

Liu YH, Lu M et al (2011b) Hydrogen sulfide and renal ischemia. Expert Rev Clin Pharmacol 4(1):49–61

Liu L, Cui J et al (2012a) H(2)S-releasing aspirin protects against aspirin-induced gastric injury via reducing oxidative stress. PLoS One 7(9):e46301

Liu YH, Lu M et al (2012b) Hydrogen sulfide in the mammalian cardiovascular system. Antioxid Redox Signal 17(1):141–185

Liu, YH, Lu M et al (2013) Hydrogen sulfide prevents heart failure development via inhibition of renin release from mast cells in the isoproterenol treated rats. Antioxid Redox Signal

Lu M, Hu LF et al (2008) Hydrogen sulfide protects astrocytes against H(2)O(2)-induced neural injury via enhancing glutamate uptake. Free Radic Biol Med 45(12):1705–1713

Lu M, Choo CH et al (2010a) Hydrogen sulfide regulates intracellular pH in rat primary cultured glia cells. Neurosci Res 66(1):92–98

Lu M, Liu YH et al (2010b) Hydrogen sulfide inhibits plasma renin activity. J Am Soc Nephrol 21(6):993–1002

Lu M, Liu YH et al (2011) Hydrogen sulfide regulates cAMP homeostasis and renin degranulation in As4.1 and rat renin-rich kidney cells. Am J Physiol Cell Physiol 302(1):C59–66

Lu M, Zhao FF et al (2012) The neuroprotection of hydrogen sulfide against MPTP-induced dopaminergic neuron degeneration involves uncoupling protein 2 rather than ATP-sensitive potassium channels. Antioxid Redox Signal 17(6):849–859

Maldonado R, Koob GF (1993) Destruction of the locus coeruleus decreases physical signs of opiate withdrawal. Brain Res 605(1):128–138

Martelli A, Testai L et al (2013) Vasorelaxation by hydrogen sulphide involves activation of K(v)7 potassium channels. Pharmacol Res 70(1):27–34

Meng QH, Yang G et al (2007) Protective effect of hydrogen sulfide on balloon injury-induced neointima hyperplasia in rat carotid arteries. Am J Pathol 170(4):1406–1414

Minamishima S, Bougaki M et al (2009) Hydrogen sulfide improves survival after cardiac arrest and cardiopulmonary resuscitation via a nitric oxide synthase 3-dependent mechanism in mice. Circulation 120(10):888–896

Mishra PK, Tyagi N et al (2010) H2S ameliorates oxidative and proteolytic stresses and protects the heart against adverse remodeling in chronic heart failure. Am J Physiol Heart Circ Physiol 298(2):H451–H456

Morrison LD, Smith DD et al (1996) Brain S-adenosylmethionine levels are severely decreased in Alzheimer's disease. J Neurochem 67(3):1328–1331

Mustafa AK, Sikka G et al (2011) Hydrogen sulfide as endothelium-derived hyperpolarizing factor sulfhydrates potassium channels. Circ Res 109(11):1259–1268

Nagai Y, Tsugane M et al (2004) Hydrogen sulfide induces calcium waves in astrocytes. FASEB J 18(3):557–559

Olson KR, Donald JA (2009) Nervous control of circulation–the role of gasotransmitters, NO, CO, and H_2S. Acta Histochem 111(3):244–256

Osipov RM, Robich MP et al (2009) Effect of hydrogen sulfide in a porcine model of myocardial ischemia-reperfusion: comparison of different administration regimens and characterization of the cellular mechanisms of protection. J Cardiovasc Pharmacol 54(4):287–297

Pan TT, Feng ZN et al (2006) Endogenous hydrogen sulfide contributes to the cardioprotection by metabolic inhibition preconditioning in the rat ventricular myocytes. J Mol Cell Cardiol 40(1):119–130

Pan TT, Neo KL et al (2008) H2S preconditioning-induced PKC activation regulates intracellular calcium handling in rat cardiomyocytes. Am J Physiol Cell Physiol 294(1):C169–C177

Pan TT, Chen YQ et al (2009) All in the timing: a comparison between the cardioprotection induced by H_2S preconditioning and post-infarction treatment. Eur J Pharmacol 616(1–3):160–165

Papapetropoulos A, Pyriochou A et al (2009) Hydrogen sulfide is an endogenous stimulator of angiogenesis. Proc Natl Acad Sci U S A 106(51):21972–21977

Perna AF, Luciano MG et al (2009) Hydrogen sulphide-generating pathways in haemodialysis patients: a study on relevant metabolites and transcriptional regulation of genes encoding for key enzymes. Nephrol Dial Transplant 24(12):3756–3763

Prathapasinghe GA, Siow YL et al (2007) Detrimental role of homocysteine in renal ischemia-reperfusion injury. Am J Physiol Renal Physiol 292(5):F1354–F1363

Prathapasinghe GA, Siow YL et al (2008) Inhibition of cystathionine-beta-synthase activity during renal ischemia-reperfusion: role of pH and nitric oxide. Am J Physiol Renal Physiol 295(4): F912–F922

Qu K, Chen CP et al (2006) Hydrogen sulfide is a mediator of cerebral ischemic damage. Stroke 37(3):889–893

Robert K, Vialard F et al (2003) Expression of the cystathionine beta synthase (CBS) gene during mouse development and immunolocalization in adult brain. J Histochem Cytochem 51(3):363–371

Rona G, Chappel CI et al (1959) An infarct-like myocardial lesion and other toxic manifestations produced by isoproterenol in the rat. AMA Arch Pathol 67(4):443–455

Savage JC, Gould DH (1990) Determination of sulfide in brain tissue and rumen fluid by ion-interaction reversed-phase high-performance liquid chromatography. J Chromatogr 526(2):540–545

Sen U, Vacek TP et al (2008) Cardioprotective role of sodium thiosulfate on chronic heart failure by modulating endogenous H_2S generation. Pharmacology 82(3):201–213

Sen U, Basu P et al (2009) Hydrogen sulfide ameliorates hyperhomocysteinemia-associated chronic renal failure. Am J Physiol Renal Physiol 297(2):F410–F419

Sen U, Munjal C et al (2010) Hydrogen sulfide regulates homocysteine-mediated glomerulosclerosis. Am J Nephrol 31(5):442–455

Shi YX, Chen Y et al (2007) Chronic sodium hydrosulfide treatment decreases medial thickening of intramyocardial coronary arterioles, interstitial fibrosis, and ROS production in spontaneously hypertensive rats. Am J Physiol Heart Circ Physiol 293(4):H2093–H2100

Shibuya N, Tanaka M et al (2009) 3-Mercaptopyruvate sulfurtransferase produces hydrogen sulfide and bound sulfane sulfur in the brain. Antioxid Redox Signal 11(4):703–714

Shibuya N, Koike S et al (2013) A novel pathway for the production of hydrogen sulfide from D-cysteine in mammalian cells. Nat Commun 4:1366

Sidhapuriwala J, Li L et al (2007) Effect of S-diclofenac, a novel hydrogen sulfide releasing derivative, on carrageenan-induced hindpaw oedema formation in the rat. Eur J Pharmacol 569 (1–2):149–154

Sivarajah A, McDonald MC et al (2006) The production of hydrogen sulfide limits myocardial ischemia and reperfusion injury and contributes to the cardioprotective effects of preconditioning with endotoxin, but not ischemia in the rat. Shock 26(2):154–161

Sodha NR, Clements RT et al (2008) The effects of therapeutic sulfide on myocardial apoptosis in response to ischemia-reperfusion injury. Eur J Cardiothorac Surg 33(5):906–913

Sodha NR, Clements RT et al (2009) Hydrogen sulfide therapy attenuates the inflammatory response in a porcine model of myocardial ischemia/reperfusion injury. J Thorac Cardiovasc Surg 138(4):977–984

Sojitra B, Bulani Y et al (2011) Nitric oxide synthase inhibition abrogates hydrogen sulfide-induced cardioprotection in mice. Mol Cell Biochem 360(1–2):61–69

Stipanuk MH, Beck PW (1982) Characterization of the enzymic capacity for cysteine desulphhydration in liver and kidney of the rat. Biochem J 206(2):267–277

Su YW, Liang C et al (2009) Hydrogen sulfide regulates cardiac function and structure in adriamycin-induced cardiomyopathy. Circ J 73(4):741–749

Sun YG, Cao YX et al (2008) Hydrogen sulphide is an inhibitor of L-type calcium channels and mechanical contraction in rat cardiomyocytes. Cardiovasc Res 79(4):632–641

Szabo C (2007) Hydrogen sulphide and its therapeutic potential. Nat Rev Drug Discov 6(11):917–935

Szabo C (2012) Roles of hydrogen sulfide in the pathogenesis of diabetes mellitus and its complications. Antioxid Redox Signal 17(1):68–80

Szabo G, Veres G et al (2011) Cardioprotective effects of hydrogen sulfide. Nitric Oxide 25(2):201–210

Tang G, Wu L et al (2005) Direct stimulation of K(ATP) channels by exogenous and endogenous hydrogen sulfide in vascular smooth muscle cells. Mol Pharmacol 68(6):1757–1764

Tang XQ, Yang CT et al (2008) Effect of hydrogen sulphide on beta-amyloid-induced damage in PC12 cells. Clin Exp Pharmacol Physiol 35(2):180–186

Tay AS, Hu LF et al (2010) Hydrogen sulfide protects neurons against hypoxic injury via stimulation of ATP-sensitive potassium channel/protein kinase C/extracellular signal-regulated kinase/heat shock protein90 pathway. Neuroscience 167(2):277–286

Tiong CX, Lu M, Bian JS (2010) Protective effect of hydrogen sulfide against 6-OHDA induced cell injury in SH-SY5Y cells involves PKC/PI3K/Akt pathway. Br J Pharmacol 161(2): 467–480

Tripatara P, Patel NS et al (2008) Generation of endogenous hydrogen sulfide by cystathionine gamma-lyase limits renal ischemia/reperfusion injury and dysfunction. Lab Invest 88 (10):1038–1048

Tripatara P, Patel NS et al (2009) Characterisation of cystathionine gamma-lyase/hydrogen sulphide pathway in ischaemia/reperfusion injury of the mouse kidney: an in vivo study. Eur J Pharmacol 606(1–3):205–209

Umemura K, Kimura H (2007) Hydrogen sulfide enhances reducing activity in neurons: neurotrophic role of H_2S in the brain? Antioxid Redox Signal 9(11):2035–2041

Vandiver MS, Paul BD et al (2013) Sulfhydration mediates neuroprotective actions of parkin. Nat Commun 4:1626

Vitvitsky V, Thomas M et al (2006) A functional transsulfuration pathway in the brain links to glutathione homeostasis. J Biol Chem 281(47):35785–35793

Wang R (2002) Two's company, three's a crowd: can H_2S be the third endogenous gaseous transmitter? FASEB J 16(13):1792–1798

Wang Y, Zhao X et al (2009) Role of hydrogen sulfide in the development of atherosclerotic lesions in apolipoprotein E knockout mice. Arterioscler Thromb Vasc Biol 29(2):173–179

Wang MJ, Cai WJ et al (2010) The hydrogen sulfide donor NaHS promotes angiogenesis in a rat model of hind limb ischemia. Antioxid Redox Signal 12(9):1065–1077

Wang X, Wang Q et al (2011) Hydrogen sulfide attenuates cardiac dysfunction in a rat model of heart failure: a mechanism through cardiac mitochondrial protection. Biosci Rep 31(2):87–98

Warenycia MW, Goodwin LR et al (1989a) Acute hydrogen sulfide poisoning. Demonstration of selective uptake of sulfide by the brainstem by measurement of brain sulfide levels. Biochem Pharmacol 38(6):973–981

Warenycia MW, Steele JA et al (1989b) Hydrogen sulfide in combination with taurine or cysteic acid reversibly abolishes sodium currents in neuroblastoma cells. Neurotoxicology 10(2):191–199

Webb GD, Lim LH et al (2008) Contractile and vasorelaxant effects of hydrogen sulfide and its biosynthesis in the human internal mammary artery. J Pharmacol Exp Ther 324(2):876–882

Wei H, Zhang R et al (2010) Hydrogen sulfide attenuates hyperhomocysteinemia-induced cardiomyocytic endoplasmic reticulum stress in rats. Antioxid Redox Signal 12(9):1079–1091

Whiteman M, Armstrong JS et al (2004) The novel neuromodulator hydrogen sulfide: an endogenous peroxynitrite 'scavenger'? J Neurochem 90(3):765–768

Whiteman M, Cheung NS et al (2005) Hydrogen sulphide: a novel inhibitor of hypochlorous acid-mediated oxidative damage in the brain? Biochem Biophys Res Commun 326(4):794–798

Whitfield NL, Kreimier EL et al (2008) Reappraisal of H_2S/sulfide concentration in vertebrate blood and its potential significance in ischemic preconditioning and vascular signaling. Am J Physiol Regul Integr Comp Physiol 294(6):R1930–R1937

Wong PT, Qu K et al (2006) High plasma cyst(e)ine level may indicate poor clinical outcome in patients with acute stroke: possible involvement of hydrogen sulfide. J Neuropathol Exp Neurol 65(2):109–115

Wu SY, Pan CS et al (2006) Hydrogen sulfide ameliorates vascular calcification induced by vitamin D3 plus nicotine in rats. Acta Pharmacol Sin 27(3):299–306

Wu N, Siow YL et al (2010) Ischemia/reperfusion reduces transcription factor Sp1-mediated cystathionine beta-synthase expression in the kidney. J Biol Chem 285(24):18225–18233

Xia M, Chen L et al (2009) Production and actions of hydrogen sulfide, a novel gaseous bioactive substance, in the kidneys. J Pharmacol Exp Ther 329(3):1056–1062

Xie L, Tiong CX et al (2012) Hydrogen sulfide protects SH-SY5Y cells against 6-hydroxydopamine-induced endoplasmic reticulum stress. Am J Physiol Cell Physiol 303(1):C81–C91

Xie L, Hu LF et al (2013) Therapeutic effect of hydrogen sulfide-releasing L-Dopa derivative ACS84 on 6-OHDA-induced Parkinson's disease rat model. PLoS One 8(4):e60200

Xu Z, Prathapasinghe G et al (2009) Ischemia-reperfusion reduces cystathionine-beta-synthase-mediated hydrogen sulfide generation in the kidney. Am J Physiol Renal Physiol 297(1): F27–F35

Xu ZS, Wang XY et al (2011) Hydrogen sulfide protects MC3T3-E1 osteoblastic cells against H_2O_2-induced oxidative damage-implications for the treatment of osteoporosis. Free Radic Biol Med 50(10):1314–1323

Xue R, Hao DD et al (2013) Hydrogen sulfide treatment promotes glucose uptake by increasing insulin receptor sensitivity and ameliorates kidney lesions in type 2 diabetes. Antioxid Redox Signal 19(1):5–23

Yan H, Du J et al (2004) The possible role of hydrogen sulfide on the pathogenesis of spontaneous hypertension in rats. Biochem Biophys Res Commun 313(1):22–27
Yang G, Wu L et al (2006) Pro-apoptotic effect of endogenous H_2S on human aorta smooth muscle cells. FASEB J 20(3):553–555
Yang G, Wu L et al (2008) H2S as a physiologic vasorelaxant: hypertension in mice with deletion of cystathionine gamma-lyase. Science 322(5901):587–590
Yang G, Tang G et al (2011) The pathogenic role of cystathionine gamma-lyase/hydrogen sulfide in streptozotocin-induced diabetes in mice. Am J Pathol 179(2):869–879
Yang HY, Wu ZY et al (2013) Hydrogen sulfide attenuates opioid dependence by suppression of adenylate cyclase/cAMP pathway. Antioxid Redox Signal (in press)
Yao LL, Huang XW et al (2010) Hydrogen sulfide protects cardiomyocytes from hypoxia/reoxygenation-induced apoptosis by preventing GSK-3beta-dependent opening of mPTP. Am J Physiol Heart Circ Physiol 298(5):H1310–H1319
Yin WL, He JQ et al (2009) Hydrogen sulfide inhibits MPP(+)-induced apoptosis in PC12 cells. Life Sci 85(7–8):269–275
Yin J, Tu C et al (2013) Exogenous hydrogen sulfide protects against global cerebral ischemia/reperfusion injury via its anti-oxidative, anti-inflammatory and anti-apoptotic effects in rats. Brain Res 1491:188–196
Yong QC, Lee SW et al (2008a) Endogenous hydrogen sulphide mediates the cardioprotection induced by ischemic postconditioning. Am J Physiol Heart Circ Physiol 295(3):H1330–H1340
Yong QC, Pan TT et al (2008b) Negative regulation of beta-adrenergic function by hydrogen sulphide in the rat hearts. J Mol Cell Cardiol 44(4):701–710
Yong QC, Choo CH et al (2010a) Effect of hydrogen sulfide on intracellular calcium homeostasis in neuronal cells. Neurochem Int 56(3):508–515
Yong QC, Hu LF et al (2010b) Hydrogen sulfide interacts with nitric oxide in the heart: possible involvement of nitroxyl. Cardiovasc Res 88(3):482–491
Yong QC, Cheong JL et al (2011) Regulation of heart function by endogenous gaseous mediators-crosstalk between nitric oxide and hydrogen sulfide. Antioxid Redox Signal 14(11):2081–2091
Zanardo RC, Brancaleone V et al (2006) Hydrogen sulfide is an endogenous modulator of leukocyte-mediated inflammation. FASEB J 20(12):2118–2120
Zhang C, Du J et al (2003) The regulatory effect of endogenous hydrogen sulfide on hypoxic pulmonary hypertension. Beijing Da Xue Xue Bao 35(5):488–493
Zhang Z, Huang H et al (2007) Hydrogen sulfide contributes to cardioprotection during ischemia-reperfusion injury by opening K ATP channels. Can J Physiol Pharmacol 85(12):1248–1253
Zhang LM, Jiang CX et al (2009) Hydrogen sulfide attenuates neuronal injury induced by vascular dementia via inhibiting apoptosis in rats. Neurochem Res 34(11):1984–1992
Zhang H, Gao Y et al (2011a) Hydrogen sulfide reduces mRNA and protein levels of beta-site amyloid precursor protein cleaving enzyme 1 in PC12 cells. Neurochem Int 58(2):169–175
Zhang H, Zhang A et al (2011b) S-diclofenac protects against doxorubicin-induced cardiomyopathy in mice via ameliorating cardiac gap junction remodeling. PLoS One 6(10):e26441
Zhao W, Wang R (2002) H(2)S-induced vasorelaxation and underlying cellular and molecular mechanisms. Am J Physiol Heart Circ Physiol 283(2):H474–H480
Zhao W, Zhang J et al (2001) The vasorelaxant effect of H(2)S as a novel endogenous gaseous K (ATP) channel opener. EMBO J 20(21):6008–6016
Zhao X, Zhang LK et al (2008) Regulatory effect of hydrogen sulfide on vascular collagen content in spontaneously hypertensive rats. Hypertens Res 31(8):1619–1630
Zhong G, Chen F et al (2003) The role of hydrogen sulfide generation in the pathogenesis of hypertension in rats induced by inhibition of nitric oxide synthase. J Hypertens 21(10):1879–1885
Zhu YZ, Wang ZJ et al (2007) Hydrogen sulfide and its possible roles in myocardial ischemia in experimental rats. J Appl Physiol 102(1):261–268

Biological Effects of H_2S Inhalation and Its Therapeutic Potential

7

Fumito Ichinose

Abstract
Hydrogen sulfide has been known as a toxic gas for many years. Following a discovery that inhaling low concentrations of H_2S induces "suspended animation" in rodents, a number of studies have been done in the hope to exploiting beneficial effects of this toxic gas. Although many intriguing results have been reported, typically, these early studies were conducted in rodents with one concentration of inhaled H_2S. Therapeutic potential of H_2S inhalation remains to be examined using large animal models.

Keywords
Suspended animation • Hypoxia • Shock • Inflammation • Reperfusion injury

Abbreviations

ATP	Adenosine triphosphate
CSE	Cystathionine γ-lyase
ECML	Extra-corporeal membrane lung
GYY-4137	Morpholin-4-ium 4 methoxyphenyl (morpholino) phosphinodithioate

F. Ichinose (✉)
Department of Anesthesia, Critical Care and Pain Medicine, Massachusetts General Hospital, Harvard Medical School, Boston, MA, USA
e-mail: fichinose@partners.org

H. Kimura (ed.), *Hydrogen Sulfide and its Therapeutic Applications*,
DOI 10.1007/978-3-7091-1550-3_7, © Springer-Verlag Wien 2013

H_2S	Hydrogen sulfide
HV_T	High tidal volume ventilation
I/R	Ischemia-reperfusion
IL-10	Interleukin-10
LPS	Lipopolysaccharide
MBB	Monobromobimane
MPTP	1-Methyl-4-phenyl-1,2,3,6-tetrahydropyridine
Na_2S	Sodium disulfide
NaHS	Sodium hydrosulfide
NO	Nitric oxide
Nrf-2	Nuclear factor (erythroid-derived 2)-like 2
PAG	DL-propagylglycine
PD	Parkinson's Disease
ppm	Parts per million
STS	Sodium thiosulfate
TH	Tyrosine hydroxylase
VCO_2	CO_2 production rate
VILI	Ventilator-induced lung injury

7.1 Introduction

Hydrogen sulfide (H_2S) is a colorless gas with a characteristic rotten-egg odor, found in volcano gas emissions, sulfur springs, bacterial decomposition of proteins, and in various sulfur-containing products (Szabo 2007a). About 10 % of total global emissions of H_2S is due to human activity, including petroleum refineries, coke ovens, paper mills, and tanneries.

Toxic effects of H_2S gas on humans have been well known for centuries (Struve et al. 2001). Our noses can detect H_2S at concentrations of 0.0047 part per million (ppm). Exposure to low (~100 ppm) concentrations of H_2S causes "gas eye", sore throat, nausea, and respiratory effects attributable to airway irritation. At 500 ppm, it impairs breathing. Exposure to 800 ppm for 5 min leads to death (Simonton and Spears 2007). Today it constitutes the number-one occupational safety hazard at oil and gas field wellheads, along pipelines, in processing plants and in refineries (Fuller and Suruda 2000). The cellular toxicity of H_2S is attributed to its capacity to inhibit cytochrome c oxidase, the terminal enzyme of oxidative phosphorylation, resulting in cellular hypoxia.

In 2005, a study published in *Science* reported that breathing low concentration of H_2S decreases both body temperature and metabolic rate inducing a "suspended animation-like state" in mice (Blackstone et al. 2005). This paper prompted a number of following experimental studies examining the effects of H_2S inhalation, expecting "beneficial" effects.

In this chapter, experimental evidence suggesting the "beneficial" effects of inhaled H_2S will be critically reviewed and the future of H_2S inhalation therapy will be discussed. Toxic effects of H_2S inhalation has been extensively studied and reviewed elsewhere and therefore not covered here.

7.1.1 Induction of "Suspended Animation-Like" State by H_2S Inhalation

Balancing cellular oxygen supply and demand is a key therapeutic approach to protecting organs such as the brain, kidneys and heart from ischemic injury. Permissive hypothermia and active cooling have been shown to reduce oxygen demands in patients experiencing stroke, cardiac arrest, cardiac surgery, severe trauma and other instances of ischemia and subsequent reperfusion (Arrich et al. 2012; Fukudome and Alam 2009; Grigore et al. 2009; Polderman 2009). However, hypothermic reduction of aerobic metabolism has been associated with adverse effects, including increased rates of infection and coagulopathy (Fries et al. 2009). Developing other methods to acutely reduce metabolism in patients could be clinically useful.

In their landmark paper, Blackstone and colleagues demonstrated that, in awake, spontaneously breathing mice, inhaling H_2S at 80 ppm induced a suspended animation-like metabolic state characterized by reduced energy expenditure and hypothermia (Blackstone et al. 2005). Subsequently, we reported that this metabolic depression was associated with bradycardia and reduced cardiac output but that blood pressure and stroke volume remained unaffected (Fig. 7.1) (Volpato et al. 2008). Consequently, given the exciting prospect of pharmacologically reducing energy expenditure to protect against ischemia by application of a gaseous drug without depressing blood pressure, the effects of inhaled H_2S were investigated in various animal models. In fact, inhaled H_2S protected rodents against otherwise lethal hypoxia (Blackstone and Roth 2007) and hemorrhage (Morrison et al. 2008b) and attenuated murine kidney, lung, and liver I/R injury (see next section) (Bos et al. 2009; Faller et al. 2010). Equivocal data, however, are available from large animals: inhaled H_2S up to 80 ppm failed to show any metabolic effect in sheep or swine (Haouzi et al. 2008; Li et al. 2008a), and the intravenous H_2S donor sodium sulfide (Na_2S) was reported either to reduce energy expenditure (Simon et al. 2008) or to have no effect at all (Drabek et al. 2011).

Struve and colleagues (Struve et al. 2001) reported that inhalation of H_2S at 200 to 400 ppm, but not at 30–80 ppm, decreased body temperature in rats. Similarly, Morrison and colleagues (2008a) showed that inhaling H_2S at 300 ppm was required to decrease VCO_2 in rats, in contrast to 80 ppm in mice. While these observations suggest that higher levels of H_2S may be required to alter metabolic rates in larger animals (Morrison et al. 2008a), it is well documented that inhalation of high concentrations of H_2S may injure the bronchial mucosa, cause pulmonary edema, and impair gas exchange (Beauchamp et al. 1984; Reiffenstein et al. 1992). In an attempt to bypass the direct pulmonary toxicity of inhaled H_2S, we used an extra-corporeal membrane lung (ECML) to directly diffuse high concentrations of H_2S gas into the blood (Fig. 7.2) (Derwall et al. 2011). The results of this proof-of-concept large animal study revealed that ventilating an ECML with up to

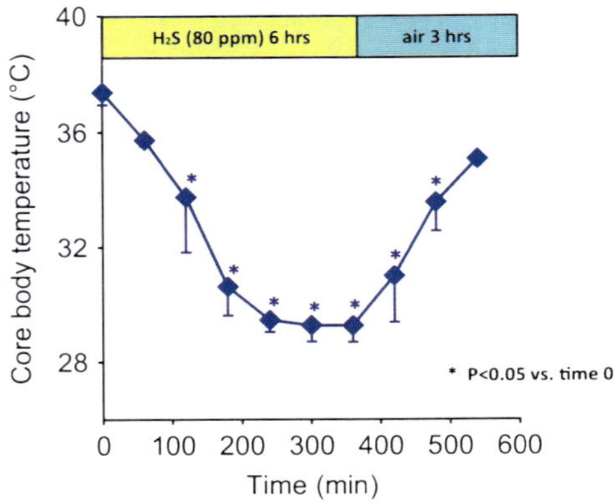

Fig. 7.1 Change of core body temperature in mice breathed H_2S at 80 ppm in air (Modified from Volpato et al. 2008)

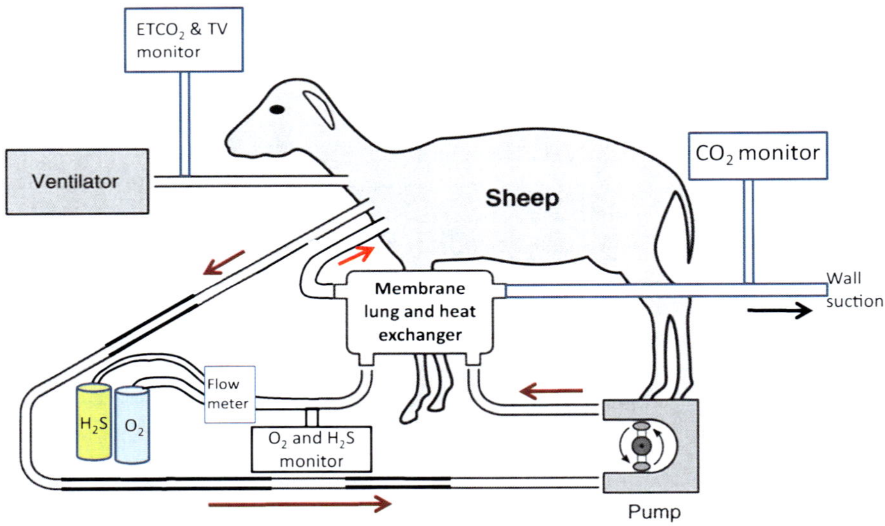

Fig. 7.2 An experimental system to administer H_2S to sheep using extra-corporeal membrane lung (*ECML*)

300 ppm H_2S in venoarterial cardiac bypass circulation does not reduce whole body CO_2 production or O_2 consumption in anesthetized sheep. In addition, we have demonstrated that administration of 300 ppm H_2S via ECML ventilation causes significant adverse effects, including pulmonary vasoconstriction, systemic vasodilation and hypoxemia. Our results showed that high concentrations of H_2S delivered via an ECML do not reduce the metabolic rate in large mammals at rest.

Why can H_2S gas reduce metabolic rate only in rodents but not in large mammals? Hydrogen sulfide may be one, but not the only, trigger for murine metabolic depression. Indeed, hypoxia, anemia and exposure to carbon monoxide have been reported to reduce aerobic metabolism in mice (Gautier and Bonora 1994; Matsuoka et al. 1994; Singer 2004), but not in large mammals (Forster et al. 1981; Frappell et al. 1992; Korducki et al. 1994). It is of note that mice are known to have a much higher specific metabolic rate (metabolic rate per unit body mass of approximately 168 kcal $kg^{-1}\cdot day^{-1}$ in a 30-g mouse) than sheep (approximately 30 kcal $kg^{-1}\cdot day^{-1}$ in a 30-kg sheep) (Schmidt-Nielsen 1984). In a previous study, we reported that H_2S inhalation reduced metabolism in awake, spontaneously breathing mice by about 40 % during normothermia, resulting in a specific metabolic rate of no more than approximately 100 $kcal\cdot kg^{-1}\cdot day^{-1}$ (Volpato et al. 2008). In contrast, it has been reported that H_2S inhalation at 100 ppm failed to reduce CO_2 production in normothermic mice that were anesthetized and mechanically ventilated (Baumgart et al. 2010). Interestingly, in anesthetized mice studied by Baumgart and colleagues (2010), the baseline CO_2 production rate before H_2S inhalation was approximately 50 % less than that in awake mice studied by Volpato et al. in our laboratory (Volpato et al. 2008). It is tempting to speculate that the ability of H_2S to reduce metabolism depends on the specific metabolic rate of animals. H_2S may reduce metabolism when the baseline rate of metabolism is high (for example, in awake mice), but not when the metabolic rate is already depressed (for example, in anesthetized mice or sheep). Nonetheless, metabolism-reducing effects of H_2S inhalation has not been reproduced in larger mammals to date.

7.1.2 Other "Beneficial Effects" of H_2S Inhalation

7.1.2.1 Protective Effects of H_2S Inhalation Against Lethal Hypoxia and Hemorrhagic Shock

An early demonstration of "protective" effects of H_2S breathing was reported in mice exposed to lethal hypoxia by Blackstone and Roth (2007). C57BL6 wild-type mice cannot survive for longer than 20 min when exposed to 5 % oxygen. However, if mice are first put into a suspended animation-like state by a 20-min pretreatment with 150 ppm H_2S and then are exposed to low oxygen, they can survive for more than 6.5 h in 5 % oxygen with no apparent detrimental effects. In addition, if mice are exposed to a 20-min pretreatment with H_2S followed by 1 h at 5 % oxygen, they can then survive for several hours at oxygen tensions as low as 3 %. Based on the reduced metabolic rates after exposure to H_2S and hypoxia, the investigators hypothesized that prior exposure to H_2S reduces oxygen demand, therefore making it possible for the mice to survive with low oxygen supply. Although similar experiments have not been attempted in larger mammals, it appears highly unlikely that pretreatment with H_2S increases resistance to lethal hypoxia in larger mammals. The "protective" effects against hypoxia appear to be solely based on the ability of H_2S to induce suspended-animation like state. As described in the previous section, H_2S inhalation does not reduce metabolic rates in larger mammals.

Reduced metabolic activity improves outcome in many clinical and experimental models of injury and diseases that result in insufficient blood supply. Morrison and colleagues hypothesized that hydrogen sulfide confers benefit in injuries and diseases related to insufficient blood supply and tested this hypothesis in rat model of lethal hemorrhagic shock (Morrison et al. 2008b). Sprague–Dawley rats were subjected to controlled hemorrhage to remove 60 % of total blood. Hydrogen sulfide was administered to rats either via airway as gas, or intravenous infusion as liquid. Using H_2S inhalation at 300 ppm, 75 % of treated and 23 % of untreated rats survived longer than 24 h. With intravenous administered sulfide, 67 % of treated and 14 % of untreated rats survived longer than 24 h. Respirometry results showed that H_2S stabilized (but not reduced) CO_2 production rates during and after hemorrhage. Protective effects of H_2S inhalation or intravenous sulfide in hemorrhagic shock may not solely depend on the ability of H_2S to reduce metabolic rates, since inhaled or IV sulfide did not reduce CO_2 production during hemorrhage in this model (Morrison et al. 2008b). While effects of H_2S inhalation in hemorrhagic shock have not been examined in larger mammals, two groups examined effects of intravenous sulfide (Na_2S) infusion on outcomes after hemorrhagic shock in pigs and reported conflicting results; one study showing no benefit (Drabek et al. 2011) while the other group showing beneficial effects of sulfide infusion (Bracht et al. 2012). The latter study even showed that sulfide infusion increased survival rate of pigs subjected to prolonged hemorrhagic shock. It is of note that beneficial effects of sulfide infusion were associated with moderate hypothermia (~35 °C). Effects of H_2S inhalation in hemorrhagic shock in larger mammals remain to be determined.

7.1.2.2 Inhibition of Ischemia-Reperfusion Injury by H_2S Inhalation

Pharmacologically reducing the demand for oxygen is a promising strategy to minimize unavoidable I/R injury during organ transplantation. Majority of studies examining the beneficial effects of H_2S on I/R injury were conducted using H_2S donor compounds (e.g., NaHS and Na_2S). A number of studies showed beneficial effects of intravenous sulfide infusion against I/R injury in small and large mammals (Elrod et al. 2007; Sodha et al. 2008). In contrast, only a few groups have studied effects of H_2S inhalation on I/R injury. Bos and colleagues have examined effects of H_2S inhalation on renal I/R injury in rats and mice (Bos et al. 2009). These investigators found that exposure to H_2S gas dose-dependently reduced mitochondrial membrane potential in a cultured proximal tubule cell line, reduced oxygen consumption and ATP production in perfused ex-vivo rat kidney, reduced CO_2 production in mice, and prevented renal I/R injury in mice. The same investigators reported protective effects of H_2S inhalation on hepatic I/R injury in mice using the similar approach (Bos et al. 2012). It is of note that these beneficial effects of inhaled H_2S were observed while body temperature of mice were clamped at 37 °C during I/R injury to separate effects of H_2S from well-known protective effects of hypothermia. Nonetheless, beneficial effects of H_2S inhalation in the kidney and liver I/R injury were associated with reduction in metabolic rates in mice. Protective effects of H_2S inhalation against I/R injury in larger mammals have not been reported to date.

7.1.2.3 Prevention of Neurodegenerative Disease by Breathing H_2S

Parkinson's disease (PD) is one of the most common neurodegenerative diseases. It is characterized by a slow and progressive degeneration of dopaminergic neurons in the substantia nigra. Although the etiology of PD is not fully understood, several mechanisms responsible for the neurodegeneration in PD have been suggested, including abnormal protein handling, oxidative stress, mitochondrial dysfunction, excitotoxicity, neuroinflammation, and apoptosis (Hirsch and Hunot 2009).

A number of animal models of PD, both toxin-induced and genetically-engineered, have been created. Although none of the animal models accurately recapitulate the pathophysiological features of PD, 1-methyl-4-phenyl-1,2,3,6-tetrahydropyridine (MPTP) currently represents the most important and most frequently used parkinsonian toxin applied in animal models. MPTP is the only known dopaminergic neurotoxin capable of causing a clinical picture indistinguishable from idiopathic PD in humans (Langston et al. 1983).

In a recent study, we sought to examine effects of H_2S inhalation in the clinically-relevant MPTP-induced PD model in mice (Kida et al. 2010). Our results revealed that inhalation of H_2S at 40 ppm for 8 h a day for 7 days prevented MPTP-induced neurodegeneration and movement disorder. The neuroprotective effects of H_2S inhalation were associated with marked attenuation of the MPTP-induced loss of TH-containing neurons in substantia nigra and striatum. Breathing H_2S prevented apoptosis and gliosis in substantia nigra 1 day after MPTP administration. The neuroprotective effects of H_2S breathing were associated with upregulation of antioxidant proteins and phase II detoxification enzymes in nigrostriatal region of the brain. Taken together, these observations suggest that inhaled H_2S confer protection against the neurotoxicity of MPTP in mice. Our results suggest that the neuroprotection afforded by inhaled H_2S is mediated at least in part via Nrf2-dependent upregulation of antioxidant defense mechanisms.

Other groups using H_2S donor compounds have confirmed the protective effects of H_2S against neurodegeneration (Xie et al. 2013). Nonetheless, potential therapeutic effects of H_2S inhalation remain to be examined in chronic genetically-induced models of PD in future studies.

7.1.2.4 Anti-inflammatory Effects of Inhaled H_2S

Similar to nitric oxide (NO), roles of H_2S in inflammation have been the focus of intensive research. To date, numerous conflicting data regarding the pro- and anti-inflammatory activity of exogenous and endogenous sulfide have been reported. For example, various H_2S donors (e.g., sodium hydrosulfide [NaHS]) have been reported to worsen (Li et al. 2005; Zhang et al. 2007) or attenuate (Li et al. 2007, 2009) inflammation, while CSE inhibitors (e.g., DL-propagylglycine [PAG]) exhibit anti-inflammatory activity in animal models of endotoxemia and sepsis (Li et al. 2005; Zhang et al. 2007). Although the reasons for these conflicting results are likely multifactorial, purity and/or specificity of chemical H_2S donors and inhibitors have been questioned. For example, solution of NaHS appears to include polysulfides and elemental sulfur (Doeller et al. 2005), whereas PAG may affect other pyridoxal 5′-phosphate-dependent enzymes as well (Li et al. 2009).

To elucidate the impact of authentic H_2S in endotoxin-induced inflammation, we examined the hypothesis that H_2S breathing prevents LPS-induced systemic inflammation and organ injury and improves survival in mice (Tokuda et al. 2012). We also sought to determine the plasma levels of H_2S and sulfide metabolites using the monobromobimane (MBB)-based HPLC method (Wintner et al. 2010). Our study revealed that breathing H_2S at 80 ppm for 6 h after LPS challenge markedly improved survival in mice. Inhalation of H_2S attenuated LPS-induced systemic nitrosative stress, liver injury, and neutrophil infiltration in the lung. The protective effects of H_2S inhalation were associated with inhibition of LPS-induced inflammatory cytokine induction and marked upregulation of anti-inflammatory cytokine IL-10 in the liver. Our study also revealed that plasma sulfide levels were reduced by LPS challenge while H_2S breathing restored sulfide levels and markedly increased thiosulfate levels during endotoxemia. The altered sulfide metabolism after LPS challenge and H_2S breathing appeared to be caused by upregulation of rhodanese. Finally, administration of sodium thiosufate (STS) dose-dependently improved survival after LPS challenge (Fig. 7.3). These observations suggest that H_2S inhalation exhibits potent anti-inflammatory effects leading to improvement of survival during murine endotoxemia in part by remodeling sulfide metabolism in mice (Fig. 7.4).

Protective effects of H_2S breathing during endotoxemia was later confirmed by Faller and colleagues who showed that LPS-induced lung injury was attenuated by breathing 80 ppm H_2S in mice (Faller et al. 2012). The same authors also reported that H_2S inhalation prevents ventilator-induced lung injury in mice (see next section) (Faller et al. 2010).

The anti-inflammatory effects of H_2S are in odds with some earlier studies. However, as in the case of NO, it is likely that H_2S dose-dependently exerts wide spectrum of effects during inflammation; low and physiological sulfide concentrations are anti-inflammatory while high sulfide levels are pro-inflammatory. Li and colleagues reported that i.p. administration of NaHS *per se* at 14 μmol/kg (=784 μg/kg) resulted in marked histological signs of lung inflammation, increased lung and liver MPO activity, and raised plasma TNF-α concentration (Li et al. 2005). Therefore, it is no surprise that high dose NaHS (10 mg/kg) aggravates inflammation in septic mice (Zhang et al. 2007). In contrast, our results are consistent with the anti-inflammatory effects of the slow-releasing H_2S donor, GYY-4137 [morpholin-4-ium 4 methoxyphenyl (morpholino) phosphinodithioate] (Li et al. 2008b), in endotoxemic rats and in macrophages incubated with LPS (Li et al. 2009). Since NaHS releases large amounts of H_2S over a period of few seconds (Li et al. 2008b), it is likely that tissues are exposed to super-physiological levels of sulfide after systemic administration of NaHS. By demonstrating the beneficial effects of sustained levels of sulfide provided by H_2S breathing for 6 h, our results support the hypothesis that low and physiological levels of sulfide confer anti-inflammatory effects during endotoxemia.

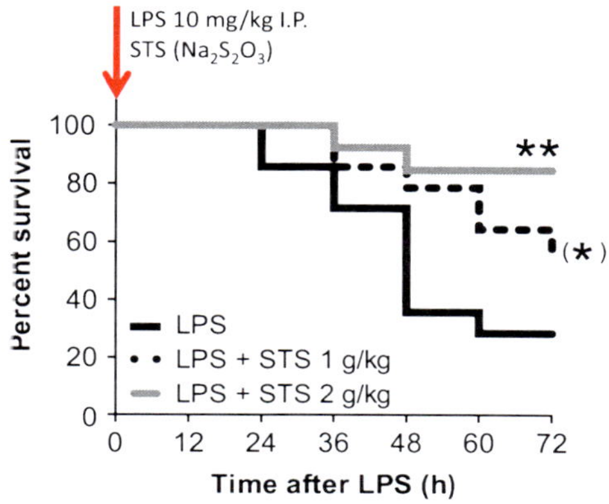

Fig. 7.3 Kaplan-Meier curve showing survival in mice challenged with LPS (LPS, n = 14), mice challenged with LPS and received 1 g/kg of sodium thiosulfate (STS) (LPS + STS 1 g/kg, n = 14), and mice challenged with LPS and received 2 g/kg of STS (LPS + STS 2 g/kg, n = 13). $*P = 0.0781$ vs LPS. $**P = 0.0047$ vs LPS (Modified from Tokuda et al. 2012)

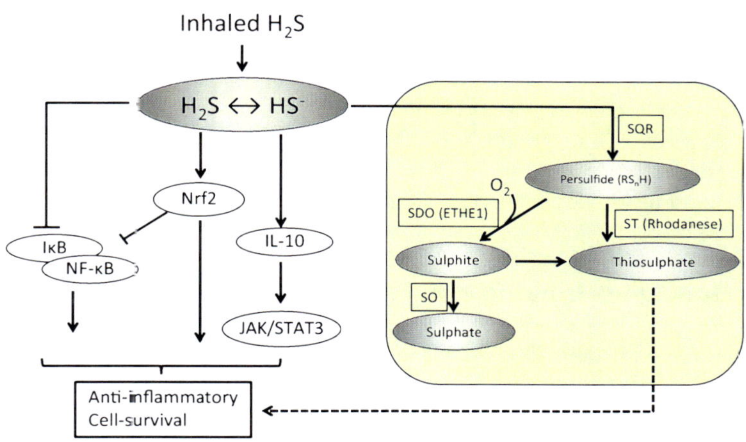

Fig. 7.4 Hypothetical mechanisms responsible for anti-inflammatory effects of inhaled H_2S

7.1.2.5 Effects of Inhaled H_2S in Acute Lung Injury

Mechanical ventilation can be life-saving in acute respiratory failure but may contribute to lung injury and increase in-hospital mortality due to its side-effects (Brower et al. 2004). Cyclic lung stretch during mechanical ventilation induces tissue disruption and activates proinflammatory pathways, thereby promoting the generation of pulmonary edema and infiltration with neutrophils (Lionetti et al. 2005). In addition, cyclic lung stretch is associated with the generation of reactive oxygen species and the production of redox imbalance in the lung (Birukov 2009; Papaiahgari et al. 2007; Vaporidi et al. 2010). Together, these events may initiate and perpetuate oxidative stress as well as local and systemic inflammatory responses, ultimately leading to enhanced lung injury, multi organ failure and death (Belperio et al. 2006; Gurkan et al. 2003; Slutsky and Tremblay 1998). Developing novel therapies targeted at oxidative stress and inflammation in ventilator-induced lung injury (VILI) could be clinically useful.

Inhalation of H_2S is known to cause airway mucosa irritation and cytotoxicity (Milby and Baselt 1999; Szabo 2007b), has been reported to exert pro-inflammatory effects in various models (reviewed in references Baumgart et al. 2009; Szabo 2007b), and has traditionally been considered a health hazard. On the other hand, Faller and colleagues recently showed that inhaled H_2S at 80 ppm prevents lung injury in mice ventilated with a tidal volume of 12 ml/kg (plateau pressure 10–13 cm H_2O, i.e. moderate stretch)(Faller et al. 2010). To elucidate the role of H_2S in VILI with increased lung stretch, such as in patients with the acute respiratory distress syndrome, we examined the effect of inhaled H_2S in a model of lung injury induced by high tidal volume (HV_T) ventilation. We found that H_2S inhalation promotes VILI and enhances the pulmonary expression of leukocyte adhesion and chemoattractant molecules. Subsequently, we hypothesized that intravascular administration of Na_2S, avoiding direct exposure of the lung to H_2S gas, could provide better protection against VILI than inhaled H_2S. We report that intravascular Na_2S both attenuates the pulmonary expression of chemoattractant and leukocyte adhesion molecules and enhances Nrf2-dependent expression of antioxidant genes, thereby attenuating VILI from HV_T ventilation.

These observations suggest a complex interaction of exogenous H_2S and innate immune response to lung stretch and subsequent inflammation. Effects of H_2S may also differ depending on cell types affected. Effects of H_2S inhalation against sepsis and acute lung injury have not been examined in larger mammals.

7.2 How Does Inhaled H_2S Exert Systemic Effects?

While the effects of inhaled H_2S on lung may be direct, the mechanisms whereby inhaled H_2S exerts effects on extra-pulmonary systemic organs are incompletely defined. H_2S is a highly reactive molecule and its plasma levels return to baseline quickly after systemic administration presumably due to rapid reaction with plasma proteins and blood-borne cells. It is conceivable that circulating cells are directly exposed to H_2S as they pass through the pulmonary capillaries and may be "pacified" before they reach the peripheral tissues (Fig. 7.5). For example, Wallace and colleagues reported that H_2S inhibits leukocyte activation thereby attenuates inflammation in gastrointestinal tract (Zanardo et al. 2006). It has also been reported that H_2S donors inhibit cytokine production from tissue macrophages (Whiteman et al. 2010). Alternatively, some H_2S, once inhaled, may be converted to relatively stable sulfide metabolites (e.g., thiosulfate) that can regenerate H_2S in the periphery (Fig. 7.5). In fact, beneficial effects of H_2S inhalation in endotoxemia in mice were associated with marked increase of plasma thiosulfate levels in our recent study (Tokuda et al. 2012). This is reminiscent of the role of nitrite in the systemic effects of inhaled NO (Cannon et al. 2001). It has been suggested that systemic effects of inhaled NO may be transferred by plasma nitrite, which can be converted back to NO in peripheral tissues.

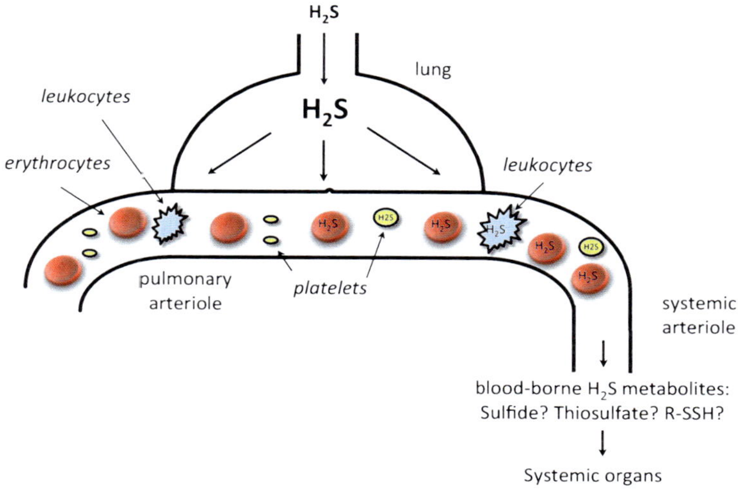

Fig. 7.5 Hypothetical mechanisms responsible for systemic effects of inhaled H_2S

7.3 Summary and Future Directions

In the past decade, H_2S has been transformed from a dreadful toxic gas to an important gaseous signaling molecule and potential therapeutic agent. One of the most illustrative parts of this transformation was the discovery of the ability of H_2S inhalation to induce "suspended animation-like" state in rodents. Although this discovery even had unusually heavy media coverage including a few TV shows and rumor of IPO of a company at one point, the initial excitement appears to be all gone when it has become clear that the suspended animation-like state cannot be induced in larger mammals. Around the same time, organ protective effects of H_2S inhalation and H_2S donating compounds were reported. These "beneficial" effects of H_2S does not appear to depend on the ability of H_2S to induce "suspended animation" since H_2S donating compounds, in general, do not reduce metabolism in mammals. While the beneficial effects of H_2S have been demonstrated many different disease models, most studies were done exclusively in rodents and very few studies examined the effects of H_2S inhalation in larger mammals. This is a stark difference between inhaled H_2S and inhaled NO. The therapeutic effects of inhaled NO were first demonstrated in a sheep model of pulmonary hypertension. For H_2S inhalation to be considered as a therapeutic modality, as in the case of inhaled NO, the efficacy of H_2S has to be clearly demonstrated in large mammal models of human disease.

Additionally, mechanisms responsible for the beneficial systemic effects of H_2S inhalation and H_2S donor compounds are incompletely understood. While H_2S induced protein S-sulfhydration has been suggested as a molecular mechanism mediating effects of H_2S (Mustafa et al. 2009), the extent and significance of this additional post-translational protein modification remains to be firmly established.

Furthermore, which molecular species (e.g., H_2S, HS-, S_2-, or other persulfide) actually mediates biological effects of H_2S remains to be determined also. If we know the active molecule, we may be able to start developing methods of administration that will provide a desired concentration of the active molecule in the active sites. For this reason, it is important to point out that H_2S inhalation at 80 ppm, which has been conveniently used since the first demonstration of the induction of suspended animation-like state in mice, has little additional scientific justification. Careful dose–response experiments should be performed in each indication in future studies.

It is probably safe to say that H_2S inhalation does not induce suspended animation in human. Nonetheless, elucidation of the "active" molecule and molecular mechanisms responsible for the "other" beneficial effects of H_2S inhalation may enable us to use this stinky gas for yet unknown therapeutic purposes in the future.

References

Arrich J, Holzer M, Herkner H, Mullner M (2012) Hypothermia for neuroprotection in adults after cardiopulmonary resuscitation. Cochrane Database Syst Rev 9:CD004128. doi:10.1002/14651858.CD004128.pub3, PMID: 22972067 [PubMed-indexed for MEDLINE]

Baumgart K, Radermacher P, Wagner F (2009) Applying gases for microcirculatory and cellular oxygenation in sepsis: effects of nitric oxide, carbon monoxide, and hydrogen sulfide. Curr Opin Anaesthesiol 22:168–176

Baumgart K, Wagner F, Groger M, Weber S, Barth E, Vogt JA et al (2010) Cardiac and metabolic effects of hypothermia and inhaled hydrogen sulfide in anesthetized and ventilated mice. Crit Care Med 38:588–595

Beauchamp RO Jr, Bus JS, Popp JA, Boreiko CJ, Andjelkovich DA (1984) A critical review of the literature on hydrogen sulfide toxicity. Crit Rev Toxicol 13:25–97

Belperio JA, Keane MP, Lynch JP 3rd, Strieter RM (2006) The role of cytokines during the pathogenesis of ventilator-associated and ventilator-induced lung injury. Semin Respir Crit Care Med 27:350–364

Birukov KG (2009) Cyclic stretch, reactive oxygen species, and vascular remodeling. Antioxid Redox Signal 11:1651–1667

Blackstone E, Roth MB (2007) Suspended animation-like state protects mice from lethal hypoxia. Shock 27:370–372

Blackstone E, Morrison M, Roth MB (2005) H_2S induces a suspended animation-like state in mice. Science 308:518

Bos EM, Leuvenink HG, Snijder PM, Kloosterhuis NJ, Hillebrands JL, Leemans JC et al (2009) Hydrogen sulfide-induced hypometabolism prevents renal ischemia/reperfusion injury. J Am Soc Nephrol 20:1901–1905

Bos EM, Snijder PM, Jekel H, Weij M, Leemans JC, van Dijk MC et al (2012) Beneficial effects of gaseous hydrogen sulfide in hepatic ischemia/reperfusion injury. Transpl Int 25:897–908

Bracht H, Scheuerle A, Groger M, Hauser B, Matallo J, McCook O et al (2012) Effects of intravenous sulfide during resuscitated porcine hemorrhagic shock*. Crit Care Med 40:2157–2167

Brower RG, Lanken PN, MacIntyre N, Matthay MA, Morris A, Ancukiewicz M et al (2004) Higher versus lower positive end-expiratory pressures in patients with the acute respiratory distress syndrome. N Engl J Med 351:327–336

Cannon RO 3rd, Schechter AN, Panza JA, Ognibene FP, Pease-Fye ME, Waclawiw MA et al (2001) Effects of inhaled nitric oxide on regional blood flow are consistent with intravascular nitric oxide delivery. J Clin Invest 108:279–287

Derwall M, Francis RC, Kida K, Bougaki M, Crimi E, Adrie C et al (2011) Administration of hydrogen sulfide via extracorporeal membrane lung ventilation in sheep with partial cardiopulmonary bypass perfusion: a proof of concept study on metabolic and vasomotor effects. Crit Care 15:R51

Doeller JE, Isbell TS, Benavides G, Koenitzer J, Patel H, Patel RP et al (2005) Polarographic measurement of hydrogen sulfide production and consumption by mammalian tissues. Anal Biochem 341:40–51

Drabek T, Kochanek PM, Stezoski J, Wu X, Bayir H, Morhard RC et al (2011) Intravenous hydrogen sulfide does not induce hypothermia or improve survival from hemorrhagic shock in pigs. Shock 35:67–73

Elrod JW, Calvert JW, Morrison J, Doeller JE, Kraus DW, Tao L et al (2007) Hydrogen sulfide attenuates myocardial ischemia-reperfusion injury by preservation of mitochondrial function. Proc Natl Acad Sci U S A 104:15560–15565

Faller S, Ryter SW, Choi AM, Loop T, Schmidt R, Hoetzel A (2010) Inhaled hydrogen sulfide protects against ventilator-induced lung injury. Anesthesiology 113:104–115

Faller S, Zimmermann KK, Strosing KM, Engelstaedter H, Buerkle H, Schmidt R et al (2012) Inhaled hydrogen sulfide protects against lipopolysaccharide-induced acute lung injury in mice. Med Gas Res 2:26

Forster HV, Bisgard GE, Klein JP (1981) Effect of peripheral chemoreceptor denervation on acclimatization of goats during hypoxia. J Appl Physiol 50:392–398

Frappell P, Lanthier C, Baudinette RV, Mortola JP (1992) Metabolism and ventilation in acute hypoxia: a comparative analysis in small mammalian species. Am J Physiol 262:R1040–R1046

Fries M, Stoppe C, Brücken D, Rossaint R, Kuhlen R (2009) Influence of mild therapeutic hypothermia on the inflammatory response after successful resuscitation from cardiac arrest. J Crit Care 24:453–457

Fukudome EY, Alam HB (2009) Hypothermia in multisystem trauma. Crit Care Med 37: S265–S272

Fuller DC, Suruda AJ (2000) Occupationally related hydrogen sulfide deaths in the United States from 1984 to 1994. J Occup Environ Med 42:939–942

Gautier H, Bonora M (1994) Ventilatory and metabolic responses to cold and CO-induced hypoxia in awake rats. Respir Physiol 97:79–91

Grigore AM, Murray CF, Ramakrishna H, Djaiani G (2009) A core review of temperature regimens and neuroprotection during cardiopulmonary bypass: does rewarming rate matter? Anesth Analg 109:1741–1751

Gurkan OU, O'Donnell C, Brower R, Ruckdeschel E, Becker PM (2003) Differential effects of mechanical ventilatory strategy on lung injury and systemic organ inflammation in mice. Am J Physiol 285:L710–L718

Haouzi P, Notet V, Chenuel B, Chalon B, Sponne I, Ogier V et al (2008) H_2S induced hypometabolism in mice is missing in sedated sheep. Respir Physiol Neurobiol 160:109–115

Hirsch EC, Hunot S (2009) Neuroinflammation in Parkinson's disease: a target for neuroprotection? Lancet Neurol 8:382–397

Kida K, Yamada M, Tokuda K, Marutani E, Kakinohana M, Kaneki M et al (2010) Inhaled hydrogen sulfide prevents neurodegeneration and movement disorder in a mouse model of Parkinson's disease. Antioxid Redox Signal 15:343–352

Korducki MJ, Forster HV, Lowry TF, Forster MM (1994) Effect of hypoxia on metabolic rate in awake ponies. J Appl Physiol 76:2380–2385

Langston JW, Ballard P, Tetrud JW, Irwin I (1983) Chronic Parkinsonism in humans due to a product of meperidine-analog synthesis. Science 219:979–980

Li L, Bhatia M, Zhu YZ, Zhu YC, Ramnath RD, Wang ZJ et al (2005) Hydrogen sulfide is a novel mediator of lipopolysaccharide-induced inflammation in the mouse. FASEB J 19:1196–1198

Li L, Rossoni G, Sparatore A, Lee LC, Del Soldato P, Moore PK (2007) Anti-inflammatory and gastrointestinal effects of a novel diclofenac derivative. Free Radic Biol Med 42:706–719
Li J, Zhang G, Cai S, Redington AN (2008a) Effect of inhaled hydrogen sulfide on metabolic responses in anesthetized, paralyzed, and mechanically ventilated piglets. Pediatr Crit Care Med 9:110–112
Li L, Whiteman M, Guan YY, Neo KL, Cheng Y, Lee SW et al (2008b) Characterization of a novel, water-soluble hydrogen sulfide-releasing molecule (GYY4137). Circulation 117:2351–2360
Li L, Salto-Tellez M, Tan CH, Whiteman M, Moore PK (2009) GYY4137, a novel hydrogen sulfide-releasing molecule, protects against endotoxic shock in the rat. Free Radic Biol Med 47:103–113
Lionetti V, Recchia FA, Ranieri VM (2005) Overview of ventilator-induced lung injury mechanisms. Curr Opin Crit Care 11:82–86
Matsuoka M, Igisu H, Tanaka I, Hori H, Koga M (1994) Effects of hypo- and hyperglycemia on brain energy metabolites in mice exposed to carbon monoxide. Toxicol Lett 73:135–143
Milby TH, Baselt RC (1999) Hydrogen sulfide poisoning: clarification of some controversial issues. Am J Ind Med 35:192–195
Morrison ML, Blackwood JE, Lockett SL, Iwata A, Winn RK, Roth MB (2008) Surviving blood loss using hydrogen sulfide. J Trauma 65:183–188
Mustafa AK, Gadalla MM, Sen N, Kim S, Mu W, Gazi SK et al (2009) H_2S signals through protein S-sulfhydration. Sci Signal 2:ra72
Papaiahgari S, Yerrapureddy A, Reddy SR, Reddy NM, Dodd OJ, Crow MT et al (2007) Genetic and pharmacologic evidence links oxidative stress to ventilator-induced lung injury in mice. Am J Respir Crit Care Med 176:1222–1235
Polderman KH (2009) Mechanisms of action, physiological effects, and complications of hypothermia. Crit Care Med 37:S186–S202
Reiffenstein RJ, Hulbert WC, Roth SH (1992) Toxicology of hydrogen sulfide. Annu Rev Pharmacol Toxicol 32:109–134
Schmidt-Nielsen K (1984) Scaling: why is animal size so important? Cambridge University Press, Cambridge, UK
Simon F, Giudici R, Duy CN, Schelzig H, Oter S, Groger M et al (2008) Hemodynamic and metabolic effects of hydrogen sulfide during porcine ischemia/reperfusion injury. Shock 30:359–364
Simonton DS, Spears M (2007) Human health effects from exposure to low-level concentrations of hydrogen sulfide. Occup Health Saf 76:102, 104
Singer D (2004) Metabolic adaptation to hypoxia: cost and benefit of being small. Respir Physiol Neurobiol 141:215–228
Slutsky AS, Tremblay LN (1998) Multiple system organ failure: is mechanical ventilation a contributing factor? Am J Respir Crit Care Med 157:1721–1725
Sodha NR, Clements RT, Feng J, Liu Y, Bianchi C, Horvath EM et al (2008) The effects of therapeutic sulfide on myocardial apoptosis in response to ischemia-reperfusion injury. Eur J Cardiothorac Surg 33:906–913
Struve MF, Brisbois JN, James RA, Marshall MW, Dorman DC (2001) Neurotoxicological effects associated with short-term exposure of Sprague–Dawley rats to hydrogen sulfide. Neurotoxicology 22:375–385
Szabo C (2007) Hydrogen sulphide and its therapeutic potential. Nat Rev Drug Discov 6:917–935
Tokuda K, Kida K, Marutani E, Crimi E, Bougaki M, Khatri A et al (2012) Inhaled hydrogen sulfide prevents endotoxin-induced systemic inflammation and improves survival by altering sulfide metabolism in mice. Antioxid Redox Signal 17:11–21
Vaporidi K, Francis RC, Bloch KD, Zapol WM (2010) Nitric oxide synthase 3 contributes to ventilator-induced lung injury. Am J Physiol 299:L150–L159

Volpato GP, Searles R, Yu B, Scherrer-Crosbie M, Bloch KD, Ichinose F et al (2008) Inhaled hydrogen sulfide: a rapidly reversible inhibitor of cardiac and metabolic function in the mouse. Anesthesiology 108:659–668

Whiteman M, Li L, Rose P, Tan CH, Parkinson DB, Moore PK (2010) The effect of hydrogen sulfide donors on lipopolysaccharide-induced formation of inflammatory mediators in macrophages. Antioxid Redox Signal 12:1147–1154

Wintner EA, Deckwerth TL, Langston W, Bengtsson A, Leviten D, Hill P et al (2010) A monobromobimane-based assay to measure the pharmacokinetic profile of reactive sulphide species in blood. Br J Pharmacol 160:941–957

Xie L, Hu LF, Teo XQ, Tiong CX, Tazzari V, Sparatore A et al (2013) Therapeutic effect of hydrogen sulfide-releasing L-Dopa derivative ACS84 on 6-OHDA-induced Parkinson's disease rat model. PLoS One 8:e60200

Zanardo RCO, Brancaleone V, Distrutti E, Fiorucci S, Cirino G, Wallace JL (2006) Hydrogen sulfide is an endogenous modulator of leukocyte-mediated inflammation. FASEB J 20:2118–2120

Zhang H, Zhi L, Moochhala S, Moore PK, Bhatia M (2007) Hydrogen sulfide acts as an inflammatory mediator in cecal ligation and puncture-induced sepsis in mice by upregulating the production of cytokines and chemokines via NF-{kappa}B. AJP – Lung Cell Mol Physiol 292:L960–L971

H_2S as a Bacterial Defense Against Antibiotics

8

Lyly Luhachack and Evgeny Nudler

Abstract

Hydrogen sulfide is a member of a group of gaseous signaling molecules, termed gasotransmitters. Like eukaryotes, most bacterial species encode a putative enzyme for the generation of hydrogen sulfide. In recent years intense attention has been given to this gasotransmitter as a signaling molecule participating in fundamental cellular processes in eukaryotes. For bacteria, hydrogen sulfide presents a novel-signaling pathway that is currently poorly understood. The only function that has been described for bacterial hydrogen sulfide is its role as a defense system against antibiotics. A better and more comprehensive understanding of the pathways and mechanisms underlying the actions of H_2S can potentiate our existing knowledge in combating infectious diseases and provide a new target for drug development.

Keywords

H_2S • Cystathionine beta-synthase • Cystathionine gamma-lyase • Nitric oxide • Bacteria • Oxidative stress • Antibiotics • Resistance

Abbreviations

3MST 3-Mercaptopyruvate sulfurtransferase
bNOS Bacterial nitric oxide synthase
CBS Cystathionine beta-synthase

L. Luhachack • E. Nudler (✉)
Department of Biochemistry, New York University School of Medicine, New York, NY 10016, USA
e-mail: evgeny.nudler@nyumc.org

H. Kimura (ed.), *Hydrogen Sulfide and its Therapeutic Applications*, DOI 10.1007/978-3-7091-1550-3_8,

CSE Cystathionine gamma-lyase
ROS Reactive oxygen species
SRB Sulfate reducing bacteria

8.1 Introduction

Hydrogen sulfide (H_2S) has long been recognized as a malodorous environmental toxin. Likewise the production of H_2S by bacteria as a by-product of sulfur metabolism was established almost a century ago (Clarke 1953). It is curious then, that the function of H_2S in bacteria had never been investigated before recent times. With the inception of gasotransmitter biology, which interestingly enough arose as a field of study in eukaryotes, the exploration for the physiological role of these gasotransmitters has now expanded to microorganisms.

The first gasotransmitter established in bacteria, was nitric oxide with the discovery of bacterial nitric oxide synthase (bNOS) (Adak et al. 2002; Gusarov and Nudler 2005). Bioinformatic analysis of sequenced bacterial genomes revealed the existence of putative bNOS homologs in a limited number of gram-positive species (Gusarov et al. 2008). This led to a search for bacterial orthologs of the mammalian enzymes that generate H_2S: cystathionine beta-synthase (CBS), cystathionine gamma-lyase (CSE) and 3-mercaptopyruvate sulfurtransferase (3MST). Unlike bNOS, H_2S producing enzymes are widely conserved as most sequenced species contain putative orthologs of at least one of its mammalian counterpart in the genome (Shatalin et al. 2011). This broad conservation is suggestive of a fundamental role for H_2S in bacteria and this chapter will serve to highlight one fascinating facet of the gasotransmitter, that of mediating resistance to antibiotics.

8.2 Sources of Hydrogen Sulfide

Long before the discovery of dedicated enzymes for endogenous hydrogen sulfide generation, it was known that H_2S could be produced through the sulfate reduction pathway by anaerobic sulfate reducing bacteria (SRB) (Wang 2012). By the same token some enteric bacteria are able to produce H_2S from the reduction of thiosulfate (Stoffels et al. 2012). On a macroscopic scale, these SRBs were of concern as an environmental pollutant as the accumulation of H_2S corrupted oil reservoirs and wastewater treatments (Nemati et al. 2001; Sublette et al. 1998). The shift from investigating H_2S not only on a macroscale but also microscale became of particular interest as it became evident that at molecular physiological concentration, this gas molecule possessed basic functions in the human body. SRBs reside in the human intestinal tract as part of the microbial community and as we come to understand the importance of our microbiome in normal physiology and pathophysiology, the exact function of sulfur metabolism by SRBs in that context will need to be defined (Wang 2012).

There are three ascertained enzymes which produce H_2S: cystathionine beta-synthase (CBS), cystathionine gamma-lyase (CSE) and 3-mercaptopyruvate sulfurtransferase (3MST). All three proteins utilize cysteine as a substrate and CBS/CSE can also metabolize homocysteine in a series of different reactions to produce H_2S and other metabolites (Mondovi et al. 1963; Wang 2012). CBS and CSE are pyridoxal 5'-phosphate dependent enzymes. 3MST first requires the catalytic activity of aspartate aminotransferase with cysteine aminotransferase activity to form a sulfane sulfur product, which can then release hydrogen sulfide in the presence of a reducing agent (Mikami et al. 2011; Singh and Banerjee 2011). In bacteria, the in vivo discovery of these enzymes was established in four different pathogenic bacteria. *Bacillus anthracis*, *Pseudomonas aeruginosa*, and *Staphylococcus aureus* encode CBS/CSE and *Escherichia coli* encode 3MST (Shatalin et al. 2011). While other protein candidates exist for the enzymatic generation of hydrogen sulfide, such as cysteine desulfurases, their overall contribution to the pool of cellular H_2S would need to be examined. For E. coli at least, 3MST is the major source of H_2S (Shatalin et al. 2011). In addition to enzymatically-generated sources of H_2S, acid-labile sulfur and reductant labile sulfur might also provide an additional supply (Shen et al. 2012; Stoffels et al. 2012).

8.3 Antibiotic Resistance

Increasingly, there has been growing concern regarding multidrug resistant pathogens in recent years. This problem is further compounded by the concurrent lack of new classes of antibiotics being developed (Cooper and Shlaes 2011). To put it into perspective, treating antibiotic resistant infections comes with an upwards $20 billion yearly price tag in the United States alone (Bush et al. 2011; Roberts et al. 2009). While the push for developing new antibiotics is certainly valid, it does not address the concern that combatting infectious diseases with antibiotics will only continually give rise to a new population that will eventually develop resistance. Therefore searching for novel antimicrobial targets and novel therapeutic intervention is an important alternative or in conjunction to existing therapies. Here, we report the existence of a universal defense system in bacteria that serves to protect against antibiotics, the gasotransmitter H_2S. Directly targeting a prevalent defensive mechanism is particularly advantageous in that the efficacy of any therapeutic measures will encompass both gram-positives and gram-negatives. Furthermore, exploiting H_2S will not require developing a new drug but augmenting current and old drugs that are already FDA approved.

In eukaryotes, hydrogen sulfide mediates diverse responses in multiple systems, from the cardiovascular system to immune system and the nervous system (Wang 2012). However, to date, the only definitive work investigating the role for H_2S in bacteria is its function as a novel and universal defense system in pathogenic bacteria against antibiotics. Inactivation of the hydrogen sulfide generating enzymes in the aforementioned pathogens increased antibiotic susceptibility to different classes of antibiotics (Shatalin et al. 2011). In accordance with an earlier study of bNOS in bacteria, nitric oxide (NO) also provided the same protective ability in gram-positives that encoded bNOS (Gusarov et al. 2009). The authors proposed that H_2S and NO imparted this defensive quality by sequestering iron that fuels the damaging Fenton reaction and augmenting the capabilities of the cells' antioxidant activity (Gusarov et al. 2009; Shatalin et al. 2011).

The first notion that hydrogen sulfide might prove to be an important signaling molecule in bacteria, stemmed from work first done with NO signaling. The breadth of roles NO played in different species ranged from stress resistance, virulence, host response modulation and cellular communication (Bowman et al. 2011; Carlson et al. 2010; Gusarov et al. 2009, 2013; Gusarov and Nudler 2005; Kers et al. 2004; Shatalin et al. 2008). In 2009, Gusarov et al., discovered that in species expressing bNOS, nitric oxide protected against a variety of antimicrobials. Interestingly, NO in *Bacillus subtilis* and *B. anthracis* also provided a small measure of protection against a toxin, pyocyanin, secreted by *P. aeruginosa* (Gusarov et al. 2009). Possibly this is indicative of an evolutionarily advantageous strategy deployed by microbes in its natural niche, that extends beyond the specific protection conferred against antibiotics in a clinical setting. From this report, we next searched for orthologs of eukaryotic H_2S generating enzymes in bacteria. When it became evident that the expression of these enzymes were much more prevalent than bNOS, the logical hypothesis constructed from that was hydrogen sulfide could also be cytoprotective in the same manner as NO, on a universal scale. Indeed, this was found to be the case when wild type and H_2S deficient mutants in *B. anthracis, S. aureus, P. aeruginosa*, and *E. coli* were exposed to antibiotics. The mutants were significantly more susceptible to antibiotic toxicity compared to the wild type. Chemical complementation of H_2S to the media could restore wild type resistant phenotype to the mutants (Shatalin et al. 2011). Recall that *B. anthracis* also possesses bNOS in addition to CBS/CSE while the other three pathogens don't express bNOS. Attempts at constructing genetic knock-outs in both bNOS and CBS/CSE in *B. anthracis* revealed that the combination resulted in a lethal mutation. Moreover chemical inhibition of one enzyme in a strain with a mutation in the other enzyme to mimic a double mutant revealed that this yielded a population that was more susceptible to antibiotics compared to a single mutant. The authors concluded that for species containing both enzymatic machineries, NO and H_2S act in synergy as a cytoprotectant against antibiotics (Shatalin et al. 2011).

How does NO and to a broader extent H_2S impart a protective effect against antibiotics? In 2005, one of the earlier reports that elucidated a role for nitric oxide in bacteria, discovered that NO conferred resistance against hydrogen peroxide induced oxidative stress (Gusarov and Nudler 2005). This posed an intriguing

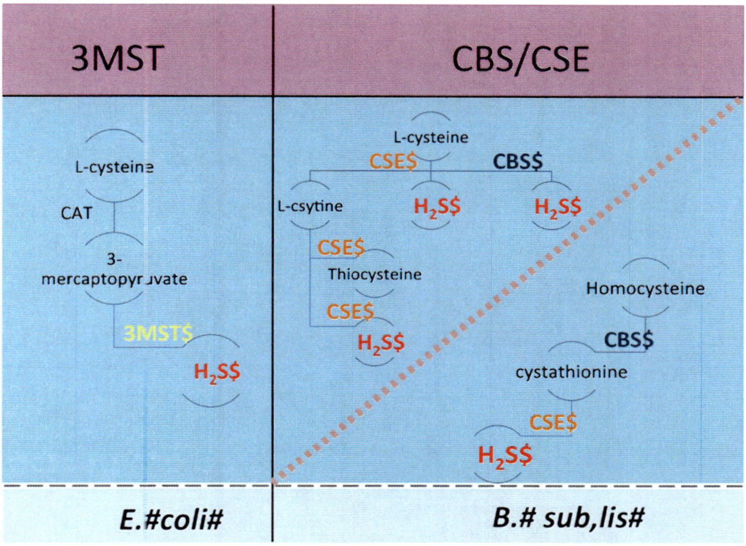

Fig. 8.1 The two pathways of endogenous H_2S production in model microorganisms. Only major products are shown. In the case of *E. coli*, H_2S generated via 3MST activity is through catabolism of L-cysteine by a cysteine aminotransferase (*CAT*) which forms a sulfane sulfur product and following reduction of this product, H_2S is produced along with pyruvate and ammonia. In the case of *B. subtilis*, three different reactions metabolize L-cysteine to generate H_2S. The CBS/CSE operon can also utilize homocysteine as a substrate to produce H_2S (Mikami et al. 2011; Singh and Banerjee 2011)

possibility that gasotransmitters mediating a defensive response against both antibiotics and oxidative stress might converge on the same pathway. Certainly, a plethora of literature has cropped up in recent years proposing a controversial new model; antibiotics induce oxidative stress leading to cellular death (Albesa et al. 2004; Kohanski et al. 2007, 2010) (Figs. 8.1 and 8.2). However the model remains incomplete and the mechanism behind such a singular pathway of antibiotics lethality is not yet understood. Efforts aimed at dissecting the pathway are hindered by the lack of a sensitive standardized assay that can directly measure reactive oxygen species (ROS) levels and the complex dynamics entailed in bacterial physiology. While there is mounting support for this oxidative stress dependency hypothesis, there are disputing reports that antibiotic killing does not involve ROS (Keren et al. 2013; Liu and Imlay 2013).

Regardless, the hydrogen sulfide (and nitric oxide) protective model convincingly establishes the fact that H_2S endows the bacteria with a novel defense system against antibiotics and oxidative stress (Shatalin et al. 2011). Whether this involves the same mechanism or not remains to be seen. It is possible that the missing link in the aforementioned pathway is H_2S. At the very least ROS is a contributing factor

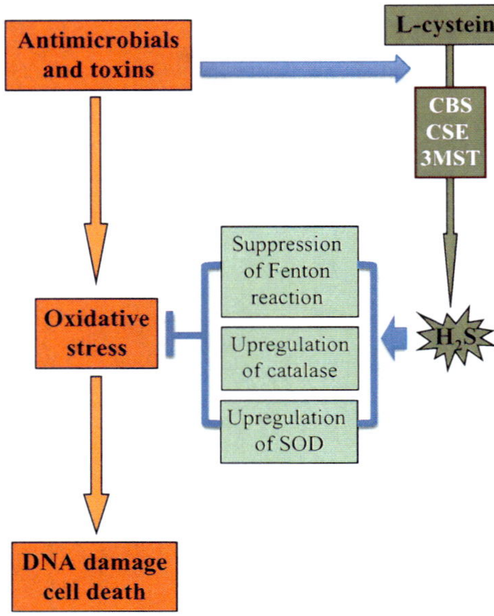

Fig. 8.2 **Model of H_2S mediated protection against antibiotics toxicity**. Antibiotics induce oxidative stress in the cell, generating ROS, which leads to DNA damage and eventual cell death. H_2S interferes in this pathway by inhibiting the Fenton reaction, which generates hydroxyl radicals. The inhibition of the Fenton reaction occurs through a direct sequestration of Fe^{2+} by H_2S and also via depleting cells of free reduced cysteine, which is a substrate for H_2S enzymes. Cysteine effectively re-reduces Fe, thus fueling the Fenton reaction (Park and Imlay 2003; Gusarov and Nudler 2005). In addition, H_2S boosts the activity of the cell's major antioxidants – superoxide dismutase (*SOD*) and catalase (Shatalin et al. 2011)

as H_2S mutants were equally as resistant to antibiotics as its wild type counterpart when grown under anaerobic conditions, suggesting a dependency on the presence of oxygen (Shatalin et al. 2011).

Conclusions

The role of hydrogen sulfide in bacteria provides an intriguing new field of research but it is obvious that much is still left to be uncovered. The duality of the gas to be both toxic and beneficial necessitates its tight regulation and how this is achieved must be clarified. The clear role of hydrogen sulfide as a unique viable target for antimicrobial therapy is especially imperative. The exact mechanism behind this cytoprotection must be defined and might unearth other targets in the pathway. Beyond its clinically valuable implications, the wide conservation of H_2S from bacteria to eukaryotes presents a new opportunity of research in the basic physiological functions of cell signaling.

References

Adak S, Aulak KS, Stuehr DJ (2002) Direct evidence for nitric oxide production by a nitric-oxide synthase-like protein from Bacillus subtilis. J Biol Chem 277(18):16167–16171. doi:10.1074/jbc.M201136200

Albesa I, Becerra MC, Battan PC, Paez PL (2004) Oxidative stress involved in the antibacterial action of different antibiotics. Biochem Biophys Res Commun 317(2):605–609. doi:10.1016/j.bbrc.2004.03.085

Bowman LA, McLean S, Poole RK, Fukuto JM (2011) The diversity of microbial responses to nitric oxide and agents of nitrosative stress close cousins but not identical twins. Adv Microb Physiol 59:135–219. doi:10.1016/B978-0-12-387661-4.00006-9

Bush K, Courvalin P, Dantas G, Davies J, Eisenstein B, Huovinen P, Jacoby GA, Kishony R, Kreiswirth BN, Kutter E, Lerner SA, Levy S, Lewis K, Lomovskaya O, Miller JH, Mobashery S, Piddock LJ, Projan S, Thomas CM, Tomasz A, Tulkens PM, Walsh TR, Watson JD, Witkowski J, Witte W, Wright G, Yeh P, Zgurskaya HI (2011) Tackling antibiotic resistance. Nat Rev Microbiol 9(12):894–896. doi:10.1038/nrmicro2693

Carlson HK, Vance RE, Marletta MA (2010) H-NOX regulation of c-di-GMP metabolism and biofilm formation in Legionella pneumophila. Mol Microbiol. doi:10.1111/j.1365-2958.2010.07259.x

Clarke PH (1953) Hydrogen sulphide production by bacteria. J Gen Microbiol 8(3):397–407

Cooper MA, Shlaes D (2011) Fix the antibiotics pipeline. Nature 472(7341):32. doi:10.1038/472032a

Gusarov I, Nudler E (2005) NO-mediated cytoprotection: instant adaptation to oxidative stress in bacteria. Proc Natl Acad Sci U S A 102(39):13855–13860. doi:10.1073/pnas.0504307102

Gusarov I, Starodubtseva M, Wang ZQ, McQuade L, Lippard SJ, Stuehr DJ, Nudler E (2008) Bacterial nitric-oxide synthases operate without a dedicated redox partner. J Biol Chem 283(19):13140–13147. doi:10.1074/jbc.M710178200

Gusarov I, Shatalin K, Starodubtseva M, Nudler E (2009) Endogenous nitric oxide protects bacteria against a wide spectrum of antibiotics. Science 325(5946):1380–1384. doi:10.1126/science.1175439

Gusarov I, Gautier L, Smolentseva O, Shamovsky I, Eremina S, Mironov A, Nudler E (2013) Bacterial nitric oxide extends the lifespan of C. elegans. Cell 152(4):818–830. doi:10.1016/j.cell.2012.12.043

Keren I, Wu Y, Inocencio J, Mulcahy LR, Lewis K (2013) Killing by bactericidal antibiotics does not depend on reactive oxygen species. Science 339(6124):1213–1216. doi:10.1126/science.1232688

Kers JA, Wach MJ, Krasnoff SB, Widom J, Cameron KD, Bukhalid RA, Gibson DM, Crane BR, Loria R (2004) Nitration of a peptide phytotoxin by bacterial nitric oxide synthase. Nature 429(6987):79–82. doi:10.1038/nature02504

Kohanski MA, Dwyer DJ, Hayete B, Lawrence CA, Collins JJ (2007) A common mechanism of cellular death induced by bactericidal antibiotics. Cell 130(5):797–810. doi:10.1016/j.cell.2007.06.049

Kohanski MA, Dwyer DJ, Collins JJ (2010) How antibiotics kill bacteria: from targets to networks. Nat Rev Microbiol 8(6):423–435. doi:10.1038/nrmicro2333

Liu Y, Imlay JA (2013) Cell death from antibiotics without the involvement of reactive oxygen species. Science 339(6124):1210–1213. doi:10.1126/science.1232751

Mikami Y, Shibuya N, Kimura Y, Nagahara N, Ogasawara Y, Kimura H (2011) Thioredoxin and dihydrolipoic acid are required for 3-mercaptopyruvate sulfurtransferase to produce hydrogen sulfide. Biochem J 439(3):479–485. doi:10.1042/BJ20110841

Mondovi B, Scioscia-Santoro A, Cavallinid D (1963) Further evidence on the identity of cystathionase and cysteine desulfhydrase. Arch Biochem Biophys 101:363–364

Nemati M, Jenneman GE, Voordouw G (2001) Mechanistic study of microbial control of hydrogen sulfide production in oil reservoirs. Biotechnol Bioeng 74(5):424–434

Park S, Imlay JA (2003) High levels of intracellular cysteine promote oxidative DNA damage by driving the fenton reaction. J Bacteriol 185:1942–1950

Roberts RR, Hota B, Ahmad I, Scott RD, Foster SD, Abbasi F, Schabowski S, Kampe LM, Ciavarella GG, Supino M, Naples J, Cordell R, Levy SB, Weinstein RA (2009) Hospital and societal costs of antimicrobial-resistant infections in a Chicago teaching hospital: implications for antibiotic stewardship. Clin Infect Dis 49(8):1175–1184. doi:10.1086/605630

Shatalin K, Gusarov I, Avetissova E, Shatalina Y, McQuade LE, Lippard SJ, Nudler E (2008) Bacillus anthracis-derived nitric oxide is essential for pathogen virulence and survival in macrophages. Proc Natl Acad Sci U S A 105(3):1009–1013. doi:10.1073/pnas.0710950105

Shatalin K, Shatalina E, Mironov A, Nudler E (2011) H2S: a universal defense against antibiotics in bacteria. Science 334(6058):986–990. doi:10.1126/science.1209855

Shen X, Peter EA, Bir S, Wang R, Kevil CG (2012) Analytical measurement of discrete hydrogen sulfide pools in biological specimens. Free Radic Biol Med 52(11–12):2276–2283. doi:10.1016/j.freeradbiomed.2012.04.007

Singh S, Banerjee R (2011) PLP-dependent H(2)S biogenesis. Biochim Biophys Acta 1814 (11):1518–1527. doi:10.1016/j.bbapap.2011.02.004

Stoffels L, Krehenbrink M, Berks BC, Unden G (2012) Thiosulfate reduction in Salmonella enterica is driven by the proton motive force. J Bacteriol 194(2):475–485. doi:10.1128/JB.06014-11

Sublette KL, Kolhatkar R, Raterman K (1998) Technological aspects of the microbial treatment of sulfide-rich wastewaters: a case study. Biodegradation 9(3–4):259–271

Wang R (2012) Physiological implications of hydrogen sulfide: a whiff exploration that blossomed. Physiol Rev 92(2):791–896. doi:10.1152/physrev.00017.2011

Hydrogen Sulfide-Mediated Cellular Signaling and Cytoprotection

9

Hideo Kimura

Abstract

Hydrogen sulfide (H_2S) is a signaling molecule in the nervous and vascular systems. It also protects various organs from oxidative stress or ischemia-reperfusion injury. H_2S is produced from L-cysteine by cystathionine β-synthase (CBS), cystathionine γ–lyase (CSE) and 3-mercaptopyruvate sulfurtransferase (3MST) along with cysteine aminotransferase (CAT). We recently found that H_2S is also produced from D-cysteine by 3MST along with D-amino acid oxidase (DAO). This pathway is mainly localized to in the cerebellum and the kidney, producing H_2S more efficiently than the pathways that utilize L-cysteine as a substrate. The administration of D-cysteine to mice ameliorates renal ischemia-reperfusion injury more effectively than that of L-cysteine, promising a therapeutic application of D-cysteine to renal diseases. We recently found that H_2S-derived polysulfides exist in the brain and induce Ca^{2+} influx by activating transient receptor potential ankyrin-1 (TRPA1) channels approximately 300 times more efficiently than H_2S in astrocytes, which surround neuronal synapses and modulate their activity. Polysulfides are possible H_2S-derived signaling molecules. This review focuses on the production of H_2S from D-cysteine and polysulfides as possible signaling molecules derived from H_2S.

Keywords

H_2S • Polysulfides • Bound sulfane sulfur • CBS • CSE • 3MST • CAT • DAO • Ca^{2+} • TRP • D-cysteine • Kidney • Cytoprotective

H. Kimura (✉)
Division of Molecular Pharmacology, National Institute of Neuroscience, Ogawa-Higashi, Kodeira 4-1-1, Tokyo 187-8502, Japan
e-mail: kimura@ncnp.go.jp

H. Kimura (ed.), *Hydrogen Sulfide and its Therapeutic Applications*,
DOI 10.1007/978-3-7091-1550-3_9,

Abbreviations

3MP	3-Mercaptopyruvate
3MST	3-Mercaptopyruvate sulfurtransferase
Ca^{2+}	Calcium ion
CAT	Cysteine aminotransferase
CBS	Cystathionine β–synthase
cGMP	Cyclic guanosine monophosphate
CHO	Chinese hamster ovary
CO	Carbon monoxide
CO_2	Carbon dioxide
CSE	Cystathionine γ–lyase
DAO	D-amino acid oxidase
DHLA	Dihydrolipoic acid
DTT	Dithiothreitol
EDHF	Endothelium-derived hyperpolarizing factor
GAPDH	Glyceraldehydes-3-phosphate dehydrogenase
Gd^{3+}	Gadolinium ion
GFAP	Glial fibrillary acidic protein
H_2O_2	Hydrogen peroxide
H_2S	Hydrogen sulfide
HEK	Human embryonic kidney
HPLC	High pressure liquid chromatography
HSNO	Thionitrous acid
I_2CA	Indol-2-carboxylate
K^+	Potassium ion
K_{ATP} channel	ATP-dependent K^+ Channel
La^{3+}	Lanthanum
LTP	Long-term potentiation
NAC	N-acetylcysteine
NAD	Nicotinamide adenine dinucleotide
NaHS	Sodium hydrosulfide
NF	Nuclear factor
NMDA	N-methyl-D-aspartate
NO	Nitric oxide
Nrf_2	Nuclear factor-E_2-related factor
O_2	Oxygen
PLP	Pyridoxal 5'-phosphate
ROS	Reactive oxygen species
SAM	*S*-adenocyl-L-methionine
siRNA	Small interfering ribonucleic acid
STAT-3	Signal transducers and activators of transcription 3

TRPA1	transient receptor potential ankyrin-1
V-H^+ ATPase	Vacuolar-type H^+ adenosine trisphosphatase
γ-GCS	γ-glutamyl-cysteine synthase

9.1 Introduction

The endogenous levels of sulfide were measured and found to be relatively abundant in the brains of humans, bovine, and rats (Goodwin et al. 1989; Warenycia et al. 1989; Savage and Gould 1990). This finding led us to the study of H_2S production and its physiological function in the brain. We found that the H_2S-producing enzymes, cystathionine β-synthase (CBS) and 3-mercaptopyruvate sulfurtransferase (3MST), are localized in the astrocytes, a type of glia, and neurons, respectively (Abe and Kimura 1996; Enokido et al. 2005; Ichinohe et al. 2005; Shibuya et al. 2009b). H_2S facilitates the induction of hippocampal long-term potentiation (LTP) by enhancing the activity of NMDA receptors (Abe and Kimura 1996). We also found that another H_2S producing enzyme, cystathionine γ–lyase (CSE), expressed vascular smooth muscle, the portal vein, and the ileum. Additionally, H_2S relaxes the smooth muscle in synergy with nitric oxide (NO) (Hosoki et al. 1997). Based on these observations, we proposed that H_2S may function as a neuromodulator in the brain and as a smooth muscle relaxant. The effect of H_2S on smooth muscle relaxation was later confirmed with the activation of K_{ATP}- and small-conductance K^+-channels to relax vascular smooth muscle (Zhao et al. 2001; Teague et al. 2002; Mustafa et al. 2011).

Others and we have discovered the cytoprotective effect of H_2S. H_2S protects neurons from oxidative stress and ischemia-reperfusion injury by enhancing the production of glutathione, an intracellular major antioxidant, as well as by scavenging reactive oxygen species, although to a lesser extent (Kimura and Kimura 2004; Whiteman et al. 2004; Kimura et al. 2006, 2010). These findings led to the identification of the protective effect on the heart and, later on, various other organs, including the kidneys, against ischemia-reperfusion injury (Elrod et al. 2007; Tripatara et al. 2008; Bos et al. 2009; Zhu et al. 2012). The third phase of clinical trials assessing the application of N-acetyl-cysteine, as a substrate for the production of H_2S, with regards to chronic renal disease has recently been completed (Clinical Trials.gov Identifier: NCT01232257). The application of H_2S to renal transplantation has also been proposed (Zhu et al. 2012; Razzak 2012).

In addition to the three pathways for the production of H_2S from L-cysteine, we recently discovered a fourth pathway utilizing D-cysteine as the substrate (Shibuya et al. 2013). This pathway is localized in the kidney and the brain, especially in the cerebellum. H_2S production from D-cysteine is 80 times greater than L-cysteine, and the oral administration of D-cysteine protects the kidney more efficiently than that of L-cysteine. Because D-cysteine is less toxic to tissues than L-cysteine, D-cysteine has greater potential for therapeutic applications (Misra 1989; Olney et al. 1990).

The levels of H_2S in tissues were recently re-evaluated and found to be much lower than initially reported, but this finding did confirm the existence of H_2S in tissues (Furne et al. 2008; Ishigami et al. 2009; Wintner et al. 2010; Levitt et al. 2011). The endogenous steady-state levels of H_2S are between 14 nM and 2 μM, while at least 10 μM H_2S is required to exert its effects. A 5 ~ 1,000-fold difference has been observed in concentrations between the endogenous and effective concentrations. This discrepancy can be partly explained by the fact that: (1) H_2S is unstable and its levels are difficult to correctly evaluate; (2) only basal steady-state levels of H_2S have been measured, but how much of an increase in its levels can be achieved in activated cells has not yet been determined; and (3) H_2S may not be a final effector molecule but an intermediate effector molecule. We recently discovered that polysulfides, which are generated from H_2S and exist at a concentration of 25 μmol/kg in wet brain tissue, induce Ca^{2+} influx by activating the transient receptor potential ankyrin-1 (TRPA1) channels in astrocytes 300 times more efficiently than H_2S. The ED_{50} of polysulfide to induce Ca^{2+} influx is approximately 90 nM, while that of H_2S is 116 μM (Nagai et al. 2006; Oosumi et al. 2010; Kimura et al. 2013). Polysulfides are thought to be possible H_2S-derived signaling molecules (Kimura et al. 2013).

9.2 H_2S-Producing Enzymes

Three pathways can produce H_2S from L-cysteine (i.e., CBS, CSE, and 3MST along with CAT) (Stipanuk and Beck 1982; Abe and Kimura 1996; Hosoki et al. 1997; Shibuya et al. 2009a, b). Another pathway with DAO and 3MST produces H_2S from D-cysteine (Shibuya et al. 2013).

9.2.1 CBS

CBS produces H_2S by a β-replacement reaction with L-cysteine and by catalyzing a β-replacement of L-cysteine with L-homocysteine (Cavallini et al. 1962; Braunstein et al. 1971; Chen et al. 2004; Singh et al. 2009). CBS has two regulatory sites, including the S-adenocyl-L-methionine (SAM) binding domain at the carboxy-terminus and a heme group at the amino terminus. SAM enhances the activity of CBS. The catalytic site, which is located at the center of the enzyme, is covered by the carboxyl-terminal SAM domain in the absence of SAM (Shan et al. 2001). The binding of SAM releases the SAM-binding domain and exposes the catalytic site, thereby activating the enzyme (Fig. 9.1).

The heme group also plays a regulatory role in CBS (Taoka and Banerjee 2001). The Fe(II) form of the enzyme is inhibited upon the binding of CO or NO, while the activity is enhanced when Fe(II) is oxidized to Fe(III). This regulation plays an important role in the microcirculation in the brain (Morikawa et al. 2012). In hypoxia, the production of CO by heme oxygenase-2 in neurons is diminished, and the suppression of CBS by CO is released. The end result is an increase in the production of H_2S from CBS located in astrocytes. Capillaries surrounded by astrocytes are relaxed by H_2S. By this mechanism, a deficiency in the oxygen supply is compensated by an increase in blood flow.

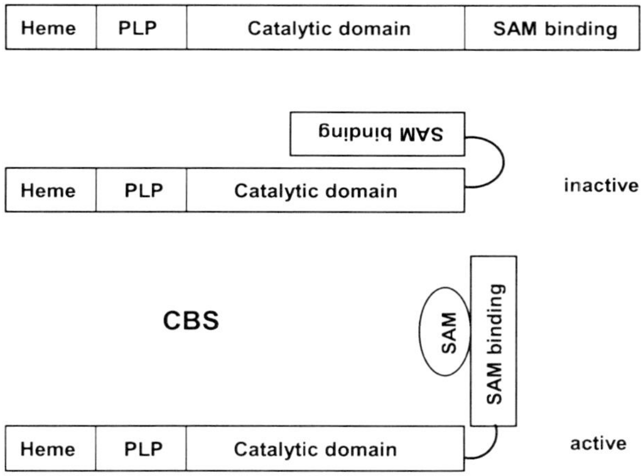

Fig. 9.1 Regulation of CBS activity by S-adenosyl methionine

9.2.2 CSE

CSE produces H_2S by the α,β-elimination reaction with L-cysteine (Singh et al. 2009; Chiku et al. 2009). CSE has approximately an 11-fold higher affinity for cystine than cysteine (Stipanuk and Beck 1982; Mikami et al. 2013). However, cystine is mainly localized in the extracellular fluid, while cysteine is the dominant form in cytosol where CSE is localized. Therefore, cysteine appears to be a preferable substrate for CSE in cells rather than cystine.

We initially reported that CSE is expressed in the vascular system and H_2S relaxes vascular smooth muscle (Hosoki et al. 1997). The relaxing effect of H_2S on vascular smooth muscle is weaker than the effect on other smooth muscles, such as the portal vain and ileum. However, we found a synergistic effect between H_2S and NO on the relaxation of vasculature (Hosoki et al. 1997). Recently, a substance known as thionitrous acid (HSNO), which is produced by the reaction between H_2S and S-nitrosothiols, has been identified (Filipovic et al. 2012). HSNO freely diffuses through the plasma membrane and may be a new signaling molecule.

We have previously described the transcriptional regulation of CSE, which has an SP1 binding site in its promoter region (Ishii et al. 2004). The SP1 binding site is activated by the multifunctional, proinflammatory cytokine tumor necrosis factor alpha (TNF-α), leading to an increase in the expression of CSE and the production of H_2S (Sen et al. 2012). TNF-α also activates the IκB kinase complex that phosphorylates IκB, resulting in IκB degradation and NF-κB translocation to the nucleus. Furthermore, H_2S produced by CSE activates NF-κB. This pathway plays an important role in the regulation of the antiapoptotic actions of NF-κB.

Although it was reported that Ca^{2+}/calmodulin regulates the H_2S-producing activity of CSE, this activity was examined in the presence of 1–2 mM Ca^{2+} in a previous study (Yang et al. 2008). The intracellular Ca^{2+} concentration was approximately 100 nM in the steady-state and increased to 2–3 μM in the activated cells. Therefore, we re-evaluated the effect of Ca^{2+} on the regulation of CSE

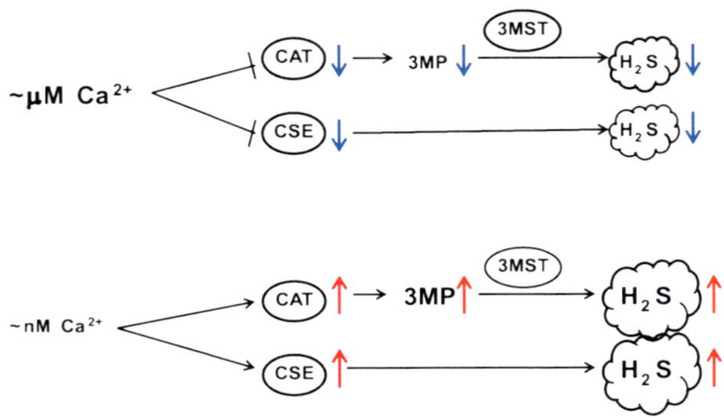

Fig. 9.2 Ca^{2+} suppresses the activity of CAT and CSE

activity (Mikami et al. 2013). The H_2S-producing activity of CSE, which is the pyridoxal 5′-phosphate (PLP)-dependent enzyme, is regulated by Ca^{2+} differently in the presence or absence of PLP. In the presence of PLP, the H_2S-producing activity is at its maximum potential in the absence of Ca^{2+}. It is suppressed by Ca^{2+} in a concentration-dependent manner up to 300 nM and the level is retained at higher Ca^{2+} concentrations (Mikami et al. 2013). In contrast, in the absence of PLP, the CSE activity is minimal in the absence of Ca^{2+} and increased by Ca^{2+} up to 300 nM. However, the maximum activity observed in the absence of PLP is lower than that in the presence of PLP. These observations suggest that H_2S may be constitutively produced by CSE in quiescent cells and that production is suppressed when Ca^{2+} influx is induced by physiological stimulations (Fig. 9.2). Calmodulin and its selective inhibitor W-7, regardless of the presence or absence of PLP, do not change CSE activity, suggesting that calmodulin is not involved in the regulation of CSE by Ca^{2+} (Mikami et al. 2013).

A possible mechanism of the regulation of CSE activity by Ca^{2+} is that PLP, which links to CSE, dissociates from CSE to form a new linkage with the substrate cysteine to produce H_2S (Mikami et al. 2013). The steady-state, low Ca^{2+} concentrations facilitate the linkage formation between cysteine and PLP, resulting in enhanced H_2S production. However, the linkage formation is suppressed when Ca^{2+} concentrations are increased, leading to the suppression of H_2S production.

9.2.3 3MST/CAT Pathway

CBS has been thought to be the major H_2S-producing enzyme in the brain. However, we found that H_2S is produced even in the brains of CBS knockout mice, suggesting that there is another enzyme or pathway in the brain. It was later identified as the 3MST and CAT pathway (Shibuya et al. 2009b). From cysteine and α–ketoglutarate, CAT produces 3-mercaptopyruvate (3MP), which is subsequently metabolized to H_2S by 3MST (Meister et al. 1954; Taniguchi and Kimura 1974; Ubuka et al. 1978; Cooper 1983; Shibuya et al. 2009b).

3MST is primarily localized in the mitochondria and secondarily in the cytosol. 3MST has a signal sequence at its amino-terminus that targets 3MST to mitochondria. Although the signal sequence of most mitochondrial proteins is removed after they are transported to the mitochondria, the signal sequence of 3MST is retained in the mitochondria without being processed (Nagahara and Nishino 1996). There are two forms of CAT, (i.e., mitochondrial- and cytosolic-CAT). Both CATs share 48 % identity in their amino acid sequences (Ubuka et al. 1978; Akagi 1982; Doyle et al. 1990). The pathway with 3MST and CAT mainly produces H_2S in the mitochondria and, to a lesser extent, cytosol (Shibuya et al. 2009b). Unlike cytosol, mitochondria contain sufficiently high concentrations of cysteine (approximately 1 mM), which is enough for the 3MST/CAT pathway to produce H_2S (Griffith 1999; Tateishi et al. 1977; Kimura et al. 2010). Mitochondria are major organelles that produce large amounts of reactive oxygen species (ROS). H_2S produced mitochondria can scavenge ROS to decrease oxidative stress (Kimura et al. 2010). H_2S produced by the 3MST/CAT pathway also plays an important role in mitochondrial electron transport and oxidative phosphorylation, and regulates cellular bioenergetics (Modis et al. 2013).

3MST requires a reducing substance, such as dithiothreitol (DTT), to produce H_2S from 3MP, while neither CBS nor CSE require such a reducing substance. Although 3MP can interact with thioredoxin, an endogenous reducing substance for the production of H_2S was not known (Westrop et al. 2009). In the current study, we found that 3MST produces H_2S in the presence of dithiols, such as thioredoxin and dihydrolipoic acid (DHLA) (Mikami et al. 2011a). Thioredoxin is found at a concentration of ~20 μM and is four times as potent as DTT in releasing H_2S from 3MST, while DHLA is found at a concentration of ~40 μM in brain mitochondria and is as potent as DTT. The observations that other reducing substances with greater reducing potential do not release H_2S from 3MST and that thioredoxin, DHLA, and DTT are dithiols suggest that dithiol is a critical factor for releasing H_2S from 3MST (Mikami et al. 2011a). A possible mechanism is that 3MST receives sulfide from 3MP to produce 3MST persulfide, which is transferred to one of the thiol residues in thioredoxin or DHLA to produce thioredoxin persulfide or DHLA persulfide. The remaining thiol in thioredoxin or DHLA attacks the persulfide to release sulfide.

DHLA also plays an important role in the transcription of genes that are involved in the production of H_2S. H_2S, which is produced by 3MST in the presence of DHLA, and α-lipoic acid, an oxidized form of DHLA, both activate nuclear factor-E2-related factor (Nrf_2), which is a nuclear basic leucine zipper transcription factor. Nrf_2, in turn, increases the expression of thioredoxin and thioredoxin reductase that regulate the production of H_2S (Suh et al. 2004). Therefore, it promotes H_2S production through Nrf_2 induction.

We found that H_2S production by the 3MST/CAT pathway is regulated by Ca^{2+}; the activity of CAT to produce 3MP is specifically suppressed by Ca^{2+} (Mikami et al. 2011b) (Fig. 9.2). Calmodulin is not involved in the regulation by Ca^{2+}. Similar regulation by Ca^{2+} without the involvement of calmodulin or any Ca^{2+}-binding domain has been reported for three mitochondrial dehydrogenases (i.e., pyruvate dehydrogenase, nicotinamide adenine dinucleotide (NAD)-isocitrate dehydrogenase, and oxoglutarate dehydrogenase) in addition to CSE, as described earlier (Denton 2009; Mikami et al. 2013). Serine racemase, which is a PLP-dependent enzyme, has a Ca^{2+}-binding site (Cook et al. 2002). The 3 mitochondrial dehydrogenases and serine racemase are activated by Ca^{2+}, while CAT and CSE are suppressed by Ca^{2+} in the presence of PLP (Mikami et al. 2011b, 2013).

9.2.4 3MST/DAO Pathway

When we examined the production of H_2S from brain homogenates, we found that H_2S was produced from D-cysteine, a negative control for L-cysteine (Shibuya et al. 2013). H_2S production from D-cysteine by bacteria and plants is well known, but its production in mammals is not well-understood (Riemenschneider et al. 2005). H_2S production from D-cysteine has properties significantly different from those of L-cysteine. For instance, optimal production occurs at pH 7.4 and is PLP-independent from D-cysteine, while the optimal production occurs at pH 8.4 and is PLP-dependent for L-cysteine (Shibuya et al. 2013). The production from D-cysteine is suppressed by indol-2-carboxylate (I_2CA), a selective inhibitor of DAO, while that from L-cysteine is not. These observations suggest the involvement of DAO in the production of H_2S from D-cysteine.

DAO, which is localized to the peroxisome, metabolizes D-cysteine to achiral 3MP, which is a substrate used by mitochondrial 3MST to produce H_2S (Gould et al. 1988) (Fig. 9.3). Mitochondria and peroxisome are essential cellular organelles in mammals and exchange various metabolites and enzymes via a specific form of vesicular trafficking. Both organelles are in close proximity to each other or have physical contact with each other (Schumann and Subramani 2008). This interaction between mitochondria and peroxisome enables cells to produce H_2S from D-cysteine (Fig. 9.4).

This pathway is restricted to the kidney and the brain, especially in the cerebellum, and the H_2S-producing activity in the kidney is seven times as potent as that in the cerebellum (Shibuya et al. 2013). The H_2S production in the kidney from D-cysteine is approximately 80 times as great as that from L-cysteine. In contrast, the enzymes, which produce H_2S from L-cysteine, such as CBS, CSE, and 3MST/CAT, are expressed in various organs.

9.3 Cytoprotective Effect of H_2S

9.3.1 Neurons

9.3.1.1 H_2S Increases Glutathione

H_2S protects neurons from oxidative stress induced by an excess amount of neurotransmitter glutamate and H_2O_2 (Kimura and Kimura 2004; Kimura et al. 2010).

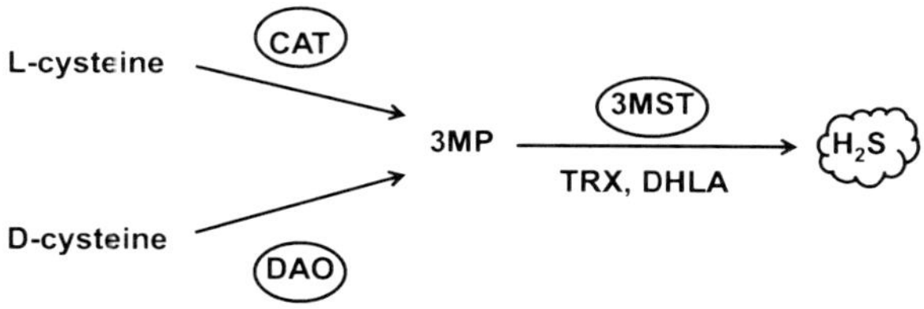

Fig. 9.3 Production of H_2S from L-cysteine and D-cysteine

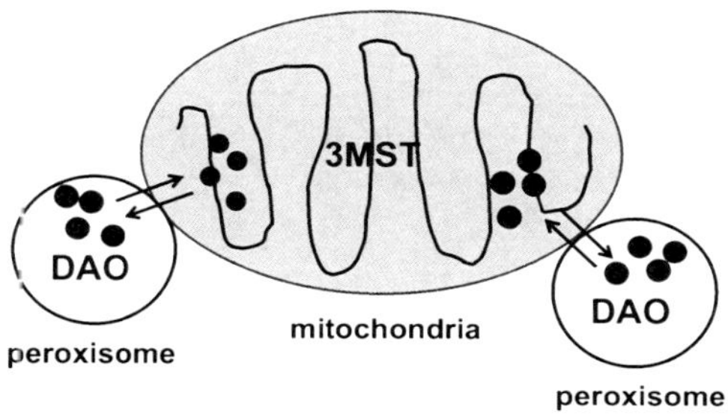

Fig. 9.4 Interaction between mitochondria and peroxisomes

Under oxidative stress, the levels of ROS increase, and it is necessary for neurons to reduce them. High concentrations of glutamate released by dead cells that can be caused by a stroke (for example) suppress the cystine/glutamate antiporter from transporting cystine into cells (Murphy et al. 1989). The lack of cystine, which is reduced to cysteine in cytosol as a substrate for glutathione production, decreases the levels of glutathione, an intracellular major antioxidant. Although the major form of cysteine in extracellular space is its oxidized form cystine, 20 μM of cysteine is also found in the blood (Richie and Lang 1987). Cysteine is more efficiently transported into cells than cystine and is used for glutathione production (Kimura et al. 2010). H_2S enhances the activity of both the cystine/glutamate antipoter xCT^- and cysteine transporter $X_{A,G}$-. It also enhances the activity of γ-glutamyl-cysteine synthase (γ-GCS), which is a limiting enzyme for glutathione synthesis (Kimura and Kimura 2004) (Fig. 9.5). By these effects, H_2S efficiently recovers glutathione levels, which are decreased by oxidative stress.

9.3.1.2 H_2S Scavenges ROS

H_2S produced by 3MST and CAT in mitochondria, which generate large amounts of ROS, protects neurons by scavenging ROS (Kimura et al. 2010). A neuroblastoma cell line was protected by H_2S from oxidative stress induced by peroxynitrite and hypochlorous acid in previous studies (Whiteman et al. 2004, 2005). Although it has not been well understood in neurons, H_2S production is triggered by oxidative stress in yeasts (Kwak et al. 2003). The steady-state levels of H_2S are approximately

Fig. 9.5 Mechanism of cytoprotection by H_2S

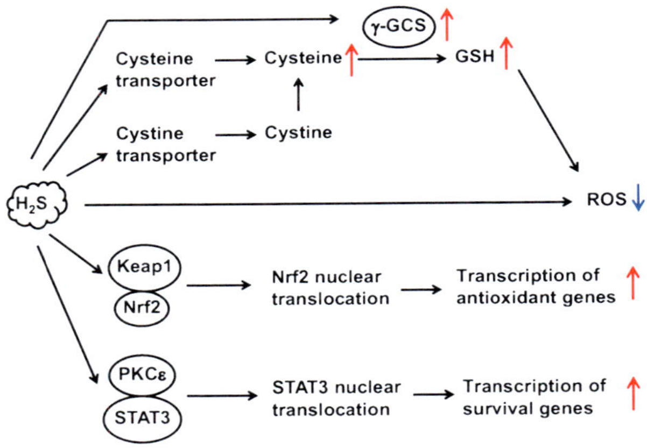

14 nM to 2 μM, while those of glutathione are in the mM order. Therefore, cytoprotection by H_2S mediated by glutathione is more efficient than the ROS scavenging effect of H_2S alone (Kimura et al. 2010).

9.3.1.3 H_2S Suppresses Ca^{2+} Influx in the Retina

The retina is susceptible to oxidative stress because of its high consumption of oxygen and daily exposure to light. Excessive light exposure leads to photoreceptor degeneration, which is an irreversible injury that is caused by various factors, such as ROS and elevated intracellular concentrations of Ca^{2+} (Wenzel et al. 2005). In photoreceptor cells, the intracellular concentrations of Ca^{2+} are maintained lower than other types of cells in which Ca^{2+} concentrations are regulated between 100 nM and a few μM. When photoreceptor cells are exposed to light, Ca^{2+} concentration is approximately 10 nM. It increases to 600 nM in darkness (Krizaj and Copenhagen 2002). 3MST and CAT are enzymes that produce H_2S in the retina (Mikami et al. 2011b). Neither CBS nor CSE are found in the mammalian retina. The production of H_2S by the 3MST/CAT pathway is greatest with a low Ca^{2+} concentration, which is achieved by exposure to light, and is suppressed by high concentrations of Ca^{2+} in darkness. When photoreceptor cells are exposed to light, cGMP-gated cation channels are closed, resulting in an abolishment of Ca^{2+} influx and suppression of the release of the neurotransmitter glutamate from photoreceptor synaptic ends to horizontal cells. Without the activation of glutamate receptors, the intracellular Ca^{2+} concentrations of horizontal cells are maintained at low levels, which activate the 3MST/CAT pathway to produce H_2S. H_2S reduces the cysteine disulfide bridge at the active site of vacuolar-type H^+ ATPase (V-ATPase) for activation in horizontal cells (Jouhou et al. 2007). V-ATPase releases H^+, which suppresses voltage-gated Ca^{2+} channels in photoreceptor cells, thereby maintaining the intracellular Ca^{2+} concentrations at lower levels in photoreceptor cells (Mikami et al. 2011b).

The effect of H_2S on suppressing intracellular Ca^{2+} concentrations involves the protection of retinal neurons from light-induced degeneration. Excessive light

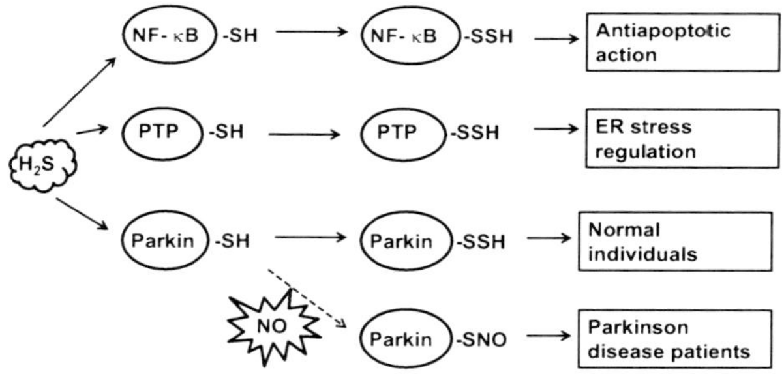

Fig. 9.6 Regulation of protein function by sulfhydration

exposure causes retinal degeneration by increasing the levels of ROS and elevating intracellular Ca^{2+} concentrations. The intraperitoneal administration of NaHS ameliorates retinal degeneration, as observed in the decreasing the number of TUNEL positive neurons and of cells containing 8-hydroxy 2′-deoxyguanosine, which is produced by the oxidation of deoxyguanosine by ROS (Mikami et al. 2011b). The intracellular concentrations of Ca^{2+} are increased during photoreceptor apoptosis, but the L-type Ca^{2+} channel blocker diltiazem prevents light-induced photoreceptor degeneration (Donovan and Cotter 2002). The lack of the V-ATPase a3 subunit causes retinal degradation in mice, suggesting that V-ATPase is involved in preventing retinal degeneration (Kawamura et al. 2010). The failure of Ca^{2+} homeostasis, which is regulated by Ca^{2+} channels and V-ATPase, causes retinal degeneration. Even under such conditions, the administration of H_2S protects neurons from light-induced degeneration.

9.3.2 Cardiovascular and Renal Systems

9.3.2.1 Cardiovascular Protection

H_2S protects cardiomyocytes from ischemia reperfusion injury (Elrod et al. 2007; Predmore and Lefer 2010). The application of H_2S at the time of reperfusion limits infarct size and preserves the left ventricular function. H_2S inhibits myocardial inflammation and caspase-3 activity, and preserves mitochondrial structure and function. The transcriptional regulation of cytoprotection was recently elucidated. H_2S dissociates Keap1 from Nrf_2 to facilitate the translocation of Nrf_2 from cytosol to the nucleus where it activates a number of antioxidants and related enzymes, including thioredoxin and heme oxygenase-1 (Calvert et al. 2009) (Fig. 9.5). H_2S also induces the phosphorylation of proteins in the protein kinase Cε/signal transducers and activators of the transcription 3 (STAT-3) signaling cascade to translocate STAT3 to the nucleus. STAT3 upregulates the expression of the survival factors Bcl-2 and Bcl-xL but downregulates the apoptotic factor Bad. H_2S induces the transcription of antiapoptotic genes by translocating NF-κB to the nucleus through sulfhydrating the p65 subunit of NF-κB (Sen et al. 2012) (Fig. 9.6).

9.3.2.2 Renal Protection

Renal dysfunction and injury caused by ischemia-reperfusion is attenuated by the application of NaHS (Tripatara et al. 2008; Bos et al. 2009; Zhu et al. 2012). Phosphorylation of mitogen-activated protein kinases (e.g., p-38, c-JUN N-terminal protein kinase ½, and extracellular signal-regulated kinase ½) is significantly suppressed by NaHS (Tripatara et al. 2008). NaHS also suppresses the activity of caspase-3 and recovers the expression of Bcl-2, which is decreased by ischemia-reperfusion injury. The administration of H_2S gas, which decreases oxygen (O_2) consumption and carbon dioxide (CO_2) production, also shows renal survival and protection against apoptosis and inflammation caused by bilateral renal ischemia reperfusion injury (Bos et al. 2009).

The third phase of the clinical trial for the effect of H_2S produced by the administration of N-acetylcysteine (NAC) on chronic kidney disease was completed in 2012 (Clinical Trials.gov Ientifier:NCT01232257). NAC readily passes through the plasma membrane and enters into cells where acetyl residue is cleaved from L-cysteine that is metabolized to produce H_2S.

Kidney homogenates produce H_2S from D-cysteine 80 times more than L-cysteine (Shibuya et al. 2013). The oral administration of D-cysteine increases the levels of bound sulfane sulfur, which is an index of produced H_2S, more efficiently than the administration of L-cysteine, resulting in the protection of kidney from ischemia-reperfusion injury by D-cysteine more effectively than L-cysteine. The pathways to produce H_2S from L-cysteine, such as CBS, CSE, and 3MST (along with CAT), are distributed in various organs, while the pathway from D-cysteine is localized specifically to the kidney and, to a lesser extent, the cerebellum (Kimura 2010, 2012). D-cysteine is less toxic than L-cysteine (Misra 1989; Olney et al. 1990). The administration of D-cysteine may provide a new therapeutic approach to protect the kidney from oxidative stress or ischemia-reperfusion injury through its conversion to H_2S via the novel 3MST and DAO pathway.

9.3.3 Bacterial Resistance to Antibiotics

The development of bacterial resistance against antibiotics was recently found to be regulated by H_2S (Shatalin et al. 2011). The cytoprotective effect of H_2S has been thought to be a universal defense mechanism that functions from bacteria to mammals. Antibiotics can induce bacterial death through oxidative stress by ROS, which cause damage to the DNA. It is known that bacteria up-regulate superoxide dismutase (SOD), catalase, and NO synthase (NOS) for protection against damage caused by ROS. Nudler and colleagues have recently found that inactivation of the bacterial homologs of CBS, CSE, and 3MST decreases the production of H_2S that causes the vulnerability of bacteria against antibiotics. Bacteria produce both NO and H_2S, and both substances show a synergistic effect against the antibiotics. Because bacterial CBS, CSE, and 3MST have evolutionally diverged from their mammalian counterparts, they may represent appropriate targets for developing a new class of antibiotics.

9.4 Bound Sulfane Sulfur and Sulfhydration

9.4.1 Bound Sulfane Sulfur

There are two forms of sulfur stores that can release H_2S in cells, including bound sulfane sulfur (which releases H_2S under reducing conditions) and acid-labile sulfur, (which releases H_2S under acidic conditions) (Ogasawara et al. 1994; Ishigami et al. 2009).

Bound sulfane sulfur, which is divalent sulfur that bounds only to other sulfurs (e.g., outer sulfur atoms of the persulfides and inner chain atoms of polysulfides). Production of bound sulfane sulfur at the cysteine residues of proteins has recently been designated as 'sulfhydration' by Snyder and colleague (Mustafa et al. 2009). Exogenously applied H_2S is immediately absorbed in tissue homogenates as bound sulfane sulfur (Ishigami et al. 2009). This observation suggests that the endogenously produced H_2S must also be immediately incorporated into bound sulfane sulfur. As expected, the expression of H_2S producing the enzymes 3MST and CAT in HEK 293F cells increased the levels of bound sulfane sulfur in cells (Shibuya et al. 2009b). The levels of bound sulfane sulfur in the kidney are greatly increased by the oral administration of D-cystein to mice or of L-cysteine, to a lesser extent (Shibuya et al. 2013). The brain contains bound sulfane sulfur, which can release approximately 1.5 μmol H_2S/g protein, and the liver contains a greater amount of bound sulfane sulfur than the brain. In addition, 25–50 % of most hepatic proteins are sulfhydrated (Ishigami et al. 2009; Mustafa et al. 2009). These observations suggest that the majority of proteins contain bound sulfane sulfur or are sulfhydrated.

Acid-labile sulfur is mainly a sulfur atom in iron-sulfur complexes, which is the active center of the enzymes, and regulates the redox reactions in the respiratory chain in mitochondria. Because H_2S is released from acid-labile sulfur at a $pH < 5.4$, acid-labile sulfur may not release H_2S under physiological conditions (Ishigami et al. 2009).

9.4.2 Sulfurhydration

Serine dehydrogenase was inhibited by sulfhydration in the presence of cysteine and CSE (Kato et al. 1966). Since then, the activity of many enzymes, such as homoserine dehydrogenase, tyrosine aminotransferase, ornithine decarboxylase, and adenylate kinase, were suppressed by sulfhydration (Toohey 1989). The inhibition is reversible by reducing agents, such as glutathione and dithiothreitol (DTT), suggesting that sulfhydration or the addition of bound sulfane sulfur is removed by reducing agents (Toohey 1989).

In contrast to the previous studies, recent observations by Snyder and colleagues show that sulfhydration activates protein activities (Paul and Snyder 2012). Sulfhydration of the glyceraldehyde-3-phosphate dehydrogenase (GAPDH) increases its glycolytic activity by sevenfold, while another modification of the same cysteine residue, S-nitrosylation, inhibits glycolytic function (Mustafa et al. 2009). H_2S also regulates endoplasmic reticulum (ER) stress through sulfhydration of the protein tyrosine phosphatase (Krishnan et al. 2011) (Fig. 9.6). Other examples are: (1) the sulfhydration of actin enhances its polymerization, and (2) that of K_{ATP} channels, and Ca^{2+}-dependent intermediate- and small conductance-K^+ channels enhance their activities (Mustafa et al. 2011). The antiapoptotic actions of NF-κB are regulated by sulfhydration mediated by H_2S (Sen et al. 2012). In contrast to sulfhydration, S-nitrosylation inhibits protein function. Recently, a typical example of the protein activity, which is reciprocally regulated by sulfhydraton and S-nitrosylation, has been reported. In the brains of patients with Parkinson's disease, the neuroprotective ubiquitin E3 ligase, parkin, is inactivated by S-nitrosylation. Physiologic modification of parkin by sulfhydration mediated by H_2S enhances its catalytic activity, and parkin sulfhydration is markedly depleted in the brains of Parkinson's patients. Based on these observations, sulfhydration is suggested to mediate the neuroprotective actions of parkin (Vandiver et al. 2013).

9.5 Polysulfides as H_2S-Derived Signaling Molecules

9.5.1 Polysulfides Induce Ca^{2+} Influx in Astrocytes

The activity of neuronal synapses are surrounded and regulated by astrocytes, a type of glia (Dani et al. 1992; Fellin et al. 2004). NaHS increases intracellular Ca^{2+} concentrations in astrocytes, and the response propagates to the neighbor astrocytes (Nagai et al. 2004). The responses are abolished in the absence of extracellular Ca^{2+} or broad spectrum blockers of transient receptor potential (TRP) channels, such as La^{3+}, Gd^{3+}, and ruthenium red, suggesting that H_2S activates TRP channels to induce Ca^{2+} influx. Glial fibrillary acidic protein (GFAP)-positive mature astrocytes respond well to H_2S, while GFAP-negative premature astrocytes and old astrocytes do not respond to H_2S (Tsugane et al. 2007).

The steady-state levels of endogenous H_2S is 14 nM to a few μM, while the ED_{50} of NaHS required to induce Ca^{2+} influx is 116 μM and maximal responses are induced at 160 μM (Nagai et al. 2004; Furne et al. 2008; Ishigami et al. 2009; Wintner et al. 2010; Levitt et al. 2011). Although it has not been measured, the levels of endogenous H_2S (when H_2S-producing pathways are activated in cells) may not be released beyond 200 μM, even considering the fact that the endogenous bound sulfane sulfur exists beyond this concentration (Ishigami et al. 2009). For these reasons, H_2S may not be enough to fully induce Ca^{2+} influx in astrocytes.

Elemental sulfur is well dissolved in a NaHS solution to produce polysulfides. Polysulfides are also produced from H_2S in the presence of oxygen and equilibrate with varying numbers of sulfurs until the number reaches 8, at which point sulfur molecules undergo cyclization and separate from polysulfides (Nagy and Winterbourn 2010; Nielsen et al. 2011; Toohey 2011).

$$2nH_2S + 1/2(2n - 1)O_2 \rightarrow H_2S_{2n} + (2n - 1)H_2O$$

$$HS^- \leftrightarrow HSS^- \leftrightarrow HSSS^- \leftrightarrow \ldots \leftrightarrow HS_7{}^- \rightarrow S_8$$

We found that polysulfides induce Ca^{2+} influx more efficiently than H_2S in astrocytes with an ED_{50} of 91 nM, and the maximal responses are induced at 0.5 μM (Nagai et al. 2006; Oosumi et al. 2010; Kimura et al. 2013). Our HPLC analysis shows that approximately 25 nmol polysulfides/gram of wet brain tissue (which corresponds to an order of μM) exist in the brain, suggesting that endogenous polysulfides are enough to induce Ca^{2+} influx in astrocytes (Kimura et al. 2013).

Responses induced by polysulfides are also suppressed in the absence of extracellular Ca^{2+} or TRP channel blockers similar to those induced by H_2S (Kimura et al. 2013; Nagai et al. 2004, 2006; Oosumi et al. 2010). Moreover, GFAP-positive astrocytes respond well to polysulfide, while GFAP-negative premature astrocytes or old astrocytes do not (Kimura et al. 2013; Tsugane et al. 2007). These observations suggest that polysulfides activate TRP channels in astrocytes, as observed with H_2S.

9.5.2 The Sensitive Molecule or Receptor of Polysulfides

TRPA1 channels are localized in unmyelinated nerve fibers in the rat urinary bladder. In addition, NaHS and TRPA1 activators initiate detrusor overactivity, suggesting that H_2S mediates sensory transduction in this organ (Streng et al. 2008). Moreover, NaHS induces Ca^{2+} responses in TRPA1-expressing Chinese hamster ovary (CHO) cells. H_2S also functions as a nociceptive messenger through the activation of TRPA1 channels in mice based on the following observations (Ogawa et al. 2012): (1) TRPA1 is expressed in the sensory neurons; (2) H_2S increases the intracellular concentrations of Ca^{2+}; (3) responses to H_2S are suppressed by TRPA1 selective inhibitors; (4) sensory neurons obtained from TRPA1 knockout mice do not respond to H_2S; (5) human embryonic kidney (HEK) 293 cells expressing TRPA1 channels respond to H_2S; (6) responses to H_2S are greater at pH 6.8 than at pH 7.4; (7) pain-related behaviors are not observed in TRPA1 knockout mice; and (8) pain-related behaviors are greatly observed at pH 6.8 compared to pH 7.4.

In both the urinary bladder and sensory neurons, responses were elicited with NaHS concentrations greater than 1 mM (Streng et al. 2008; Ogawa et al. 2012). Considering the fact that the steady-state levels of endogenous H_2S are between 14 nM and 2 μM, H_2S may not be an endogenous ligand for TRPA1 channels. As described earlier, polysulfides induce Ca^{2+} influx in rat astrocytes with endogenous concentrations, and the effect of polysulfides is approximately 300 times as efficient as H_2S, suggesting that polysulfides may be endogenous ligands to TRPA1 channels (Nagai et al. 2006; Oosumi et al. 2010; Kimura et al. 2013).

Until recently, TRPA1 was not thought to exist in astrocytes. A transcriptome database shows that TRPA1 mRNA is not expressed in astrocytes, and immunohistochemistry with an antibody against TRPA1 fails to show the existence of TRPA1 in astrocytes (Cahoy et al. 2008; Shigetomi et al. 2012). This could be a reflection of the fact that TRPA1 channels function at expression levels below the reach of immunocytochemistry (Shigetomi et al. 2012). We found that TRPA1-selective agonists, allyl isothiocyanate and cinnamaldehyde, induce Ca^{2+} influx and responses to polysulfides are suppressed by TRPA1-selective inhibitors

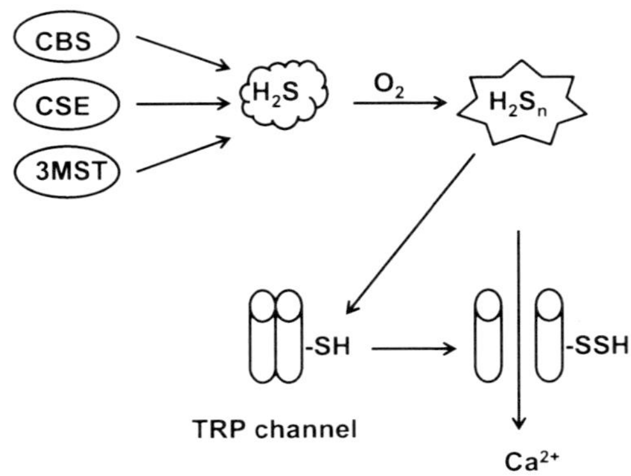

Fig. 9.7 A possible mechanism of the regulation of TRPA1 channels by polysulfides

(e.g., HC-030031 and AP-18) as well as by siRNAs selective to TRPA1 (Kimura et al. 2013). Considering the fact that the H_2S-producing enzymes 3MST and CBS are localized in neurons and astrocytes, respectively (Enokido et al. 2005; Shibuya et al. 2009b), and that polysulfides activate TRP channels more potently than H_2S, polysulfides may be H_2S-derived bioactive molecules that stimulate TRPA1 channels (Kimura et al. 2013).

9.5.3 A Possible Mechanism for the Function of H_2S-derived Polysulfides

The sulfane sulfur of polysulfides is a reactive electrophile and is readily transferred to a nucleophilic protein thiolate to generate the protein persulfide (sulfhydration) (Kato et al. 1966; Mustafa et al. 2009; Toohey 2011). The sulfhydration activity of polysulfides is greater than their parental molecule, H_2S (Toohey 2011). This mechanism explains our initial observation on the activation of NMDA receptors by H_2S (Abe and Kimura 1996). NMDA receptors have a cysteine disulfide bridge, whose reduction by DTT enhances the activity of NMDA receptors (Aizenman et al. 1989). H_2S, which is a reducing substance, reduces the disulfide bridge of receptors and enhances their activity. However, the activity of H_2S is not fully explained by the reduction of the disulfide bridge. DTT enhances the activity of NMDA receptors, but 10-fold decrease in the concentration of H_2S, which is a weaker reductant, further enhances the activity of the receptors. Polysulfides produced by the oxidation of H_2S may effectively add sulfane sulfur to the cysteine thiolate, which is produced by the reduction of the cysteine disulfide bridge by DTT, to further activate NMDA receptors.

The amino-terminus of TRPA1 plays an important role in gating because many agonists are electrophiles that activate the channel through reversible covalent modification of cysteine residues within this domain. Polysulfides may transfer sulfane sulfur to the cysteine residues in the modulatory domain of the channel, resulting in conformational changes that lead to the activation of the channel (Fig. 9.7).

9.6 Concluding Remarks

Four pathways that produce H_2S (i.e., CBS, CSE, CAT/3MST, or DAO/3MST). A common substrate for CBS, CSE, and CAT/3MST is L-cysteine, while DAO/3MST uses D-cysteine as the substrate. Since racemace, which converts L-cysteine to D-cysteine, has not been identified, an understanding of the endogenous production of D-cysteine is awaited. L-Amino acids are nonenzymatically racemized by heat and alkaline conditions applied during food processing, and almost a half of L-cysteine is changed to D-cysteine by alkaline treatment (Liardon and Ledermann 1986; Friedman 2010). Therefore, the DAO/3MST pathway may function under physiological conditions. This pathway is restricted to the cerebellum and the kidney compared to the L-cysteine pathways, which are expressed in various organs. In particular, the production of H_2S from D-cysteine is 80 times greater than from L-cysteine in the kidney. In addition, H_2S exerts significantly more efficient protection in the kidney. The therapeutic application of D-cysteine for kidney diseases or transplantation is promising.

Polysulfides, which are produced by oxidizing H_2S, more efficiently activate TRPA1 channels than its parental molecule, H_2S (Nagai et al. 2006; Oosumi et al. 2010; Kimura et al. 2013). The mechanism for the activation of the channels may be the modification of channel proteins through the transfer of sulfane sulfur to the cysteine residues (sulfurhydration) localized to the amino-terminal modulatory domain of the channels. Whether H_2S is able to sulfurhydrate the cysteine residues is still a controversial topic of discussion (Toohey 2011). Because Km values of H_2S-producing enzymes are higher than the endogenous concentration of their substrate (i.e., cysteine), the production rate must be very slow and may not be enough to exert its effect. Therefore, it is necessary to change H_2S to a more stable and potent form (i.e., polysulfides). Since sulfane sulfur in polysulfides more efficiently sulfurhydrates cysteine residues than H_2S, endogenously produced H_2S immediately generates polysulfides that sulfurhydrate target proteins to modify their function. It is necessary to clarify where, when, and how H_2S is converted into polysulfides.

Acknowledgements This work was supported by a grant from National Institute of Neuroscience and KAKENHI (23659089) from Grant-in-Aid for Challenging Exploratory Research to H.K.

References

Abe K, Kimura H (1996) The possible role of hydrogen sulfide as an endogenous neuromodulator. J Neurosci 16:1066–1071

Aizenman E, Lipton DA, Loring RH (1989) Selective modulation of NMDA responses by reduction and oxidation. Neuron 2:1257–1263

Akagi R (1982) Purification and characterization of cysteine aminotransferase from rat liver cytosol. Acta Med Okayama 36:187–197

Bos EM, Leuvenink HGD, Snijder PM, Koosterhuis NJ, Hillebrands J-L, Leemans JC, Florquin S, Goor H (2009) Hydrogen sulfide-induced hypometabolism prevents renal ischemia/reperfusion injury. J Am Soc Nephrol 20:1901–1905

Braunstein AE, Goryachenkowa EV, Tolosa EA, Willhardt IH, Yefremova LL (1971) Specificity and some other properties of liver serine sulphhydrase: evidence for its identity with cystathionineβ-synthase. Biochim Biophys Acta 242:247–260

Cahoy JD, Emery B, KaushalA FLC, Zamanian JL, Christopherson KS, Xing Y, Lubischer JL, Krieg PA, Krupenko SA, Thompson WJ, Barres BA (2008) A transcriptome database for astrocytes, neurons, and oligodendrocytes: a new resource for understanding brain development and function. J Neurosci 28:264–278

Calvert JW, Jha S, Gundewar S, Elrod JW, Ramachandran A, Pattillo CB, Kevil CG, Lefer DJ (2009) Hydrogen sulfide mediates cardioprotection through Nrf_2 signaling. Circ Res 105:365–374

Cavallini D, Mondovi B, De Marco C, Scioscia-Santoro A (1962) The mechanism of desulphhydration of cysteine. Enzymologia 24:253–266

Chen X, Jhee KH, Kruger WD (2004) Production of the neuromodulator H_2S by cystathionine beta-synthase via the condensation of cysteine and homocystein. J Biol Chem 279:52082–52086

Chiku T, Padovani D, Zhu W, Singh S, Vitvitsky V, Banerjee R (2009) H_2S biogenesis by human cystathionine γ–lyase leads to the novel sulfur metabolites lanthionine and homolanthionine and is responsive to the grade of hyperhomocysteinemia. J Biol Chem 284:11601–11612

Cook SP, Galve-Roperh I, Martinez del Pozo A, Rodriguez-Crespo I (2002) Direct calcium binding results in activation of brain serine racemase. J Biol Chem 277:27782–27792

Cooper AJL (1983) Biochemistry of sulfur-containing amino acids. Annu Rev Biochem 52:187–222

Dani JW, Chernjavsky A, Smith SJ (1992) Neuronal activity triggers calcium waves in hippocampal astrocyte networks. Neuron 8:429–440

Denton RM (2009) Regulation of mitochondrial dehydrogenases by calcium ions. Biochim Biophys Acta 1787:1309–1316

Donovan M, Cotter TG (2002) Caspase-independent photoreceptor apoptosis in vivo and differential expression of apoptotic protease activating factor-1 and caspase-3 during retinal development. Cell Death Differ 9:1220–1231

Doyle JM, Schinina ME, Bossa F, Doonan S (1990) The amino acid sequence of cytosolic aspartate aminotransferase from human liver. Biochem J 270:651–657

Elrod JW, Calvert JW, Morrison J, Doeller JE, Kraus DW, Tao L, Jiao X, Scalia R, Kiss L, Szabo C (2007) Hydrogen sulfide attenuates myocardial ischemia-reperfusion injury by preservation of mitochondrial function. Proc Natl Acad Sci U S A 104:15560–15565

Enokido Y, Suzuki E, Iwasawa K, Namekata K, Okazawa H, Kimura H (2005) Cysstathionine beta-synthase, a key enzyme for homocysteine metabolism, is preferentially expressed in the radial glia/astrocyte lineage of developing mouse CNS. FASEB J 19:1854–1856

Fellin T, Pascual O, Gobbo S, Pozzan T, Haydon PG, Carmignoto G (2004) Neuronal synchrony mediated by astrocytic glutamate through activation of extrasynaptic NMDA receptors. Neuron 43:729–743

Filipovic MR, Miljkovic JL, Nauser T, Royzen M, Klos K, Shubina T, Koppenol WH, Lippard SJ, Ivanovic-Burmazovic I (2012) Chemical characterization of the smallest S-nitrosothiol, HSNO; cellular cross-talk of H_2S and S-nitrosothiols. J Am Chem Soc 134:12016–12027

Friedman M (2010) Origin, microbiology, nutrition and pharmacology of D-amino acids. Chem Biodivers 7:1491–1530

Furne J, Saeed A, Levitt MD (2008) Whole tissue hydrogen sulfide concentrations are orders of magnitude lower than presently accepted values. Am J Physiol Regul Integr Comp Physiol 295:R1479–R1498

Goodwin LR, Francom D, Dieken FP, Taylor JD, Warenycia MW, Reiffenstein RJ, Dowling G (1989) Determination of sulfide in brain tissue by gas dialysis/ion chromatography: postmortem studies and two case reports. J Anal Toxicol 13:105–109

Gould SJ, Keller GA, Subramani S (1988) Identification of peroxisomal targeting signals located at the carboxy terminus of four peroxisomal proteins. J Cell Biol 107:897–905

Griffith OW (1999) Biologic and pharmacologic regulation of mammalian glutathione synthesis. Free Radic Biol Med 27:922–935

Hosoki R, Matsuki N, Kimura H (1997) The possible role of hydrogen sulfide as an endogenous smooth muscle relaxant in synergy with nitric oxide. Biochem Biophys Res Commun 237:527–531

Ichinohe A, Kanaumi T, Takashima S, Enokido Y, Naai Y, Kimura H (2005) Cystathionine beta-synthase is enriched in the brains of Down's patients. Biochem Biophys Res Commun 338:1547–1550

Ishigami M, Hiraki K, Umemura K, Ogasawara Y, Ishii K, Kimura H (2009) A source of hydrogen sulfide and a mechanism of its release in the brain. Antioxid Redox Signal 11:205–214

Ishii I, Akahoshi N, Yu X-N, Kobayashi Y, Namekata K, Komaki G, Kimura H (2004) Murine cystathionine γ-lyase: complete cDNA and genomic sequences, promoter activity, tissue distribution and developmental expression. Biochem J 381:113–123

Jouhou H, Yamamoto K, Homma A, Hara M, Kaneko A, Yamada M (2007) Depolarization of isolated horizontal cells of fish acidifies their immediate surrounding by activating V-ATPase. J Physiol 585:401–412

Kato A, Ogura M, Suda M (1966) Control mechanism in the rat liver enzyme system converting L-methionine to L-cystine. 3. Noncompetitive inhibition of cystathionine synthetase-serine dehydratase by elemental sulfur and competitive inhibition of cystathionine-homoserine dehydratase by L-cysteine and L-cystine. J Biochem 59:40–48

Kawamura N, Tabata H, Sun-Wada GH, Wada Y (2010) Optic nerve compression and retinal degeneration in Tcirg 1 mutant mice lacking the vacuolar type H^+-ATP ase a3 subunit. PLoS One 5:e12086

Kimura H (2010) Hydrogen sulfide: from brain to gut. Antioxid Redox Signal 12:1111–1123

Kimura Y, Kimura H (2004) Hydrogen sulfide protects neurons from oxidative stress. FASEB J 18:1165–1167

Kimura Y, Dargusch R, Schubert D, Kimura H (2006) Hydrogen sulfide protects HT22 neuronal cells from oxidative stress. Antioxid Redox Signal 8:661–670

Kimura Y, Goto Y-I, Kimura H (2010) Hydrogen sulfide increases glutathione production and suppresses oxidative stress in mitochondria. Antioxid Redox Signal 12:1–13

Kimura H, Shibuya N, Kimura Y (2012) Hydrogen sulfide is a signaling molecule and a cytoprotectant. Antioxid Redox Signal 17:45–57

Kimura Y, Mikami Y, Osumi K, Tsugane M, Oka J-I, Kimura H (2013) Polysulfides are possible H_2S-derived signaling molecules in rat brain. FASEB J 27:2451–2457

Krishnan N, Fu C, Pappin DJ, Tonks NK (2011) H_2S-induced sulfhydration of the phosphatase PTP1B and its role in the endoplasmic reticulum stress response. Sci Signal 4:ra86

Krizaj D, Copenhagen DR (2002) Calcium regulation in photoreceptors. Front Biosci 7: d2023–d2044

Kwak WJ, Kwon GS, Jin I, Kuriyama H, Sohn HY (2003) Involvement of oxidative stress in the regulation of H2S production during ultradian metabolic oscillation of Saccharomyces cerevisiae. FEMS Microbiol Lett 219:99–104

Levitt MD, Abdel-Rehim MS, Furne J (2011) Free and acid-labile hydrogen sulfide concentrations in mouse tissues: anomalously high free hydrogen sulfide in aortic tissue. Antioxid Redox Signal 15:373–378

Liardon R, Ledermann S (1986) Racemization kinetics of free and protein-bound amino acids under moderate alkaline treatment. J Agric Food Chem 34:557–565

Meister A, Fraser PE, Tice SV (1954) Enzymatic desulfuration of β-mercaptopyruvate to pyruvate. J Biol Chem 206:561–575

Mikami Y, Shibuya N, Kimura Y, Nagahara N, Ogasawara Y, Kimura H (2011a) Thioredoxin and dihydrolipoic acid are required for 3-mercaptopyruvate sulfurtransferase to produce hydrogen sulfide. Biochem J 439:479–485

Mikami Y, Shibuya N, Kimura Y, Nagahara N, Yamada M, Kimura H (2011b) Hydrogen sulfide protects the retina from light-induced degeneration by the modulation of Ca^{2+} influx. J Biol Chem 286:39379–39386

Mikami Y, Shibuya N, Ogasawara Y, Kimura H (2013) Hydrogen sulfide is produced by cystathionine γ–lyase at the steady-state low intracellular Ca^{2+} concentrations. Biochem Biophys Res Commun 431:131–135

Misra CH (1989) Is a certain amount of cysteine prerequisite to produce brain damage in neonatal rats? Neurochem Res 14:253–257

Modis K, Coletta C, Erdelyi K, Papapetropoulos A, Szabo C (2013) Intramitochondrial hydrogen sulfide production by 3-mercaptopyruvate sulfurtransferase maintains mitochondrial electron flow and supports cellular bioenergetics. FASEB J 27:601–611

Morikawa T, Kajimura M, Nakamura T, Hishiki T, Nakanishi T, Yukutake Y, Nagahata Y, Ishikawa M, Hattori K, Takenouchi T, Takahashi T, Ishii I, Matsubara K, Kabe Y, Uchiyama S, Nagata E, Gadalla MM, Snyder SH, Suematsu M (2012) Hypoxic regulation of the cerebral microcirculation is mediated by a carbon monoxide-sensitive hydrogen sulfide pathway. Proc Natl Acad Sci U S A 109:1293–1298

Murphy TH, Miyamoto M, Sastre A, Schnaar RL, Coyle JT (1989) Glutamate toxicity in a neuronal cell line involves inhibition of cystine transport leading to oxidative stress. Neuron 2:1547–1558

Mustafa AK, Gadalla MM, Sen N, Kim S, Mu W, Gazi SK, Barrow RK, Yang G, Wang R, Snyder SH (2009) H_2S signals through protein S-sulfhydration. Sci Signal 2:ra72

Mustafa AK, Sikka G, Gazi SK, Steppan J, Jung SM, Bhunia AK, Barodka VM, Gazi FK, Barrow RK, Wang R, Amzel LM, Berkowitz DE, Snyder SH (2011) Hydrogen fulfide as endothelium-derived hyperpolarizing factor sulfhydrates potassium channels. Circ Res 109:1259–1268

Nagahara N, Nishino T (1996) Role of amino acid residues in the active site of rat liver mercaptopyruvate sulfurtransferase. cDNA cloning, overexpression, and site-directed mutagenesis. J Biol Chem 271:27395–27401

Nagai Y, Tsugane M, Oka J, Kimura H (2004) Hydrogen sulfide induces calcium waves in astrocytes. FASEB J 18:557–559

Nagai Y, Tsugane M, Oka J, Kimura H (2006) Polysulfides induce calcium waves in rat hippocampal astrocytes. J Pharmacol Sci 100(Suppl I):200, ISSN 1347–8613

Nagy P, Winterbourn CC (2010) Rapid reaction of hydrogen sulfide with the neutrophil oxidant hypochlorous acid to generate polysulfides. Chem Res Toxicol 23:1541–1543

Nielsen RW, Tachibana C, Hansen NE, Winther JR (2011) Trisulfides in proteins. Antioxid Redox Signal 15:67–75

Ogasawara Y, Isoda S, Tanabe S (1994) Tissue and subcellular distribution of bound and acid-labile sulfur, and the enzymic capacity for sulfide production in the rat. Biol Pharm Bull 17: 1535–1542

Ogawa H, Takahashi K, Miura S, Imagawa T, Saito S, Tominaga M, Ohta T (2012) H2S functions as a nociceptive messenger through transient receptor potential ankyrin 1 (TRPA1) activation. Neuroscience 218:335–343

Olney JW, Zorumski C, Price MT, Labruyere J (1990) L-cysteine, a bicarbonate-sensitive endogenous excitotoxin. Science 248:596–599

Oosumi K, Tsugane M, Ishimami M, Nagai Y, Iwai T, Oka J, Kimura H (2010) Polysulfide activates TRP channels and increases intracellular Ca^{2+} in astrocytes. Bull Japanese Soc Neurochem 49:517, ISSN: 0037–3796

Paul BD, Snyder SH (2012) H_2S signalling through protein sulfhydration and beyond. Nat Rev Mol Cell Biol 13:499–507

Predmore BL, Lefer DJ (2010) Development of hydrogen sulfide-based therapeutics for cardiovascular disease. J Cardiovasc Transl Res 3:487–498

Razzak M (2012) Transplantation: hydrogen sulphide reduces warm renal ischaemic injury. Nat Rev Urol 9:670

Richie JP Jr, Lang CA (1987) The determination of glutathione, cyst(e)ine, and other thiols and disulfides in biological samples using high-performance liquid chromatography with dual electrochemical detection. Anal Biochem 163:9–15

Riemenschneider A, Wegele R, Schmidt A, Papenbrock J (2005) Isolation and characterization of a D-cysteine desulfhydrase protein from *Arabidopsis thaliana*. FEBS J 272:1291–1304

Savage JC, Gould DH (1990) Determination of sulfide in brain tissue and rumen fluid by ion-interaction reversed-phase high-performance liquid chromatography. J Chromatogr 526:540–545

Schumann U, Subramani S (2008) Special delivery from mitochondria to peroxisomes. Trends Cell Biol 18:253–256

Sen N, Paul BD, Gadalla MM, Mustafa AK, Sen T, Xu R, Kim S, Snyder SH (2012) Hydrogen sulfide-linked sulfhydration of NF-κB mediates its antiapoptotic actions. Mol Cell 45:13–24

Shan X, Dunbrack RLJ, Christopher SA, Kruger WD (2001) Mutation in the regulatory domain of cystathionine β–synthase can functionally suppress patient-derived mutations in cis. Human Mol Genet 10:635–643

Shatalin K, Shatalina E, Mironov A, Nudler E (2011) H_2S: a universal defense against antibiotics in bacteria. Science 334:986–990

Shibuya N, Mikami Y, Kimura Y, Nagahara N, Kimura H (2009a) Vascular endothelium expresses 3-mercaptopyruvate sulfurtransferase and produces hydrogen sulfide. J Biochem 146:623–626

Shibuya N, Tanaka M, Yoshida M, Ogasawara Y, Togawa T, Ishii K, Kimura H (2009b) 3-Mercaptopyruvate sulfurtransferease produces hydrogen sulfide and bound sulfane sulfur in the brain. Antioxid Redox Signal 11:703–714

Shibuya N, Koike S, Tanaka M, Ishigami-Yuasa M, Kimura Y, Ogasawara Y, Fukui K, Nagahara N, Kimura H (2013) A novel pathway for the production of hydrogen sulfide from D-cysteine in mammalian cells. Nat Commun 2371

Shigetomi E, Tong X, Kwan KY, Corey DP, Khakh BS (2012) TRPA1 channels regulate astrocyte resting calcium and inhibitory synapse efficacy through GAT-3. Nat Neurosci 15:70–80

Singh S, Padovani D, Leslie RA, Chiku T, Banerjee R (2009) Relative contributions of cystathionine beta-synthase and gamma-cystathionase to H_2S biogenesis via alternative trans-sulfuration reactions. J Biol Chem 284:22457–22466

Stipanuk MH, Beck PW (1982) Characterization of the enzymic capacity for cysteine desulphhydration in liver and kidney of the rat. Biochem J 206:267–277

Streng T, Axelsson HE, Hedlund P, Andersson DA, Jordt SE, Bevan S, Andersson KE, Hogestatt ED, Zygmunt PM (2008) Distribution and function of the hydrogen sulfide-sensitive TRPA1 ion channel in rat urinary bladder. Eur Urol 53:391–399

Suh JH, Shenvi SV, Dixon BM, Liu H, Jaiswal AKI, Liu RM, Hagen TM (2004) Decline in transcriptional activity of Nrf_2 causes age-related loss of glutathione synthesis, which is reversible with lipoic acid. Proc Natl Acad Sci U S A 101:3381–3386

Taniguchi T, Kimura T (1974) Role of 3-mercaptopyruvate sulfurtransferase in the formation of the iron-sulfur chromophore of adrenal ferredoxin. Biochim Biophys Acta 364:284–295

Taoka S, Banerjee R (2001) Characterization of NO binding to human cystathionine beta-synthase: possible implications of the effects of CO and NO binding to the human enzyme. J Inorg Biochem 87:245–251

Tateishi N, Higashi T, Naruse A, Nakashima K, Shiozaki H (1977) Rat liver glutathione: possible role as a reservoir of cysteine. J Nutr 107:51–60

Teague B, Asiedu S, Moore PK (2002) The smooth muscle relaxant effect of hydrogen sulphide *in vitro*: evidence for a physiological role to control intestinal contractility. Br J Pharmacol 137:139–145

Toohey JI (1989) Sulphane sulphur in biological systems: a possible regulatory role. Biochem J 264:625–632

Toohey JI (2011) Sulfur signaling: is the agent sulfide or sulfane? Anal Biochem 413:1–7

Tripatara P, Patel NSA, Collino M, Gallicchio M, Kieswich J, Castiglia S, Benetti E, Stewart KN, Brown PAJ, Yaqoob MM, Fantozzi R, Thiemermann C (2008) Generation of endogenous hydrogen sulfide by cystathionine γ–lyase limits renal ischemia/reperfusion injury and dysfunction. Lab Invest 88:1038–1048

Tsugane M, Nagai Y, Kimura Y, Oka J-I, Kimura H (2007) Differentiated astrocytes acquire sensitivity to hydrogen sulfide that is diminished by the transformation into reactive astrocytes. Antioxid Redox Signal 9:257–269

Ubuka T, Umemura S, Yuasa S, Kinuta M, Watanabe K (1978) Purification and characterization of mitochondrial cysteine aminotransferase from rat liver. Physiol Chem Phys 10:483–500

Vandiver MS, Paul BD, Xu R, Karuppagounder S, Rao F, Snowman AM, Ko HS, Lee YII, Dawson VL, Dawson TM, Sen N, Snyder SH (2013) Sulfhydration mediates neuroprotective actions of parkin. Nat Commun 4:1626

Warenycia MW, Goodwin LR, Benishin CG, Reiffenstein RJ, Grancom DM, Taylor JD, Dieken FP (1989) Acute hydrogen sulfide poisoning. Demonstration of selective uptake of sulfide by the brainstem by measurement of brain sulfide levels. Biochem Pharmacol 38:973–981

Wenzel A, Grimm C, Samardzija M, Reme CE (2005) Molecular mechanisms of light-induced photoreceptor apoptosis and neuroprotection for retinal degeneration. Prog Retin Eye Res 24:275–306

Westrop GD, Georg I, Coombs GH (2009) The mercaptopyruvate sulfurtransferase of Trichomonas vaginalis links cysteine catabolism to the production of thioredoxin persulfide. J Biol Chem 284:33485–33494

Whiteman M, Armstrong JS, Chu SH, Jia-Ling S, Wong BS, Hheung NS, Halliwell B, Moore PK (2004) The novel neuromodulator hydrogen sulfide: an endogenous peroxynitrite 'scavenger'? J Neurochem 90:765–768

Whiteman M, Cheung NS, Zhu YZ, Chu SH, Siau JL, Wong BS, Armstrong JS, Moore PK (2005) Hydrogen sulphide: a novel inhibitor of hypochlorous acid-mediated oxidative damage in the brain? Biochem Biophys Res Commun 326:794–798

Wintner EA, Deckwerth TL, Langston W, Bengtsson A, Leviten D, Hill P, Insko MA, Dumpit R, VandenEkart E, Toombs CF, Szabo C (2010) A monobromobimane-based assay to measure the pharmacokinetic profile of reactive sulphide species in blood. Br J Pharmacol 160:941–957

Yang G, Wu L, Jiang B, Yang W, Qi J, Cao K, Meng Q, Mustafa AK, Mu W, Zhang S, Snyder SH, Wang R (2008) H_2S as a physiologic vasorelaxant: hypertension in mice with deletion of cystathionine γ-lyase. Science 322:587–590

Zhao W, Zhang J, Lu Y, Wang R (2001) The vasorelaxant effect of H_2S as a novel endogenous gaseous K_{ATP} channel opener. EMBO J 20:6008–6016

Zhu JXG, Kalbfleisch M, Yang YX, Bihari R, Lobb I, Davison M, Mok A, Cepinskas G, Lawendy A-R, Sener A. (2012) Detrimental effects of prolonged warm renal ischaemia-reperfusion injury are abrogated by supplemental hydrogen sulphide: an analysis using real-time intravital microscopy and polymerase chain reaction. BJU 110:E1218–E1227

Index

DOI 10.1007/978-3-7091-1550-3,

Printed by Publishers' Graphics LLC
ICISO130906.15.15.5